PARADOXES AND PHYSICAL LIMITS OF INFORMATION THEORY

World Scientific Series on Quantum Algorithms, Information, and Learning

Series Editor: Wojciech Szpankowski (*Purdue University, USA*)

Published

Vol. 1   *Paradoxes and Physical Limits of Information Theory*
         by Philippe Jacquet

PARADOXES AND PHYSICAL LIMITS OF INFORMATION THEORY

Philippe Jacquet

The National Institute for Research in Digital Science
and Technology (INRIA), France

World Scientific

NEW JERSEY · LONDON · SINGAPORE · BEIJING · SHANGHAI · TAIPEI · CHENNAI

Published by

World Scientific Publishing Co. Pte. Ltd.

5 Toh Tuck Link, Singapore 596224

USA office: 27 Warren Street, Suite 401-402, Hackensack, NJ 07601

UK office: 57 Shelton Street, Covent Garden, London WC2H 9HE

Library of Congress Cataloging-in-Publication Data

Names: Jacquet, P. (Philippe), author.
Title: Paradoxes and physical limits of information theory / Philippe Jacquet,
    The National Institute for Research in Digital Science and Technology (INRIA), France,
    Univeristé Paris Sud (Paris XI), France.
Description: New Jersey : World Scientific, [2025] | Series: World scientific series on quantum
    algorithms, information, and learning ; vol. 1 | Includes bibliographical references and index.
Identifiers: LCCN 2024028347 | ISBN 9789811293597 (hardcover) |
    ISBN 9789811293603 (ebook for institutions) | ISBN 9789811293610 (ebook for individuals)
Subjects: LCSH: Information theory. | Quantum computing. | Quantum theory.
Classification: LCC Q360 .J33 2025 | DDC 003/.54--dc23/eng20241001
LC record available at https://lccn.loc.gov/2024028347

British Library Cataloguing-in-Publication Data
A catalogue record for this book is available from the British Library.

For any available supplementary material, please visit
https://www.worldscientific.com/worldscibooks/10.1142/13850#t=suppl

Desk Editors: Soundararajan Raghuramans/Steven Patt

Typeset by Stallion Press
Email: enquiries@stallionpress.com

Preface

Information has become the defining currency of our age, fueling revolutions in communication, science, and technology. From the birth of information theory to its profound implications in artificial intelligence, quantum information physics, and even black hole physics, the study of information challenges our deepest intuitions about reality. This book, *Paradoxes and Physical Limits of Information Theory*, embarks on a journey through the fundamental questions, paradoxes, and constraints that shape our understanding of information itself.

We begin by questioning the very nature of information: What distinguishes raw data from knowledge? How do signals and entropy shape our ability to communicate, predict, and process information? We tried to build a structured approach to these questions, beginning with a historical perspective and simple everyday examples before delving into the rigorous mathematical foundations that underpin information theory.

One of the most striking aspects of this exploration is the paradoxical nature of information. How can we compress data without losing meaning? Shannon's answer is in the measure of information quantity via entropy, but how can the pattern-matching algorithms extended by Lempel and Ziv truly help build predictors? How did the information theory lead to the creation of the internet? How internet got wireless? Why increasing user density and user mobility can paradoxically amplify the Gupta–Kumar and Grossglauser–Tse transport capacity of wireless networks? How can the fractal nature of natural or man-made landscapes can also contribute to the capacity amplification? These counterintuitive phenomena reveal the intricate and sometimes contradictory behavior of information in different contexts.

Beyond classical theory, this book ventures into uncharted territories where information theory intersects with artificial intelligence and quantum computing. Can information theory define the boundaries of learnability in artificial intelligence and clarify the vision of Turing when he was talking about intelligent machines? Does quantum mechanics offer an escape from the computational limits of classical computers as envisioned by Richard Feynman? What is the von Neumann zero entropy paradox? These are not mere theoretical exercises but questions with profound implications for the future of technology and scientific discovery.

Finally, we confront one ultimate paradox: the fate of information in black holes as hinted by Stephen Hawking. If information cannot be lost, as quantum mechanics suggests, yet black holes seem to destroy it, what does this mean for the fundamental laws of physics and for which impact in information theory? Could non-unitary physics open doors to faster-than-light communication or even the possibility of transmitting information backward in time?

This book is not merely a technical treatise but an invitation to engage with the deepest mysteries of information. It challenges readers to think critically, question assumptions, and appreciate the elegance and complexity of one of the most essential forces shaping our universe. Whether you are a student, researcher, or curious thinker, *Paradoxes and Physical Limits of Information Theory* aims to spark intellectual curiosity and open new pathways of understanding.

About the Author

Philippe Jacquet has been with the National Research Institute on Computer Science (Inria), France since 1984. Between 1984 and 2011, he has worked on the analysis of algorithms. He got his PhD in 1989 in this domain from the University of Paris, Orsay, under the supervision of Philippe Flajolet (1948–2011), the guru of computational complexity and combinatorics. Philippe Jacquet's initial main topics at this time were the data structures and the protocols of communication which naturally led him to the information theory. In 1998, he got his "accreditation to lead research" (HDR) dedicated to algorithmic implications of information theory. In 2012, he spent a long stay with Nokia Bell Labs as the head of the Math Department. In 2019, he was back in Inria as a research director at the Saclay-Ile-de-France center. Meanwhile, he has been a professeur chargé de cours in computer science at Ecole Polytechnique (France) from 2006 to 2016.

His current scientific interests are the analysis of algorithms, information theory, wireless telecommunication, artificial intelligence and quantum information theory. He was lucky enough to witness the rise of wireless local area networks, with WiFi and HIPERLAN, and during the advent of mobile *ad hoc* networks, he initiated the OLSR protocol (the original draft totalizes more than 9,000 citations). Since 2019, he is an IEEE Fellow in the Information Theory Society.

Contents

Preface   v

About the Author   vii

1. What is Information?   1

  1.1   The Information Age   1
  1.2   What is Information? Support of Information   5
  1.3   The Information Pyramid   6
  1.4   Basic Information Paradoxes Play Ground   9
      1.4.1   The paradox of the halted clock   9
      1.4.2   The exam week paradox 10
  1.5   The Blue Eyes Paradox 12
  1.6   Exercises 14
      1.6.1   The 100 prisoners problem 14
      1.6.2   The 100 prisoners again 15

2. The Basic Mathematics Inside Information Theory   17

  2.1   Information Theory 17
  2.2   The Entropy and Its Properties 18
  2.3   Dependent Components 19
  2.4   Example 1: The Additive Noise 20
  2.5   Example 2: The Seven Error Game 20
  2.6   Conditional Entropy, Channel Entropy 22
  2.7   Mutual Information, Channel Capacity 23
  2.8   Information Theory and Telecommunication 24
      2.8.1   Perfect codes 26

2.9   Exercises 28
     2.9.1   Shannon half theorem: Counterexample . . . 28
     2.9.2   Second half theorem: Counterexample . . . 28
     2.9.3   Second half theorem: Generalities 28
     2.9.4   Seven errors game 29
     2.9.5   One error game 29
     2.9.6   Error correction 29
     2.9.7   Unperfect codes 29

**3. Probabilistic Information Theory,
the Paradoxes of Data Compression
and Event Prediction**   **31**

3.1   The Probabilistic Entropy $h(Z)$ 31
     3.1.1   Dependent components 33
3.2   Channel Coding, Error Correction 34
     3.2.1   Seven error game revisited,
               Bernoulli channel 35
     3.2.2   Conditional entropy 35
     3.2.3   Typical sequences and Shannon
               theorem 36
     3.2.4   Typical set 38
3.3   Source Coding, Data Compression 40
     3.3.1   Lossless compression 41
     3.3.2   Symbol compression, Huffman
               algorithm 43
     3.3.3   Pattern matching lossless compression
               algorithm, Ziv–Lempel algorithm 47
3.4   Pattern Matching Predictor 51
3.5   Lossy Compression, Rate Distortion 56
3.6   Exercises 58
     3.6.1   Bernoulli channels superposition 58
     3.6.2   Entropy of Markov chain 59
     3.6.3   Huffman code length 59

**4. The Challenge of Information Networks, the
Triumph of the Algorithms over the Complexity**   **61**

4.1   Building the Most Complex and Distributed
        System 62

4.1.1   Arpanet, Internet 64

4.2   The Routing Internet Protocol 66

4.2.1   The organization of Internet 66

4.2.2   Routing internet protocol (RIP) 70

4.2.3   The border gateway protocol (BGP) 74

4.2.4   And the winner is the Arpanet accident 76

4.2.5   Why has the Internet become the universal digital telecommunications medium? 79

4.3   The Wireless Internet, Topology Compression 83

4.3.1   The topology compression 87

4.3.2   The multipoint relay selection of OLSR . . . 88

4.3.3   The Erdos–Renyi graph model for indoor wireless network graph model 91

4.3.4   The unit disk graph model, outdoor wireless network graph model 93

4.3.5   Computing OLSR routing tables 94

4.3.6   Flooding optimization 96

4.3.7   Hello packets 99

4.4   Exercises 100

4.4.1   RIP count to infinity 100

4.4.2   MPR selection in OLSR 101

4.4.3   Remote spanner and topology compression 101

5. The Performance Paradoxes of Wireless Networks Caused by Physics   **103**

5.1   Wired versus Wireless Networking 104

5.1.1   The "brave old world" of wired networking 104

5.1.2   The physics of wireless networks 105

5.2   Space Adds Capacity 106

5.3   Time Adds Capacity 111

5.4   Geometry Can Add Capacity 118

5.4.1   The Poisson shot model 118

5.4.2   The Shannon capacity of wireless networks 124

5.5   Fractal Geometries 129
5.6   Hyperfractals 140
       5.6.1   Urban Canyon effect and fixed relays 145
5.7   Exercises 148
       5.7.1   Laplace transform of signal level in
                Poisson shot model 149
       5.7.2   Probability of correct reception 150
       5.7.3   The field differentiation theorem 150

**6. The Limit of Artificial Intelligence Imposed
      by Information Theory   153**
6.1   Artificial Intelligence versus Information
       Theory 153
       6.1.1   Shannon point of view 154
       6.1.2   Solving a problem with AI 161
       6.1.3   Taxonomy of learning strategies 162
       6.1.4   Regret MinMax analysis 166
6.2   The Point of View of Turing 169
       6.2.1   Inside the machinery 170
       6.2.2   A school example: Quantum
                tomography 173
       6.2.3   How universal are neural networks? 177
       6.2.4   Perspectives for a theory of learning 188
       6.2.5   Temporary conclusion about autonomous
                self-optimizing AI 190
6.3   Problems and Projects 191
       6.3.1   The mind reading machine 191
       6.3.2   Joint complexity 193

7. Quantum Information Theory   197
7.1   Physics and Information 198
7.2   The Time Arrow 202
7.3   Information and Quantum Physics 204
       7.3.1   The algebra of quantum physics 207
       7.3.2   Theory of quantum measurement,
                quantum information 209
       7.3.3   Spin algebra, photon polarization 212

7.3.4 Density operator, von Neumann
entropy 215

7.4 Quantum Computers 217

7.4.1 The Grover algorithm 218

7.4.2 Peter Shor's algorithm 219

7.4.3 Quantum computer versus quantum
simulator 221

7.5 Entanglement, the End of the Hidden Variables and
the Paradox of the Non-locality 222

7.5.1 Causality and information theory 223

7.5.2 Bell inequalities versus experiment 225

7.5.3 Entanglement against eavesdropping 229

7.6 Quantum Teleportation 232

7.7 Exercises 235

7.7.1 Bell inequality 235

7.7.2 Entanglement 236

8. Non-unitary Quantum Information   **237**

8.1 The Quantum Accident 237

8.1.1 The many-worlds interpretation 239

8.1.2 The zero entropy paradox 242

8.1.3 Divorce, quantum style 243

8.2 Black Hole and Information 246

8.2.1 The limit of information storage:
The Bekenstein bound 248

8.3 Time Travel and Information 249

8.3.1 Dialogue in a classroom 249

8.3.2 Unitarity and retro-information 254

8.3.3 Retro-information, time paradoxes,
and causality 261

8.4 The Black Hole Information Loss Paradox 266

8.5 Programming a Non-unitary Computer 269

8.5.1 $P = NP$ with non-unitary computers 272

8.5.2 Hard time for non-unitary computers on
NP-hard problems 275

8.6 Exercises 279

8.6.1 Lorentzian referential 279

8.6.2   Closed time curve and
         retro-information 279
8.6.3   Ascending order for minimal dominating set
         with non-unitary computer 280
8.6.4   Project: Message from the future 280

9.   **Answer to Exercises**   **283**

Bibliography   307

Index   317

Chapter 1

What is Information?

Abstract

We begin with a brief history of the information age. Then, we pose
the fundamental question, as a preamble to information theory: what is
information? We then present the tetrarchy of the main elements that
make up the information chain: physical signals, raw data, information
and knowledge. We illustrate this construction with several games and
paradoxes.

1.1 The Information Age

Since the dawn of time, humanity has evolved around communication. But spoken language appeared relatively recently (dated
between 50,000 and 20,000 years ago) compared to the birth of prim-
itive humanity (around 4 million years ago). If you think about it,
our ancestors over less than 1,000 generations were probably mute.
However, the use of signs, language sounds and graphic expressions
(found on cave walls) have guided mankind on the long and winding
road across the continent of information. This journey greatly accelerated since the invention of writing and printing and, more recently,
by the advent of digital information. What we mean today by the
(late) *Information Age* hardly spans the last two centuries.

In fact, the major events on which this book is based form the
information explosion that took place during the 20th century which
witnessed the increase by more than a dozen orders of magnitude in
the performance of the telecommunication media. We can start with

1

the first experiments in wireless telecommunication around 1900. Edison and above all Marconi invented the first fundamental components of radio telecommunication. Marconi carried out a famous first experiment over the Strait of Messina and then over the English Channel. On the eve of the First World War, Férier installed the first permanent radio telecommunication station at the top of the Eiffel Tower. The antenna covered one-fifth of the surface area of France $(100,000 \, \text{km}^2)$ with a data rate of just a few bits per second via Morse code. The code, operating on short electric pulses, covered the entire electromagnetic spectrum with a power consumption of 1,000 watts.

One hundred years later, the same territory is now covered by a constellation of cellular base stations and WiFi relays, each consuming 0.01 watts and delivering 10 million bits per second per hectare. The communication flows between the terminals and the bases are distinct, furthermore the bases operate on separate bands when they are adjacent, so their data rates add up, leading to an increase in overall traffic of 10^{15} in a century.

There are few situations in history where a technology has experienced such an explosive increase in usage and performance in such a short time interval. By comparison, the explosive power of atomic weapons, compared with the explosive power of dynamite per kilo, has increased by a "mere" factor of 10^5. More peacefully, automobile traffic has also increased by a factor of 10^5 since 1900 (based on the number of cars per road kilometer), see Figure 1.1. However, this increase is significant enough to affect daily life and seriously damage the Earth's climate for centuries to come.

Fig. 1.1.   Car traffic: (left) in Paris in 1900 and (right) traffic jam on the US–Mexican border circa 2000.
Courtesy: Wikimedia Commons https://upload.wikimedia.org/wikipedia/commons/a/a0/75-Paris-Avenue-des-Champs-Elysés-et-les-Chevaux-de-Marly-ND.JPG and Wikipedia.

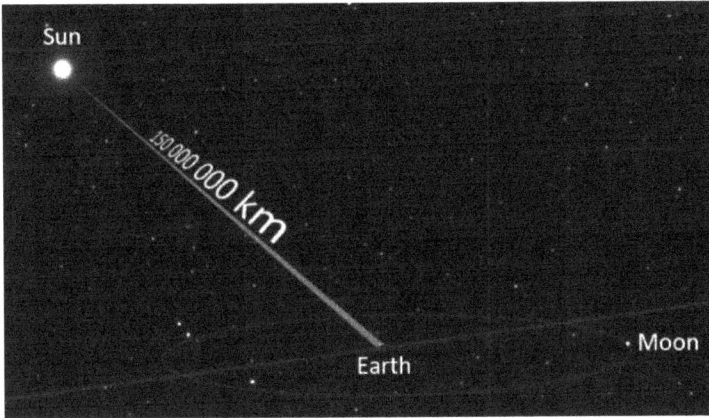

Fig. 1.2. The ratio of 10^{15} is equivalent to the ratio between the thickness of this page of the book and distance from the sun to Earth (the distance to the Moon is not to scale: only 450,000 km).
Courtesy: Wikipedia.

This enormous 10^{15} factor has few equivalents to be compared with. It is equivalent to the ratio between the thickness of a thin piece of paper and the Earth's distance from the Sun (the astronomical unit, see Figure 1.2). Figure 1.3 shows the history of the explosive evolution of this technology since 1900. This history can be divided into three main periods of almost equal impact. The first period, from 1900 to 1948, i.e. the part of the curve between Marconi's and Shannon's portraits, corresponds to the period of the *triumphs over the matter*, when most progresses are essentially due to the understanding of the physics of electromagnetism and in the advances in the electrical technologies. This period ended with the discovery of information theory and with the invention of the transistor, both led to the mathematization of telecommunications and electronics. The development of the transistor, for example, made it possible to organize circuits and chips as graphs and diagrams, and to rationalize printed circuits. We might call this period the *triumph over the numbers* which set a new field of mathematics: the applied mathematics. Progress in physics did not stop meanwhile and continued as indicated by the dashed curve. The imaginary dashed curve tries to figure out what progresses in communication would have been like without the contribution of information theory. The aim of the newborn theory was to optimize the flow of information, in a time where

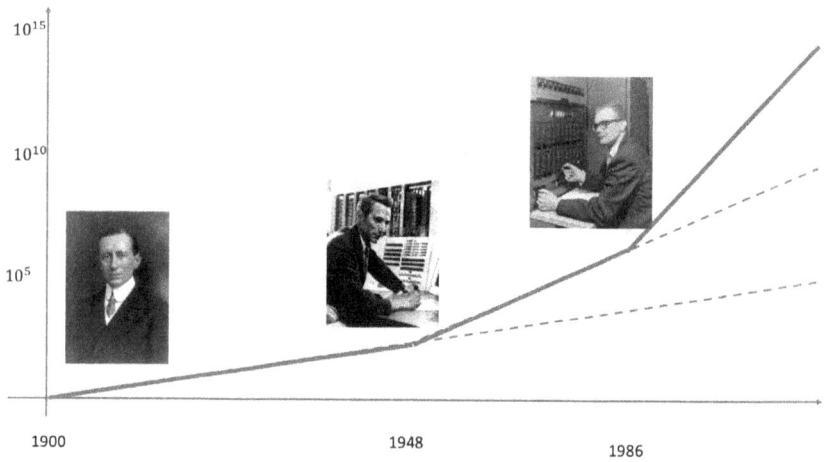

Fig. 1.3.   The information age explosion, wireless throughput capacity increase from 1900 to 2000 in logarithmic scale.

most telecommunications, especially wireless, were based on broadcasting (radio and television), with few sources and lot many more receivers. This period came to an end in 1986 with the rise of the Internet. We've chosen 1986 as the date of the Internet's official birth, with the publication of the first Internet protocol standard, but the revolution took place a little earlier with Arpanet and the introduction of computers into the telephone network. It is illustrated by the portrait of Don Knuth, one of the fathers of the pursuit of complexity in computing, universally known for his masterwork "The Art of Programming" ([70] and still in production). The existing and new networks federated by the Internet have given rise to the most complex human artifact, with multiple sources of information and multiple receivers (terminals). Sources and terminals constantly exchange roles. Until then, only the simple telephone network had anything approaching this complexity, but it remained several orders of magnitude behind until the computerization of telephone technologies allowed their eventual merger with the Internet. We'll call this period the *triumph over the complexity*: it wasn't enough to have a good control over the individual components of a telecommunication network involving the appropriate physics and mathematics, but there was a crucial need to have also the control over the whole set and to let it functioning without being overwhelmed by its complexity.

Figure 1.3 is of course displaying a very rough abstraction. For example, the increase between 1900 and 1948 was not linear as so depicted, there have been several bumps, in particular a sharp rise by a factor of 10^3 in the 1920s when phonic radio telecommunications came to the public to replace the more confidential application of Morse code, and another bump when television spread after WW2. Also we stress again that the dashed continuation is just hypothetical, since most progresses needed the combination of physics, information theory and later computer science in order to succeed.

1.2 What is Information? Support of Information

Information is an elusive term. It is difficult to define precisely what information is as such. For example, an image is information, a page of music is information. However, we won't dwell on the meaning behind a piece of information. The fact that the photo is of a famous actress or actor will not make them to take up more or less space on your hard disk.

Many old-timers at the early time of information were reluctant to understand that information and computing actually have value, since their apparent vocation seemed to restrict them to being purely immaterial goods. We see in the final chapters that this idea is misleading because information does indeed have a physical weight, or at least we can say that information must be supported by a physical medium and carry a minimum mass or energy actually proportional to the quantity of information. This point was made by Brillouin in the 1930s to resolve the famous Maxwell's paradox (see Chapter 6). But the same old-timers might have wondered about the fact that money is even more immaterial than information, since if two pictures of one megabyte each will take up two megabytes in your hard disk, an accounting line of two million dollars will take up exactly the same space as an accounting line of one million dollars in the banker's hard disk. To terminate this remark about money, just imagine the skeptics' faces when their monthly salary is delayed by just a few days.

Defining information provides the opportunity for a little game. At the start of my course, I always ask my students to give examples of what is information, and examples of what is not.

In the 2016 session, I got the following answers (they were nothing special, but they survived in some of my notes):

What is information? A visa card PIN code, a prediction, a sentence in a text, the last slide of this class, a sequence of numbers, a weather report, an image, a Master's course.

What is not information? Chance, a potato field, a student, emptiness, a T-shirt, a number, a character, the wind.

It's immediately apparent that while information is mostly immaterial, its medium is not, and almost all the elements listed as non-information could serve as a medium or support for information, even including the potato field.

1.3   The Information Pyramid

In order to discuss the above assertion, we must look at the following pyramid, see Figure 1.4. From bottom to top:

- **The physical signals** which are space-time events and constitute the ground foundations of the theory, they are the physical supports of information, for example, the signals associated with the variations of the electromagnetic field;
- **The data**, a numerical measure or a collection of symbols carried by the physical signal, for example, the intensity and phase of the measured electromagnetic field;
- **The information** *per se* which connect data, for example, the codewords extracted from the data;
- **The knowledge**, which connects several sources of information to create language, and entitles generalizations and predictions, for example, several sequences of codewords can describe the theory of gravity.

Now we can review each of the item in the list "**what is information?**":

> **a visa card PIN:** This is data, unless we go to more specific "this sequence is the PIN of this card";

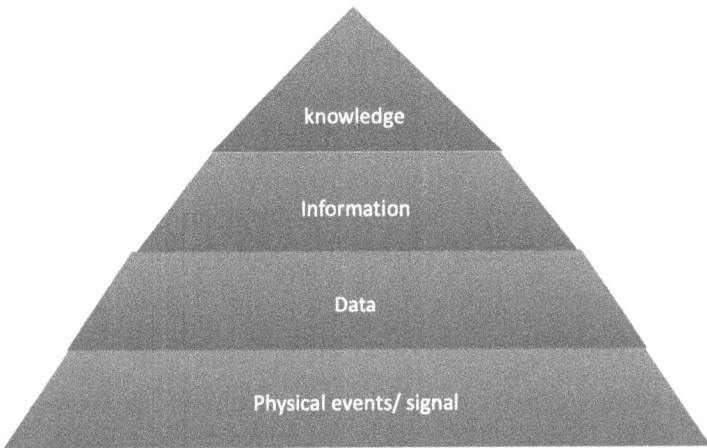

Fig. 1.4. The information pyramid.

a prediction: this is an information if we limit our considerations on the text which describes it. However, this is also an element of knowledge which enables the possibility of prediction;

a sentence in a text: an information;

the last slide: this is a space time event (but can support information);

a sequence of numbers: these are data, unless we add, "this sequence of numbers is written on this wall";

a weather bulletin: an information;

a picture: a physical event;

a Master class: a physical event;

the zero site: no idea what it can be.

From the list "**what is not information?**":

randomness: knowledge, being is a mathematical concept;

a potato field: an extended physical event;

a student: a physical event, but "this person is a student" is an information;

the vacuum: knowledge, this is a physical concept;

a T-shirt: a physical event;

a number: data;

a character: data, but information if we complement by "this character is in this text";

the wind: knowledge issued from meteorology, but "there is wind today" is information.

Let's take the example of a teacher dropping a chalk on the floor (see Figure 1.5). The situation can be viewed from three angles:

Data: The chalk is on the floor; it's simply the image of a spatio-temporal event.

Information: What happened to the chalk? It's now on the floor; this links the fact that the teacher dropped the chalk and the fact that the chalk hit the floor.

Knowledge: Law of gravity; it can be predicted that if the teacher or someone else drops a piece of chalk or something else, on Earth or any other planet, the object will fall to the ground.

In general, a specific piece of information can be seen as the answer to a sequence of questions. For example, information about a falling chalk can be described via the sequence of questions listed in Figure 1.6. Since any question can be divided into a sequence of binary questions, the answers to which are either "yes", "no" or, by

Fig. 1.5.   The chalk falls on the floor.

Fig. 1.6. The chalk information decision tree.

convention, "0" or "1", information can be described by a sequence of bits.

1.4 Basic Information Paradoxes Play Ground

In this section, we give a number of examples to illustrate the different states of information elements. They can be used to entertain students.

1.4.1 *The paradox of the halted clock*

Simply show a stopped clock and imagine the following dialog:

WHO1: Your watch is useless, why do you keep it?
WHO2: My watch is accurate twice a day, yours is constantly early or late and will never be accurate.
WHO1: But you'll never know when your watch will show the right time!
WHO2: That's not true: I can tell when my watch is accurate. It's exactly at 5:32! [see Figure 1.7]

This far-fetched dialog reflects an obvious ambiguity. When who1 says that the watch is accurate at 5:32, who1 is simply saying that the data displayed by the watch is 5:32 but not that the watch will

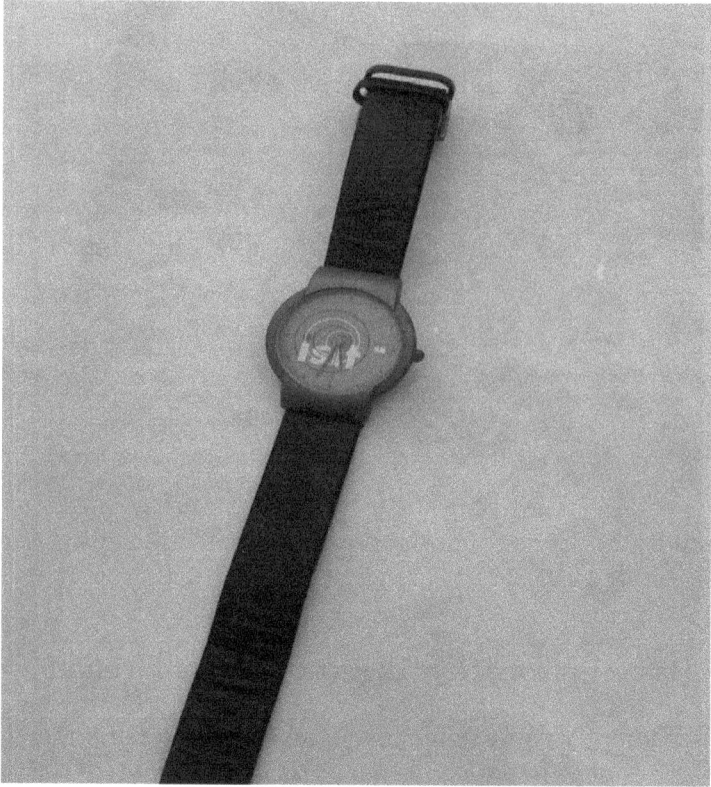

Fig. 1.7.   This halted watch is accurate at 5:32 twice a day.

send a particular signal at 5:32. This is the difference between a signal, i.e. a spatio-temporal event, and a datum. The data 5:32 is not information because it is not linked to any other data or event. An actual information could be "it's exactly 5:32 now".

This consideration is rather hollow in itself, but it could be entangled with many other non-trivial situations, described as follows.

1.4.2   *The exam week paradox*

In a classroom, the teacher tells his students: "There will be an exam day next week, scheduled in such a way that you'll never see it coming even the day before". A student raises his hand: "Madam/Sir, there will be no exam day. In fact, it can't be on Friday because, being

Fig. 1.8. The exam week paradox ("une salle d'examen").
Courtesy: Wikipedia.

the last day of the week, we could predict that it's exam day on Thursday evening. It can't be on Thursday either, because as Friday is forbidden, we could say that Thursday is the last possible date. And so on, it can't be Wednesday, Tuesday or Monday. So no exam day next week".

But the exam took place on Tuesday, and none of the students could predict it (see Figure 1.8). This is a classic paradox. It seems to be based on an apparent confusion of language: How can a prediction exist if its existence cancels out the object of the prediction? This would be equivalent to the self-contradictory phrase "I'm lying". If I am a liar, the statement "I lie" would mean the truth, which is not possible. If I'm telling the truth, then "I'm lying" would be a lie.

Information theory does not deal with semantic contradictions. So, to remove the ambiguity, we replace the verb "to predict" with the verb "to say". You can say, but it doesn't necessarily mean that a prediction will prevent the review, it just tells that you can say it. With this change, the student's reasoning leads to the following sequence of events: On Thursday, the student will say "the exam is for tomorrow", on Wednesday, (s)he will say "the exam is for tomorrow", etc. At each evening of the week, the student will say "the exam is for tomorrow". In this way, the student acts like a stopped clock: Every day, (s)he shows the same data "the exam is for tomorrow", but this is not information.

1.5   The Blue Eyes Paradox

It's again a classic paradox. On a tiny, isolated island, the native population has no mirrors. There's an absolute taboo: people's eye color (see Figure 1.9). To have blue eyes is such a disgrace that the island's powerful cultural tradition dictates that anyone who discovers one day that (s)he have blue eyes must commit suicide that very evening by jumping from the top of the island's highest cliff, called the Blue Eyes Cliff.

As there are no mirrors on the island and no one will dare tell someone that they have blue eyes, there's no way for them to realize that they actually have blue eyes.

It turns out that actually 12 people on the island have blue eyes. That's how many they are on the day a missionary lands on the island. As this is a matter of curiosity, the entire population of the island gathers in the village square to welcome him. Little aware of the island's customs, at one point the missionary exclaims: "There are blue eyes among you!". As this statement falls into the category of the island's highest taboos, the missionary is immediately eaten.

Fig. 1.9.   The blue eyes paradox.

Twelve days later, all the blue-eyed inhabitants jump off the blue-eyed cliff.

How did the blue-eyed people come to this conclusion? We assume that all the island's inhabitants know each other (so they know who has blue eyes and who doesn't, except for themselves).

At this point, we can play the following game with the students. We take a collection of red hats and blue hats. An innocent person (the teacher) places a cap of an arbitrary color on each student's head so that none can see his or her own color. Students observe each other, and the teacher plays the role of missionary (careful not to get eaten). The teacher begins to list the virtual days, each time asking students who feel they are wearing a blue hat to raise their hand. Repeat several times, changing hats. Generally speaking, all students quickly understand the trick.

The case where there is only one blue hat can be a key moment. The student wearing the only blue hat sees only red hats in front of him. As the teacher asserts that there are blue hats, the student immediately understands that it's him and immediately raises his hand. The case of the two blue hats helps advance their thinking. Each student with a blue hat sees only one other blue hat. After the first virtual day, the student realizes that the assumption that the other blue hat is unique turns out to be wrong, since he didn't immediately raise his hand, so he must have a blue hat too. So, both blue hats raise their hands on the second day.

The general solution to the problem is recursion. For a total population of $n - 1$ blue-cap holders, there will be $n - 1$ days of silence, and on the nth day, the blue-cap holders will jump off the cliff, as they understand that each of the $n - 1$ other blue-cap holders should have raised their hand the previous day if they indeed constituted the entire blue-cap population of the classroom.

This brings two important remarks:

Note 1. Silence is information. The fact that nothing happened for the first 11 days brings important information to the 12 blue-eyed ones.

Note 2. The missionary's statement brings no new information, since everyone on the island knows there are blue eyes, and everyone knows that everyone knows.

What, then, is the exact characterization of the missionary's statement? In fact, the missionary's declaration is not a piece of information but a *signal*, a spatio-temporal event that triggers the fatal countdown for blue-eye owners.

The fact that such a countdown was not previously possible stems from the fact that no one on the island had been able to establish an unspoken consensus value for the minimum number of blue-eyed people (i.e. a number known to all as indisputable common knowledge).

An interesting alternative to the missionary's statement would be "there are at least 12 blue-eyed people on this island". In this case, it's a different initial signal: All the blue-eyed people would jump the Blue-Eyed Cliff that very first evening. Alternatively, the missionary could have said "there are at least k blue-eye holders". In that case, all n ($n \geq k$) blue-eyed people would jump on the $n - k + 1$th night.

If the missionary were a liar, he could have declared a number k greater than n. If $k = 1$ and $n = 0$, then the entire population of the island would jump off the cliff. If $n \geq 1$ and $k = n + 1$, then everyone without blue eyes would jump off the cliff, but not the blue-eyed ones, who would realize that the missionary is a liar, because even adding themselves individually wouldn't compensate for the k declared by the blue-eyed ones, and they would have the compensation of eating the missionary. If $k > n + 1$, no one would move and we'd eat the missionary, who would ultimately have to stay at home.

1.6   Exercises

The main problem is in the fact that students are generally very clever and they quickly exhaust the games explained above. Following are exercises which look similar but need some more thinking.

1.6.1   *The 100 prisoners problem*

A prison holds 100 prisoners. One day, the bored warden comes up with an amusing idea. He gathers the 100 prisoners in a single room, one wall of which is fitted with a lever. The lever doesn't do anything, it just moves up and down. The guard explains that, from now on, he will take the prisoners one by one in an arbitrary order (with arbitrary repetitions) and at an arbitrarily variable frequency to show them around the room. Each visit to the room is secret from the other prisoners; the only trace the prisoner can leave by visiting the room is whether or not to activate the lever. The process can take an eternity, but it's fair, so every prisoner is guaranteed to pass through at least once, in fact an infinite number of times if the process never stops.

The guard issues the following challenge: Any prisoner can stop the process if he or she claims that all prisoners have visited the room at least once. If no prisoner stops the process, the prisoners are detained for life. If a prisoner stops the process and is right, all the prisoners are released. If he's wrong, they stay in the prison forever.

The guard leaves the 100 prisoners in the room for 30 minutes to agree on a strategy, before escorting them back to their cells (individual, isolated cells). Thereafter, the prisoners cannot communicate with each other, except by activating the lever when they return to the room individually.

Exercise 1

What is the prisoners' strategy for stopping the process without failing? See the Answer of Exercise 1 in Chapter 9.

1.6.2 *The 100 prisoners again*

The same 100 prisoners. They may have failed the previous test, or worse, passed it, and their keeper is still resourceful in cruelty. He gathers the prisoners back into the room and announces: "You will be released, but I have not said whether you will leave the prison alive. As a final challenge, I'll take you back to your cells, and to each of you I'll stick a card bearing an arbitrary number between 1 and 100 on your forehead so that you can't see it (no mirror). Then I'll bring you back into the room so you can see the numbers of the other inmates, but you won't be able to communicate. Then I'll take you back to your cell and ask you separately to tell me a number. If at least one of you tells me the number on your forehead, then you're all released unharmed, otherwise you're all executed. But first,

I'll leave you alone for 30 minutes to discuss a strategy amongst yourselves".

Exercise 2

What is the strategy to ensure that the prisoners are released alive? The numbers they carry on their foreheads don't all have to be different, they can be repeated or even all be equal to each other.

See the Answer of Exercise 2 in Chapter 9.

Chapter 2

The Basic Mathematics Inside Information Theory

Abstract

This chapter introduces the first steps in the "mathematization" of information theory by discussing the concept of entropy. In its simplest definition, without any reference to probability theory, the entropy of a system is nothing more than the binary logarithm of the number of distinct states of that system, i.e. the number of bits needed to name each of the system's state and to differentiate from the other states. We call it "deterministic entropy" because it doesn't refer to any probabilistic framework, which usually makes it very difficult for beginners to approach the theory. Indeed, deterministic entropy assumes that all system states have the same weight. Nevertheless, this didactic approach already allows us to go very far into the theory, right up to the concept of the capacity of a communication channel.

2.1 Information Theory

Information theory is an elegant theory whose aim is to describe the quantity of information that a given medium can carry. Examples include a newspaper, a memory chip in a computer or a telecommunications system. It can also be used to describe the internal complexity of a system, such as a machine or a file system. In other words, it answers the following question: "What is the minimum number of bits I need to describe a piece of information?". This theory was initiated in 1948 by Shannon and then a researcher at Bell Laboratories [105]. There is a more dynamic version of information theory, called Kolmogorov complexity [16], which essentially answers the question "what is the size of the minimal binary automaton that can mimic the behavior of a system?". When the system consists of a simple information carrier with no special computer processing, Kolmogorov's and Shannon's theories coincide. In what follows, we only deal with Shannon's theory about static information.

2.2   The Entropy and Its Properties

The entropy is a way to measure the quantity of information contained by a system Z. Let $\mathcal{S}(Z)$ be the set of all possible states of Z.

The deterministic entropy $h^D(Z)$: We set $h^D(Z) = \log_2 |\mathcal{S}(Z)|$, where $|\mathcal{S}(Z)|$ is the number of elements in $\mathcal{S}(Z)$. If we enumerate the various states from 1 to $\mathcal{S}(Z)|$, then to identify an arbitrary state, say state number i, we can write the number i in a binary representation that has $\lceil \log_2 |\mathcal{S}(Z)| \rceil$ bits.

If we want to write i with tribits (bits in base 3), the number of tribits needed to identify the state i will be $\lceil \log_3 |\mathcal{S}(Z)| \rceil$. In other words, if we want to identify a state of Z using a finite alphabet $\mathcal{A}$ of size $|\mathcal{A}|$, then the minimum number of symbols to describe a state is closely less than $\frac{h^D(Z)}{\log_2 |\mathcal{A}|}$.

If, for example, Z is the set of all numbers between 0 and $N-1$, then $h^D(Z) = \log_2 N$. If Z is the set X^n of all binary words of length n, then $|\mathcal{S}(X^n)| = 2^n$ and $h^D(X^n) = n$: exactly n bits are needed to describe an arbitrary sequence of n binary symbols. If X^n is the set of all sequences written with n symbols taken from the alphabet $\mathcal{A}$, then $h^D(X^n) = n \log_2 |\mathcal{A}|$.

Let n_0 and n_1 be fixed integers, such that $n_0 + n_1 = n$. Let X^{n_0,n_1} be the set of balanced binary sequences x^n that have exactly n_0 "0" and n_1 "1". There are exactly $\binom{n}{n_0} = \frac{n!}{n_0!n_1!}$ of such sequences, so

$$h^D(X^{n_0,n_1}) = \log_2 \binom{n}{n_0} = \log_2(n!) - \log_2(n_0!) - \log_2(n_1!). \quad (2.1)$$

Theorem 2.1. *If the system $f(Z)$ is an image of the system Z by some function f, then $h^D(f(Z)) \le h^D(Z)$; we have equality when f is bijective.*

Proof.   It comes from the fact that $|\mathcal{S}(f(Z))| \le |\mathcal{S}(Z)|$. $\square$

2.3 Dependent Components

Let Z be a system composed of two components X and Y: $Z = (X, Y)$. By this we mean that each state z is of the form (x, y), where x is a state of X and y is a state of Y. If Z is a watch, X can be the internal mechanism and Y the display. In other words, $\mathcal{S}(Z) \subset \mathcal{S}(X) \times \mathcal{S}(Y)$.

Theorem 2.2. *We have the inequality*

$$\max\{h^D(X), h^D(Y)\} \leq h^D(X, Y) \leq h^D(X) + h^D(Y). \quad (2.2)$$

Proof. Since the system X is indeed the projection of the system (X, Y) on its first component, thus the result of the mapping $f(X, Y) = X$, we have $h^D(X) \leq h^D(X, Y)$ and similarly $h^D(Y) \leq h^D(X, Y)$. In other words, the first inequality comes from the fact that $\max\{|\mathcal{S}(X)|, |\mathcal{S}(Y)|\| \leq |\mathcal{S}(X, Y)|$.

Similarly, since $\mathcal{S}(X, Y) \subset \mathcal{S}(X) \times \mathcal{S}(Y)$ and that $|\mathcal{S}(X) \times \mathcal{S}(Y)| = |\mathcal{S}(X)|.|\mathcal{S}(Y)|$ and therefore $\log_2(|\mathcal{S}(X)|.|\mathcal{S}(Y)|) = h^D(X) + h^D(Y)$, we have the inequality

$$h^D(X, Y) \leq h^D(X) + h^D(Y). \qquad \square$$

The second inequality becomes an identity only when $\mathcal{S}(Z) = \mathcal{S}(X) \times \mathcal{S}(Y)$. In this case, the components are said to be independent. For example, in a digital watch, the hour and minute indications are two independent components. In an analog watch, this is no longer the case where the two hands move in continuity. When the components are not independent, there are $x \in \mathcal{S}(X)$ and $y \in \mathcal{S}(Y)$ such that $(x, y) \notin \mathcal{S}(Z)$. For example, when it's 10:50, the short hand (indicating the hour) is closer to the 11th marker than to the 10th marker, and the opposite is impossible unless the long hand (indicating the minutes) is before marker 30. Figure 2.1 shows the breakdown of a mechanical watch. The Z system consists of a short hand, a long hand and a display. If we consider only discrete moves for the hands, we have a display with 12 states for the short hand and 60 states for the long hand. If we include the internal mechanism, which would have around 1,000 states, we get $|S(Z)| = 720,000$. If Z is a car, the car's front wheel and steering wheel are another example of a dependent component, since there is no global state where the front

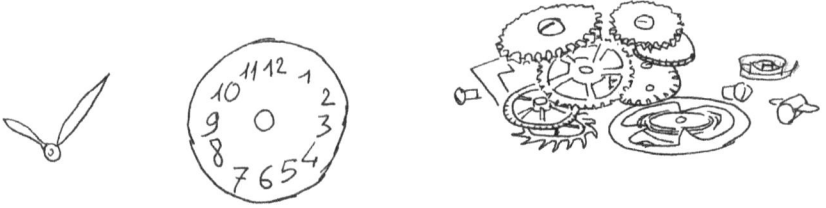

Fig. 2.1.   The watch system with 720,000 states.

wheel is oriented to the left with the steering wheel turned to the right (in normal operating mode).

2.4   Example 1: The Additive Noise

Suppose $Z = (X, Y)$ where X and Y are two systems made of integers: X are integers between 0 and $N - 1$ and Y are integers between 0 and $N + B - 1$. To simplify, we identify X and Y with the integers. We suppose the constraint that $Y = X + \beta$ with β between 0 and $B - 1$. Clearly, X and Y are dependent and $h^D(X) = \log_2 N$, $h^D(Y) = \log_2(N + B - 1)$.

Note that X and $Y - X$ are independent since $S(X) = \{0, \ldots, N - 1\}$ and $S(Y - X) = \{0, \ldots, B - 1\}$ and $S(X, Y - X) = \{0, \ldots, N - 1\} \times \{0, \ldots, B - 1\}$, so $h^D(X, Y - X) = \log_2 N + \log_2 B$. We can consider the variable $Y - X$ as additive noise on a transmitted signal X which gives a received signal Y. Since the two systems $(X, Y - X)$ and (X, Y) are in bijective correspondence, $h^D(X, Y) = h^D(X, Y - X) = \log_2 N + \log_2 B < h^D(X) + h^D(Y) = \log_2 N + \log_2(N + B - 1)$.

2.5   Example 2: The Seven Error Game

We consider that a system $Z = (X^n, Y^n)$ is made of two components X^n and Y^n, where X^n and Y^n are binary sequences. We assume that $(x^n, y^n) \in \mathcal{S}(Z)$ if and only if x^n and y^n are binary sequences which differ on exactly n_1 bits. In other words, if $x^n \oplus y^n$ denotes the bitwise addition *modulo* 2 of both sequences (XOR $\oplus$), $x^n \oplus y^n \in X^{n_0, n_1}$

Fig. 2.2.   The seven error game ($n_1 = 7$).

($n_0 = n - n_1$). See Figure 2.2 for a free illustration with $n_1 = 7$ (but errors are not bitwise, they are just classical drawing errors).

We can see X^n as a codeword sent by a source and Y^n received by a destination through a channel which corrupts arbitrarily exactly

n_1 bits. It is easy to prove that $h^D(X^n) = h^D(Y^n) = n$. But we also have $h^D(X^n, Y^n) = h^D(X^n) + h^D(X^{n0,n1}) = n + \log_2\binom{n}{n_1}$. Indeed, we have a bijection between (X^n, Y^n) and $(X^n, X^n \oplus Y^n)$, therefore $h^D(X^n, Y^n) = h^D(X^n, X^n \oplus Y^n)$, but in this case, $X^n \oplus Y^n = X^{n0,n1}$ and the components $(X^n, X^{n0,n1})$ are independent which leads to $h^D(X^n, X^{n0,n1}) = h^D(X^n) + h^D(X^{n0,n1}) = n + \log_2\binom{n}{n_1}$.

2.6 Conditional Entropy, Channel Entropy

The following quantity $h^D(X, Y) - h^D(X)$ is always positive and is called the conditional entropy of Y with respect to X and is denoted $h^D(Y|X)$:

$$h^D(Y|X) = \log_2 \frac{|\mathcal{S}(X,Y)|}{|\mathcal{S}(X)|}.$$

This is the logarithm of the average number of states (x, y) in $\mathcal{S}(X, Y)$ for a given state x in $\mathcal{S}(X)$ when averaged over all x. In other words, it measures the degree of freedom of the variable Y for a fixed value X. For example, if X is the aiming vector of a firearm and Y is the location of the bullet's impact in a paper target, the quantity $h^D(Y|X)$ measures the dispersion of the bullet's impact for a given aim. Dispersion can be due to the quality of the powder, the mobility of the air through which the bullet travels, etc. In other words, $h^D(Y|X)$ can be thought of as the entropy added by the propagation medium from an arbitrary, fixed initial state of the transmitter X to the receiving state Y. The quantity $h^D(Y|X)$ is sometimes called the channel entropy, if we consider the states X and Y to be the two ends of a telecommunication channel.

We have $h^D(Y|X) \leq h^D(Y)$ since $h^D(Y)$ measures the global dispersion of the bullet impact Y considering all the aiming positions X. We leave as an exercise a more formal proof of the inequality.

If we take again the example of additive noise where $Z = (X, Y)$ with X being all numbers between 0 and $N-1$ and $Y - X$ being all numbers between 0 and $B-1$, then $h^D(Y|X) = h^D(X,Y) - h^D(X) = \log_2 B$: there are B different Y for a given X.

We can also apply this formula to the seven error games, namely the system $Z = (X^n, Y^n)$, where $(x^n, y^n) \in \mathcal{S}(Z)$ if and only if

$x^n \oplus y^n \in X^{n_0,n_1}$ the number of possible y^n for a given x^n is exactly $|\mathcal{S}(X^{n_0,n_1})|$ we can conclude that

$$h^D(Y^n|X^n) = \log_2 \binom{n}{n_1}.$$

2.7 Mutual Information, Channel Capacity

When X and Y are non-independent components of a system, then the quantity $h^D(X) + h^D(Y) - h^D(X,Y)$ is strictly positive. It is essentially the difference between the system with independent components and the system with dependent components. This quantity is called mutual information and is denoted $I(X,Y)$. It measures the quantity of dependencies (measured in information quantities) shared by systems X and Y:

$$I(X,Y) = h^D(X) + h^D(Y) - h^D(X,Y).$$

It is also referred to as channel capacity, where X is the element (codeword/signal) to be transmitted at the transmitter end and Y is the physical effect measured at the receiver end. In fact, we have

$$I(X,Y) = h^D(Y) - h^D(Y|X).$$

Conceptually, it's the difference in *net* information received when we subtract from the total entropy $h^D(Y)$ received by Y, the entropy $h^D(Y|X)$ created by the channel (considered as noise).

Returning to the example of additive noise where $Z = (X,Y)$ with X representing all numbers between 0 and $N-1$ and Y all numbers between X and $X + B - 1$, the passage through the communication channel consists in the simple addition to X of an arbitrary number between 0 and $B-1$ and thus $I(X,Y) = \log_2(N+B-1) - \log_2(B) = \log_2\left(1 + \frac{N-1}{B}\right)$.

Remark. The mutual information is symmetrical, i.e. $I(X,Y) = I(Y,X)$ since $h(Y,X) = h(X,Y)$. This can lead to apparent reversals of causality if care is not taken (a specific chapter is devoted to this subject). However, the channel entropy $h(Y|X)$ is *not* symmetrical. In general, $h^D(Y|X) \neq h^D(X|Y)$, but when $h^D(X) = h^D(Y)$, we have the identity $h^D(Y|X) = h^D(X,Y)$, then we say that the channel is *reversible*.

2.8　Information Theory and Telecommunication

In the previous section, we used the words "channel capacity" to name certain parameters built with the abstract quantities we have defined in the context of information theory. But this gives no clue as to how this quantity relates to telecommunications, i.e. to the practical problem of sending and receiving information. We see that, in fact, under certain general conditions, channel capacity measures the greatest amount of information the channel can carry. We first focus on the deterministic framework of information theory, which makes this chapter readable without any prerequisites on probability theory. We use the following chapter to tackle the richer aspects of probabilistic information theory.

We return to the two-component system $Z = (X, Y)$, where X is the signal to be sent by the transmitter and Y is the signal received by the receiver. From now on, X will be called the *codeword*. The Z system is a telecommunication system (see Figure 2.3).

Suppose the source wants to send information Q, which may be contained in a binary sequence or in an integer. The source will translate Q into an encoded word X via the encoding function $e(.)$: $X = e(Q)$. The channel distorts X due to interference or noise into Y. The destination decodes Y by applying a decoding function $d(.)$. Transmission is successful when $d(Y) = Q$ and fails when $d(Y) \neq Q$.

In the deterministic information theory framework, we want a perfect code, i.e. one that never fails. To obtain a perfect code, we need a set $\mathcal{W}$ consisting of k codewords $x_i \in \mathcal{S}(X)$ such that the sets $\mathcal{S}(Y(x_i))$ are disjoint. We denote by $Y(x)$ the system defined by $\{y, (x, y) \in \mathcal{S}(X, Y)\}$, and $\mathcal{S}(Y(x))$ is the dispersion set of x. In this case, encoding consists in mapping the information Q to a given x_i and, assuming that y is the receive code, decoding consists in identifying the codeword x_i such that $\mathcal{S}(Y(x_i))$ contains y.

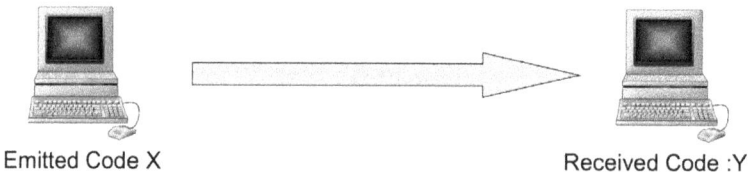

Emitted Code X　　　　　　　　　　　　　　　Received Code :Y

Fig. 2.3.　The information pair $Z = (X, Y)$ for a telecommunication system.

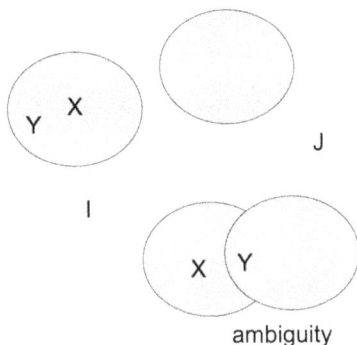

Fig. 2.4.   Decoding ambiguity when the codeword I and J dispersion area overlap.

Since the dispersal sets $\mathcal{S}(Y(x_i))$ are disjoint, there is no ambiguity (see Figure 2.4). A necessary condition is that

$$\sum_{i=1}^{i=k} \mathcal{S}(Y(x_i))| \leq |\mathcal{S}(Y)|.$$

The number of distinct pieces of information is k, or in other words $h^D(\mathcal{W}) = \log_2 k$. The "decodable" capacity of the channel is at least $h^D(\mathcal{W})$.

Theorem 2.3 (Shannon half theorem). *Under some conditions, the maximum decodable capacity of the channel $h^D(\mathcal{W})$ is upper-bounded by $I(X, Y)$.*

We name this theorem the Shannon *first half* theorem, since it addresses the upper bound of the channel decodable capacity.

Figure 2.5 shows three illustrations of the theorem. From top to bottom: The first illustration is an image X made up of 12,000 pixels (so $h(X) = 12k$) transmitted on a channel and received as a gray square Y has a value of zero pixels ($h(Y) = 0$) whatever the image sent X ($h(Y|X) = 0$). The channel capacity $I(X, Y)$ is therefore zero.

In the second illustration, the received image Y is the exact opposite of the sent image X: $Y = \bar{X}$. Thus, $h(Y) = 12k$ and $h(Y|X) = 0$ since there is no dispersion of the received image Y for a fixed transmitted image X. So $I(X, Y) = 12k$, and the perfect code is to invert the received image.

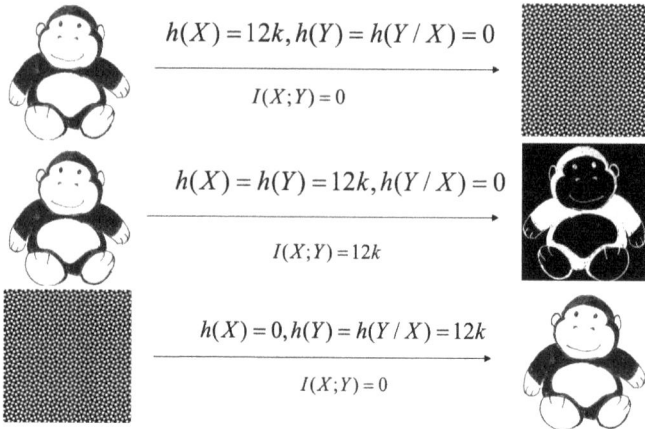

Fig. 2.5. Three illustrations of the Shannon theorem.

The third illustration is the inversion of the first: The transmitter transmits nothing, symbolized by a gray square ($h(X) = 0$). The receiver displays an image of 12,000 pixels ($h(Y) = 12k$). Despite a very nice image at the receiver, the capacity $I(X, Y)$ is zero. It's like a screen saver on a computer, which in no way reflects the device's internal activity.

2.8.1 *Perfect codes*

In the deterministic framework, the theorem is not always true if perfect codes are required, but it is nevertheless valid in certain fairly general cases. Let's take a look at some of these cases. For example, the theorem remains valid when the dispersion sets are uniform over all codewords. By uniform dispersion, we assume that for any $x \in \mathcal{S}(X)$ the quantities $|\mathcal{S}(Y(x))|$ are identical and all equal to $\frac{|\mathcal{S}(X,Y)|}{|\mathcal{S}(X)|} = h^D(Y|X)$. This is the case with the seven-error games, when $|\mathcal{S}(Y(x))| = |\mathcal{S}(X^{n_0,n_1})|$. The maximum value of k is therefore $|\mathcal{S}(Y)|$ divided by $\frac{|\mathcal{S}(X,Y)|}{|\mathcal{S}(X)|}$ and therefore

$$h^D(\mathcal{W}) \leq \log_2 \frac{|\mathcal{S}(X)| \times |\mathcal{S}(Y)|}{|\mathcal{S}(X,Y)|} = I(X,Y).$$

For example, in the case of additive noise, the codeword x is an integer between 0 and $N - 1$ and the channel adds to x an integer between 0 and $B - 1$, then an obvious encoding is to take the codewords $x_i = e(i) = (i - 1)B$ for i from 1 to $k = \lceil \frac{N}{B} \rceil$. All codewords are spaced at least B units apart, and the decoding function is $d(y) = \lfloor \frac{y}{B} \rfloor$ which is a perfect code. The code is perfect, but the maximum value of the decodable capacity is $\log_2 k = \log_2 \lceil \frac{N}{B} \rceil$, which is smaller (but not by much) than the maximum value of the decodable capacity, which is $\log_2 \lceil (1 + \frac{N-1}{B}) \rceil) = I(X, Y)$. Note, however, that $\frac{k}{1 + \frac{N-1}{B}} \to 1$ when $N, B \to \infty$.

This example is interesting because it can be linked to the world of wireless communication technologies (see Figure 2.6). Indeed, X can be the amplitude of a signal sent by the transmitter and Y the amplitude of the signal received by the receiver after an arbitrary noise whose maximum amplitude is B. Our previous analysis shows that capacity is close to $\Phi \log_2(1 + \frac{N}{B})$ expressed in bits per second, if Φ is the frequency at which signals are sent, or the *bandwidth*. In simplified terms, this is the Shannon–Hartley wireless capacity [50]. In this case, $\frac{N}{B}$ is exactly the signal-to-noise ratio (SNR). Remarkably, although Φ and the SNR are physical quantities, the expression of the resulting capacity borrows nothing from physics and is merely the consequence of simple mathematical considerations.

Theorem 2.4 (Shannon second half theorem). *Under certain mild conditions, when $I(X, Y)$ tends toward infinity, the decodable capacity of the channel tends to be equivalent to $I(X, Y)$.*

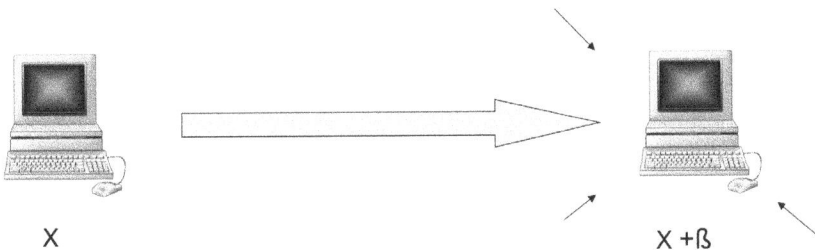

X   X +ß

Fig. 2.6.   The additive noise for wireless transmission.

This theorem is the "Holy Grail" of telecommunications, as error-correcting code experts exhaust themselves trying to scrape a few percent off the Shannon limit. We've already proved the theorem for additive noise, since when $\frac{N}{B} \to \infty$, then $\log_2(\lceil \frac{N-1}{B-1} \rceil) \sim \log_2(1 + \frac{N-1}{B})$. In fact, to obtain a perfect code, all that's needed is for the two-hop dispersion sets $\mathcal{S}(X(Y(x_i)))$, which we call $\mathcal{S}(X^2(x_i))$ for simplicity, to be pairwise disjoint.

2.9   Exercises

2.9.1   *Shannon half theorem: Counterexample*

$X = \{x_1, \ldots, x_{2n}\}$ and $Y = \{y_1, \ldots, y_{2n}\}$. $\forall i \leq n$: $Y(x_i) = \{y_i\}$, $\forall i > n$ $Y(x_i) = \{y_{n+1}, \ldots, y_{2n}\}$.

Exercise 3

Evaluate $I(X, Y)$. See the Answer of Exercise 3 in Chapter 9.

2.9.2   *Second half theorem: Counterexample*

Let $X = \{x_1, \ldots, x_n\}$ and $Y = \{y_1, \ldots, y_{n+1}\}$. $\forall i \leq n$: $Y(x_i) = \{y_i, y_{n+1}\}$.

Exercise 4

Evaluate $I(X, Y)$.
   See the Answer of Exercise 4 in Chapter 9.

2.9.3   *Second half theorem: Generalities*

Definition for $x \in X$: $X^2(x) = X(Y(x))$ is the two-hop neighborhood of x.

   Assume $\{x_1, \ldots, x_k\}$ is a maximal set of correct codewords. Show that $\sum_{i=1}^{i=k} |X^2(x_i)| \geq N(X)$.

   Assume A such that $\forall x \in X$: $|X^2(x)| = A$. Then $\log_2 k \geq h(X) - \log_2(A)$.

Exercise 5

Evaluation of A. Show that $\sum_{x \in X} |X^2(x)| = \sum_{y \in Y} |X(y)|^2$.

2.9.4 *Seven errors game*

Exercise 6

By using the Stirling formula valid on each natural number $n! \sim \sqrt{2\pi n} n^n e^{-n}$ shows that $I(X^n, Y^n) \sim (1 + p_0 \log_2 p_0 + p_1 \log_2 p_1)$.

Exercise 7

Show that n_1 and $n - n_1$ are symmetric. Consider that $\theta \in Z_k^n$, where Z_k^n is the set of binary sequences of length n with at most k 1s. Show that when $n_1 \geq n/2$ we only have $k = 1$. See the Answer of Exercise 7 in Chapter 9.

2.9.5 *One error game*

Exercise 8

Compute $I(X, Y)$. See the Answer of Exercise 8 in Chapter 9.

2.9.6 *Error correction*

Exercise 9

Describe an error correcting algorithm. See the Answer of Exercise 9 in Chapter 9.

2.9.7 *Unperfect codes*

This exercise requires deeper knowledge about probability theory and can be skipped, since this chapter is supposed not to depend on such a prerequisite. We assume that codewords are binary sequences of length n. We still use the fact that the set $\mathcal{W}$ is given but we don't make mandatory to have the $\mathcal{S}(Y(x_i)$ pairwise disjoint and it may happen that some dispersion sets overlap, if $\mathcal{S}(Y(x_i)) \cap \mathcal{S}(Y(x_j)) \neq \emptyset$

for some $i \neq j$. Therefore, the decoding function will fail if $y \in \mathcal{S}(Y(x_i)) \cap \mathcal{S}(Y(x_j))$ since there is no way to distinguish codeword x_i with codeword y_j. The goal is to minimize the decoding failure.

Assume we select the code set $\mathcal{W}$ at random. Let us assume that we want to send a codeword x_i and the dispersion set is $x_i \oplus \mathcal{S}(X^{n_0,n_1})$. We denote $x \oplus \mathcal{S}(X^{n_0,n_1}) = \mathcal{S}(x, n_1)$, the Hamming sphere of center x and radius n_1 (the set of sequence that differs from sequence x by exactly n_1 bits). Our goal is to give an upper bound of the probability of decoding failure, i.e. the probability that for an element $y \in \mathcal{S}(x_i, n_1)$ there exists an other integer j such that $y \in \mathcal{S}(x_j, n_1)$.

Exercise 10

We assume that $n_1 < pn$ for some $p < 1/2$.

Using the inequality

$$P(\exists j \neq i :: y \in \mathcal{S}(x_j, n_1)) \leq \sum_{j \neq i} P(y \in \mathcal{S}(x_j, n_1))$$

and the identity for all x

$$P(y \in \mathcal{S}(x, n_1)) = \frac{\binom{n}{n_1}}{2^n} = 2^{-I(X,Y)},$$

show that taking $k = 2^{I(X,Y)(1-\epsilon)}$ for any arbitrary $\epsilon < 1$ makes the failure probability tending to zero (in fact exponentially fast) when $n \to \infty$.

See the Answer of Exercise 10 in Chapter 9.

Chapter 3

Probabilistic Information Theory, the Paradoxes of Data Compression and Event Prediction

Abstract

This chapter extends information theory to probabilistic systems which, by being more general, allow a complete definition of entropy and capacity with much richer consequences. This leads to one of the most paradoxical applications of information theory: data compression and event predictors. The aim of compression is to reduce the medium of information to a strictly smaller medium without loss of information. This seems contradictory, as one might imagine that the process can be repeated until the medium is completely annihilated. We see how the probabilistic framework eliminates the paradox. Most compression algorithms exploit repetition patterns in the medium. Prediction algorithms work in the opposite way: From a visible fraction of the information carrier, called the *past information*, the algorithm is able to generate the missing part, called the *next information*.

3.1 The Probabilistic Entropy $h(Z)$

We now assume that the system Z described in the previous chapter still has $S(Z)$ as its set of feasible states, but with the difference that the actual states are obtained via a random processes. Let $z \in S(Z)$, and let $p(z)$ be the probability that "Z is in state z". Let $h(Z)$ be

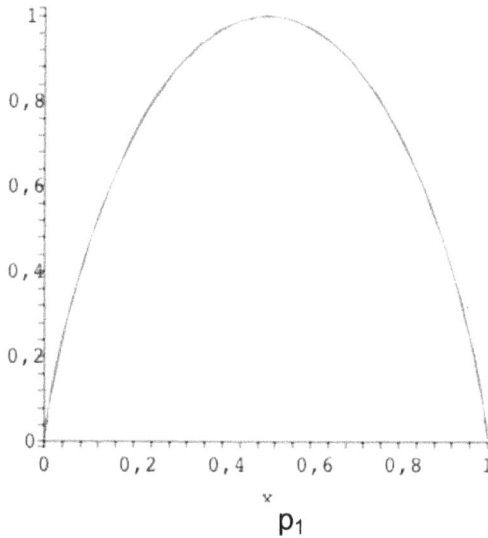

Fig. 3.1.   The entropy value of a two-atom probability distribution versus p_1.

the probabilistic entropy as follows:

$$h(Z) = \sum_{z \in S(Z)} p(z) \log_2 \frac{1}{p(z)}. \tag{3.1}$$

Figure 3.1 shows the value of entropy on a system with two states numbered 0 and 1. Let p_0 be the probability of state 0 and p_1 that of state 1 ($p_0 + p_1 = 1$). The probabilistic entropy is $p_0 \log_2 \frac{1}{p_0} + p_1 \log_2 \frac{1}{p_1}$. By convention, if there are states z such that $p(z) = 0$, we fix $p(z) \log \frac{1}{p(z)} = 0$, by continuity of the function $\log x$ where $\lim_{x \to 0} x \log x = 0$. However, these points are singularities and the slope of the entropy curve is steep at these points. This property is exploited in the lossy compression application known as rate distortion.

We note that $h^D(Z) = h(Z)$ when (and only when) the probability distribution on the states of Z is uniform: that is, $p(z) = \frac{1}{|S(Z)|}$ for all $z \in S(Z)$: $p(z) = \frac{1}{|S(Z)|}$. By the convexity argument of the function $x \log x$, we always have $h(Z) < h^D(Z)$ except in the uniform case where we have $h^D(Z) = h(Z)$.

Let $Z = X^n(p_0, p_1)$ be a binary Bernoulli source, i.e. the system consisting of binary sequences of length n where the bits are distributed independently and identically, with p_0 the probability of obtaining 0 and p_1 the probability of obtaining 1. Let x^n be one of its sequences with n_0 0 and n_1 1 ($n_0 + n_1 = n$), we have $p(x_n) = p_0^{n_0} p_1^{n_1}$. Since we have $\binom{n}{n_0} = \frac{n!}{n_0! n_1!}$, we obtain

$$
\begin{aligned}
h\left(X^n(p_0, p_1)\right) &= \sum_{n_0} \binom{n}{n_0} p_0^{n_0} p_1^{n_1} \left(n_0 \log_2 \frac{1}{p_0} + n_1 \log_2 \frac{1}{p_1} \right) \\
&= \sum_{n_0} \binom{n}{n_0} p_0^{n_0} p_1^{n_1} n_0 \log_2 \frac{1}{p_0} \\
&\quad + \sum_{n_0} \binom{n}{n_0} p_0^{n_0} p_1^{n_1} n_1 \log_2 \frac{1}{p_1} \\
&= n p_0 \log_2 \frac{1}{p_0} + n p_1 \log_2 \frac{1}{p_1}.
\end{aligned}
$$

Meanwhile, we still have the deterministic entropy which gives $h^D(X^n(p_0, p_1)) = n$ and therefore differs significantly from $h(X^n(p_0, p_1))$ when $p_0 \neq p_1$ because, although all possible sequences are feasible, some have a greater probability weight than others. Note that in general $h(X^n(p_0, p_1)) < h^D(X^n(p_0, p_1))$ except when $p_0 = p_1$, where we have equality.

Remark.   Generally speaking, information theory is strongly associated with probability theory and is mainly expressed in the probabilistic framework. For this reason, we have dropped any superscript in the "h" of the expression for the probabilistic entropy of Z. In any case, some results may be easier to understand when translated into the deterministic framework as we did in the previous chapter.

3.1.1   *Dependent components*

Let Z be a system made up of two components X and Y: $Z = (X, Y)$. In the probabilistic framework, we assume that X and Y are independent when for all $(x, y) \in \mathcal{S}(X, Y)$: $p(x, y) = p_X(x) p_Y(y)$, where p_X and p_Y are the probability distributions of states on $\mathcal{S}(X)$

and $S(Y)$, respectively. In this case, we have

$$h(X,Y) = \sum_{(x,y)\in S(X)\times S(Y)} p(x,y)\log_2 \frac{1}{p(x,y)}$$

$$= \sum_{(x,y)\in S(X)\times S(Y)} p_X(x)p_Y(y)\left(\log_2 \frac{1}{p_X(x)} + \log_2 \frac{1}{p_Y(y)}\right)$$

$$= \sum_{x\in S(X)} p_X(x)\log_2 \frac{1}{p_X(x)} + \sum_{y\in S(Y)} p_Y(y)\log_2 \frac{1}{p_Y(y)}$$

$$= h(X) + h(Y)$$

using the fact that $\sum_{x\in S(X)} p_X(x) = \sum_{y\in S(Y)} p_Y(y) = 1$.

It is easy to show that when the random variables X and Y are not independent, then $h(X,Y) < h(X) + h(Y)$. The components X and Y are said to be dependent.

3.2 Channel Coding, Error Correction

Channel coding is an important branch of information theory. Its aim is to design efficient coding and decoding functions on error-prone channels. The miracle of channel coding is to enable virtually error-free mass telecommunications over channels which, in routine operations, suffer substantial error rates. The system is so efficient that no judge on earth would accept a plaintiff's excuse that an e-mail arrived tainted by transmission errors. Reliable communication over an unreliable medium is at the root of the Internet's universal success.

Channel coding has two main components:

- **Protocol retransmission:** Data is transmitted in packets with sequence numbers. An unacknowledged packet triggers retransmission by the source via protocols integrated into several protocol layers. We deal with this part in the presentation of the Internet.
- **Forward error correction:** Data is transmitted in packets (of size n), and the packet contains redundant fields that enable errors

to be corrected without requiring the packet to be retransmitted. This is important for wireless transmission, where systematic errors would trigger an infinite cascade of retransmissions.

In this section, we focus on Forward Error Correction, leaving the protocol for packet acknowledgement and retransmission to other chapters.

3.2.1 *Seven error game revisited, Bernoulli channel*

A Bernoulli channel is a channel whose errors are reflected in a random change of bit values. Assume that the bit changes its value with probability p_1 or its value remains unchanged with probability p_0, independent of the original bit value. We consider that the transmitter sends a Bernoulli sequence of length n with uniform distribution $X^n(\frac{1}{2}, \frac{1}{2})$ and that the code received through the Bernoulli channel is $Y^n = X^n \oplus X^n(p_0, p_1)$ with $X^n(p_0, p_1)$ independent of X^n and $\oplus$ is the XOR operation.

We consider the system $Z = (X^n, Y^n)$ which we call the *Bernoulli channel*. We have $h(X^n) = n$, leaving as an exercise the proof that Y^n has the same distribution as $X^n(\frac{1}{2}, \frac{1}{2})$ (this does not mean that $X^n = Y^n$). Since the probabilities are uniform, the probabilistic and determinist entropies coincide and $h(Y^n) = n$. Following the previous analysis for the deterministic case, we have (X^n, Y^n) bijective with the system $(X^n, X^n \oplus Y^n)$ and the components are now independent: $h(X^n, Y^n) = h(X^n) + h(X^n \oplus Y^n) = n + h(X^n(p_0, p_1))$.

3.2.2 *Conditional entropy*

The probabilistic expression of $h(Y|X)$ is a little more complex than the deterministic expression. It involves the weight of conditional probabilities $p(y|x) = p(x, y)/p(x)$:

$$h(Y|X) = \sum_{x \in \mathcal{S}(X)} p(x) \sum_{y \in \mathcal{S}(Y)} p(y|x) \log_2 \frac{1}{p(y|x)}.$$

It is also equal to $\sum_{x,y \in \mathcal{S}(X,Y)} p(x, y) \log_2 \frac{1}{p(y|x)}$, but the first expression is the most interesting. It means that $h(Y|X)$ is in fact the average value over all x values of the entropy of (X, Y) when X is fixed at x. We could say $h(Y|X) = \sum_{x \in S(X)} p(x)h(Y|X = x)$ with the convention that $(Y|X = x)$ is the random variable Y seen under the condition $X = x$. Note that if $h(Y|X) = 0$, then for any $x \in S(X)$, such that $p(x) \neq 0$: $h(Y|X = x) = 0$, which means that the variable Y has no measurable statistic variations when X varies. In other words, there is a function f such that $Y = f(X)$, except perhaps for a non-measurable set of states.

By simple algebraic manipulations, it turns out that $h(Y|X) = h(X, Y) - h(X)$, which is the same expression as in the deterministic case.

In the Bernoulli channel example where $X^n = X^n(\frac{1}{2}, \frac{1}{2})$ and $Y^n = X^n \oplus X^n(p_0, p_1)$, we have for all x^n sequences $(Y^n|X^n = x^n) = x^n \oplus X^n(p_0, p_1)$, and therefore $h(Y^n|X^n = x^n) = h(X^n(p_0, p_1)) = np_0 \log_2 \frac{1}{p_0} + np_1 \log_2 \frac{1}{p_1}$.

3.2.3   *Typical sequences and Shannon theorem*

In this subsection, we assume that X and Y describe infinite random sequences over a finite alphabet and that X^n (resp. Y^n) is the portion of the sequence X (resp. Y) up to index n. Let $R_n = \frac{I(X^n, Y^n)}{n}$. Note that $R_n = 1$ when $p_1 = 0$.

Theorem 3.1. *Under certain mild conditions, if $R < \lim_{n \to \infty} R_n$, then there exists an encoding and decoding scheme that sends information at a rate nR, and the probability of decoding error tends to 0 when $n \to \infty$.*

If $R > \lim_{n \to \infty} R_n$, then any encoding–decoding scheme will fail with a probability tending toward at least $1 - \lim_{n \to \infty} R_n/R$ when $n \to \infty$.

Proof.   We're going to prove the theorem in the simplified situation of the binary Bernoulli channel with $p_1 < 0.5$ (if $p_1 > 0.5$, it's enough to invert the 0s and 1s in the received codewords). We choose the set of $\mathcal{W} = \{x_i\}$ codes at random from X^n. Suppose we want to send a codeword x_i, passing through the channel will result in the reception of $y = x_i \oplus X^n(p_0, p_1)$ at the other end.

Let's take $r = \lceil np_1(1 + \epsilon') \rceil$ for a certain $\epsilon' > 0$. If $x \in S(X^n)$ and $r > 0$, we denote $\mathcal{B}(x, r)$ the Hamming ball of the sequence of

X^n which are the sequences that differ by at most r bits from the sequence x.

The receiver receives $y = x_i \oplus X^n(p_0, p_1)$ and executes the decoding function $\mathrm{d}(y)$, which returns the identity of x_j such that $y \in \mathcal{B}(x_j, r)$. The decoding function can fail for two reasons:

- when $y \notin \mathcal{B}(x_i, r)$,
- when there exists $j \neq i$ such that $y \in \mathcal{B}(x_i, r) \cap \mathcal{B}(x_j, r)$.

The probability of the first failure case is equal to the probability that $|X^n(p_0, p_1)|_{"1"}$ is greater than r, where $|x|_{"1"}$ denotes the number of 1's in an x sequence. Since $r = np_1(1 + \epsilon')$ and np_1 is the average number of 1's in $X^n(p_0, p_1)$, by application of the law of large numbers, the probability tends to zero. This is left as an exercise.

Regarding the second case of failure, denoting k the size of the code set, $|\mathcal{W}| = k$, the probability of decoding failure is upper bounded by the probability that y belongs to $\cup_{j \neq i} \mathcal{B}(x_j, r)$, i.e. by $\frac{|\cup_i \mathcal{B}(x_i, r)|}{2^n}$ which is smaller than $\frac{|\sum_i \mathcal{B}(x_i, r)|}{2^n}$. Since the sizes of all the balls are the same, and smaller than $\sum_{d=0}^{d=r} \binom{n}{d} \leq r\binom{n}{r}$. Using the Stirling formula, which tells that for all integers r tending to infinity:

$$r! \sim r^r e^{-r} \sqrt{2\pi r},$$

we have $\binom{n}{r} \sim 2^{-h(X^n(p_0,p_1)(1+\epsilon))}$ for any $\epsilon > 0$ which tends to 0 when $\epsilon' \to 0$, and $h(X^n(p_0, p_1)) = O(I(X^n, Y^n))$, then the probability of the second failure case is bounded by:

$$np_1(1 + \epsilon')k2^{-I(X^n, Y^n)(1+\epsilon)}$$

which tends to 0 as soon as $\log_2 k - I(X^n, Y^n)(1 + \epsilon) \to -\infty$.

To prove the opposite result, we supplement the $\mathcal{W}$ codebook with other random codewords so that $\log_2 k - I(X^n, Y)(1 + \epsilon) \to \infty$. The probability of successful decoding is limited by the probability that the Hamming ball $\mathcal{B}(y, r)$ around the output sequence y is devoid of any further codewords x_j with $j \neq i$. The probability of such an event for a given j is less than $1 - \frac{\binom{n}{r}}{2^n}$. Taking all other $k - 1$ different codewords, we obtain the upper bound of the decoding probability

$$\left(1 - \frac{\binom{n}{r}}{2^n}\right)^{k-1}$$

which turns out to be smaller than $\exp\left(-(k-1)\frac{\binom{n}{r}}{2^n}\right)$ or, in other words, $\exp\left(-2^{\log_2(k-1)-I(X^n,Y^n)(1+\epsilon)}\right)$ which tends toward zero.

This result is very severe, as the probability of decoding tends either toward 1 or toward 0. We can imagine that even when $\mathcal{B}((1+\epsilon)p_1n, x_i)$ contains several codewords x_j, the receiver can randomly choose which codeword is the decoded version of the received signal y and fall on the right value. If m is the number of such sequences, assuming $m \geq 1$, then the probability of correct decoding would be $\frac{1}{m}$. Let $p(r)$ be the probability that a codeword x_j, for $j \neq i$, belongs to $\mathcal{B}(r, y)$. We know that $p(r) = \frac{\binom{n}{r}}{2^n}$.

If $r_n(\epsilon) = (1+\epsilon)p_1n$, we know that $p(r_n(\epsilon)) = \frac{1+\epsilon'}{R_n}$ for a certain ϵ' which tends to zero when $\epsilon \to 0$. We know that the original codeword x_i belongs to $\mathcal{B}(y, r_n(\epsilon))$ with a high probability. The probability that the number $m - 1$ of other codewords captured in $\mathcal{B}(y, r_n(\epsilon))$ is exactly equal to $\binom{k}{m}p(r_n(\epsilon))^{m-1}(1-p(r_n(\epsilon)))^{k-m}$ which, as an exercise, can be shown to be asymptotically equivalent $\frac{p(r_n(\epsilon))^{m-1}}{(m-1)!}e^{-kp(r_n(\epsilon))}$. The mean value of $\frac{1}{m}$ is asymptotically equal to

$$\frac{1-e^{-kp(r_n(\epsilon))}}{kp(r_n(\epsilon))} \leq \frac{I(X_n, Y_n)(1+\epsilon')}{R} = (1+\epsilon')\frac{R_n}{R}$$

which gives the expected probability of failure when $\epsilon \to 0$. $\square$

3.2.4 *Typical set*

There is a more general proof of Shannon's theorem using typical sequences. If $n_1 = \lfloor np_1 \rfloor$ and $n_0 = n - n_1$, we consider $\mathcal{B}(X^{n_0,n_1}, n\epsilon)$, i.e. the set of sequences located at a Hamming distance less than $n\epsilon$ from at least one of the sequences of $\mathcal{S}(X^{n_0,n_1})$. Each element of $\mathcal{B}(X^{n_0,n_1}, n\epsilon)$, considered as an element of $\mathcal{S}(X^n(p_0, p_1))$, has in the latter set an individual probability between $2^{-H(X^n(p_0,p_1)(1+\epsilon)}$ and $2^{-h(X^n(p_0,p_1)(1-\epsilon)}$ so that all sequences in the set tend to have the same individual probability when $\epsilon \to 0$. Furthermore, the complementary set of $\mathcal{B}(X^{n_0,n_1}, n\epsilon)$ has a weight in $\mathcal{S}(X^n(p_0, p_1))$ that tends to zero when $n \to \infty$ for any fixed $\epsilon > 0$. This property is called the typical set property. All random sequences X_n with i.i.d. bits have this property, and the property can be extended to sets that are not bit sequences, proving Shannon's theorem in a more general framework.

Definition 3.1. A sequence of random systems X^n is typical if for all $\epsilon > 0$ there exists a covering set $A_\epsilon(X^n) \subset \mathcal{S}(X_n)$ such that

- (i) for all $x^n \in A_\epsilon(X^n)$: $2^{-h(X_n)(1+\epsilon)} < P(x^n) < 2^{-h(X_n)(1-\epsilon)}$,
- (ii) $P(\mathcal{S}(X_n) \setminus A_\epsilon(X^n)) \to 0$ when $n \to \infty$.

We note that, as with i.i.d. bit sequences, the elements in $A_\epsilon(X^n)$ tend to have the same uniform probability. Thus, the number of elements in $A_\epsilon(X^n)$ tends to be equivalent to $2^{h(X^n)}$ when $n \to \infty$ and $\epsilon \to 0$.

Let $Z^n = (X^n, Y^n)$ be a sequence of two-component systems (transmit and receive states) that enjoys the typical property. We denote $B_\epsilon(X^n)$ (resp. $B_\epsilon(Y^n)$) the system consisting of the projection of $A_\epsilon(X^n, Y^n)$ onto $\mathcal{S}(X^n)$ (resp. onto $\mathcal{S}(Y^n)$). Shannon's theorem is as follows.

Theorem 3.2 (Shannon theorem on typical sequences). *Let* (X^n, Y^n) *be a typical sequence with covering set* $A_\epsilon(X^n, Y^n)$. *If* X^n *and* Y^n *are both typical with respective covering sets* $B_\epsilon(X^n)$ *and* $B_\epsilon(Y^n)$, *then the decodable capacity of system* (X^n, Y^n) *is* $I(X^n, Y^n)$.

Proof.   This is only a sketch of the proof. Condition (ii), on asymptotically zero weight, clearly applies to the sets $B_\epsilon(X^n)$ and $B_\epsilon(Y^n)$; the only new information provided by the conditions of the theorem is that the individual probabilities on these sets are asymptotically uniform. Therefore, if X_ϵ^n (resp. Y_ϵ^n) is the system X^n (resp. Y^n) bounded by $B_\epsilon(X^n)$ (resp. $B_\epsilon(Y^n)$), we are under the conditions of deterministic entropy where all weights are uniform and equal. In this case, we can refer to Section 2.9.7 to obtain the result. By the way, for x being a state of $B_\epsilon(X^n)$, the set $\mathcal{S}(Y_\epsilon^n(x))$ is just the projection into $B_\epsilon(Y^n)$ of all the tuples (x, y) included in $A_\epsilon(X^n, Y^n)$.   □

We note that the proofs of Shannon's theorem are not constructive, in the sense that they give no practical indication of how to construct an efficient algorithm for encoding and decoding information. The scheme of selecting k codewords at random is impractical, as it would require a codebook of size k which is exponential of order 2^{R_n}. The field that deals with the science of building such algorithms is called error correction. Practical solutions that bypass the obstacle of the exponential book of codes do indeed exist. These include convolutional codes, block codes, turbo codes, sparse codes and polar codes [57,72]. Most of them operate on a block basis, using algebraic operations on finite fields, the elements of which are obtained by grouping bits in the message. It should be noted that the closer the codes are to the Shannon limit, the heavier they are to implement, but there is no theoretical estimate of the increase in implementation costs as a function of distance from the Shannon limit.

In many cases (but this is not a universal rule), a codeword is made up of two parts: the clear message area and the redundancy area. The redundancy zone is a function of the plaintext zone and turns the whole message into a codeword. It is used to recover potential errors in the plaintext area (bearing in mind that errors can also propagate in the redundancy area). The ratio between the clear area and n is the information rate, denoted R. The size of the redundant part will be at least $n - I(X^n, Y^n) = h(Y^n|X^n)$.

If the redundant part is obtained from an algebraic manipulation of the plaintext part, the analysis of the entire received sequence theoretically makes it possible to detect the existence of errors in the collected sequence with a very high probability. This is a costly method of error detection, since the redundant part *a fortiori* makes it possible to locate and correct errors. In the case of error detection, a simple parity check would suffice.

There's a particular aspect of error correction called "erasure correction", in which errors are located by some external, perhaps physical, means. It's exactly as if the supposedly erroneous bits were given a specific gray value (0.5, for example). In this case, the parity check is sufficient to find the hidden value. In the case of a single erasure, one-bit redundancy is sufficient. In the case of conventional single error recovery, without localization by gray value, the size of the redundant part should be $h(Y^n|X^n) = h(X^{n-1,1}) = \log_2 n$.

3.3   Source Coding, Data Compression

The second foundation of information theory is source coding. It mainly consists in designing bandwidth-saving algorithms through data compression. Data compression is an important tool in telecommunications. It consists in taking information X, supposed to be *a priori* a binary sequence, and inserting it into a binary code $C(X)$ which is supposed to take up less space: $|C(X)| < |X|$. The ratio $\frac{|C(X)|}{|X|}$ is the compression ratio. There are two types of data compression algorithms:

- **Lossless compression:** There is a decompression function D which allows us to retrieve the original data without error: $D(C(X)) = X$.
- **Lossy compression:** The decompression function does not retrieve the original data but a copy with error so that $D(C(X)) = X + \epsilon$ for a small ϵ in a given metric.

Lossy compression allows extremely low compression ratios, e.g. for images and videos, always with a quality acceptable to the user. In fact, the search metric is adapted to the user's perception of quality. This is the basis of the main use of the Internet around streaming, which makes up 99% of Internet usage.

3.3.1 *Lossless compression*

The paradoxical aspect of data compression is that it gives the impression that repeating the compression process on a file would lead to an arbitrarily small support. As if you could compress an elephant into a mouse (see Figure 3.2).

In fact, data compression does not work in the special case of deterministic information theory. The best we can do is to leave the data intact when, for example, the file set is the set of all binary sequences X^n of length n, all having the same weight. To obtain interesting results, we need to switch to probabilistic information theory and see how this resolves the above paradox.

Theorem 3.3. *If the original file and the compressed file are both binary, then the average lossless compression rate cannot be smaller than $\frac{h(X)}{|X|}$.*

Fig. 3.2.   Compressing the elephant into a mouse.

Proof.   We give a partial but easy proof. We assume that all sequences proposed for compression X are of length n. The average compression ratio is then $\frac{E[|C(X)|]}{n}$, where $E[|C(X)|]$ is the expectation of the compressed file length. Without loss of generality, since we're looking for a lower bound, we assume that for any X: $|C(X)| \leq n$, i.e. in the event that the compressed file is larger than the original file, we keep the latter intact. The key to the theorem is the fact that $h(C(X)) = h(X)$ since X and $C(X)$ are in bijective correspondence. But

$$h(C(X)) = h(|C(X)|) + h(C(X)||C(X)|).$$

Since $|C(X)|$ is a number between 1 and n, we have $h(|C(X)|) \leq \log_2 n$. Similarly, $h(C(X)||C(X)|) = \sum_{k=1}^{n} P(|C(X)| = k)h(C(X)||C(X) = k)$. But if $|C(X)| = k$, then $h(C(X)||C(X)| = k) \leq k$ because it is a binary file. Thus

$$h(C(X)) \leq \log_2 n + E[|C(X)|].$$

The identity $h(C(X)) = h(X)$ allows us to conclude that

$$\frac{E[|C(X)|]}{n} \geq \frac{h(X)}{n} - \frac{\log_2 n}{n}. \qquad \square$$

We note that we haven't managed to completely prove the theorem because we have the additional term $\frac{\log_2 n}{n}$, but this tends to be negligible when $n \to \infty$. But the presence of this last term is interesting, as it underlines the fact that the length of the compressed code can also carry information that also contributes to the compression ratio. This is reminiscent of the blue-eyed paradox (see Section 1.5), where silences are also information.

3.3.2   *Symbol compression, Huffman algorithm*

The first wireless telegraphic telecommunication network was invented by Claude Chappe (1763–1805) during the French Revolution. The system sends each symbol of the message separately, giving its rank in the alphabet, via optical signals. The optical signal is created from the wing configuration of a *moulin*, as shown in Figure 3.3.

Fig. 3.3.   The Moulin of Chappe.
Courtesy: Wikipedia.

Fig. 3.4.   Left: Samuel Morse. Right: Table of Morse symbols.
Courtesy: Wikipedia.

The Morse system, invented fifty years later by Samuel Morse (1791–1872), but more than a century before information theory, was a great improvement, assigning shorter signals to the most frequent symbols. For example, the dot corresponds to the letter "e", the most frequent character in English, as shown in Figure 3.4.

In 1952, David Huffman ([56] and see Figure 3.5) invented a practical way of coding symbols in a file according to an arbitrary statistic of occurrence, e.g. the occurrence in a given text to be compressed without language pre-requisites.

The algorithm is designed to be near-optimal in terms of symbol entropy. Each symbol is compressed separately, regardless of its order of appearance. The algorithm works as follows. Suppose we want to compress the following text "aaaaaaaabbbbbbccccdd". We sort the symbols according to their frequency:

- "a": 11 times;
- "b": 6 times;
- "c": 6 times;
- "d": 2 times.

Figure 3.6 illustrates the frequency decomposition.

The algorithm takes the two least frequent symbols "c" and "d" and replaces them with a merged symbol "c" + "d", "cd" by abuse of notation, whose frequency is the sum of the frequencies of the two

Fig. 3.5.   David Huffman.

Fig. 3.6.   Symbol of "aaaaaaaaaabbbbbbcccccccdd" sorted by frequency.

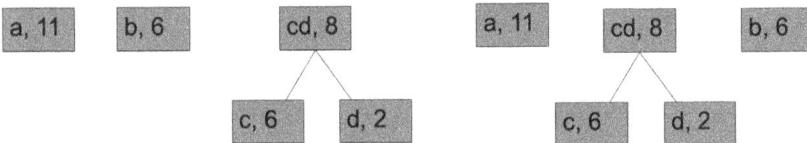

Fig. 3.7.   Left: Merging the two less frequent symbols "c" and "d". Right: Sort new symbols.

omitted symbols, i.e. 8. In doing so, it creates the root of a tree with two hanging leaves "c" and "d". He sorts the remaining symbols by frequency: Since the new symbol "cd" has a frequency of 8, it must be placed between "a" (12) and "b" (6): see Figure 3.7.

In the new list, "cd" and "b" are the two least frequent symbols and are merged into the new symbol "bcd" ("b"+"c"+"d") of frequency 14, now at the top, see Figure 3.8.

In the final operation, "bcd" and "a" are merged into a final symbol "abcd": There is now just one tree, the Huffman tree, and the original symbols are the leaves of the tree, see Figure 3.9. By assigning the label "0" to one sub-tree and "1" to the other sub-tree

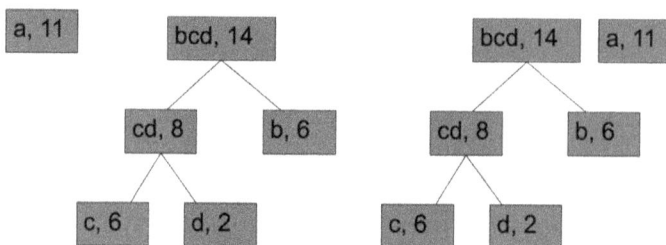

Fig. 3.8. Left: Merging the two less frequent symbols "cd" and "b". Right: Sort new symbols.

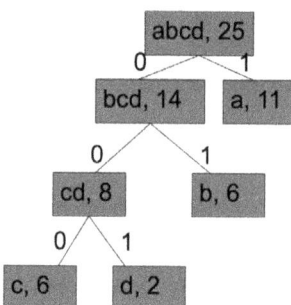

Fig. 3.9.   The Huffman tree of "aaaaaaaaaaabbbbbbccccccdd".

starting from the root (the order can be arbitrary), and repeating the operation up to the leaves of the tree, we find that the symbol "c" is reached by the label sequence "000", "d" by the sequence "001", "b" by "01" and finally "a" (originally the most frequent symbol) by "1".

These sequences are indeed the encoding of the symbols. Namely in the text "aaaaaaaaaaabbbbbbccccccdd":

- Each "a" is replaced by "1",
- each "b" is replaced by "01",
- each "c" is replaced by "001",
- each "d" is replaced by "000".

Thus the text "aaaaaaaaaaabbbbbbccccccdd" is compressed into

"((a,1), (b,01), (c,001), (d,000))

(1111111111101010101010100100100100100100000000)".

The first tuple indicates the symbol mapping table, the second tuple the compressed sequence itself. The interesting fact is that coding can only be decoded in one way because codes are naturally prefix-free: No sequence of symbols is the prefix of another, larger sequence of symbols. This is a direct consequence of the tree structure.

An inattentive reading would give the impression that the new coding doesn't resemble efficient compression. But in fact, each letter of the alphabet appears as a byte in ASCI and the text "aaaaaaaabbbbbbcccccdd" is in fact the binary file: "01100001 01100001 01100001 01100001 01100001 01100001 01100001 01100001 01100001 01100001 01100010 01100010 01100010 01100010 01100010 01100010 01100011 01100011 01100011 01100011 01100011 01100011 01100100 01100100".

There is a relatively easy proof about the quasi optimality of the Huffman code but we partially leave it as exercise. Instead, we show the following theorem.

Theorem 3.4 (Huffman code length). *Let $\mathcal{A}$ be the finite alphabet from which the symbols are drawn. For each $a \in \mathcal{A}$, let $P(a)$ be the occurrence probability of symbol a. The length of the code encoding a is not greater than $\lceil \log_2 \frac{1}{P(a)} \rceil$.*

As a direct corollary of the theorem, we note that an upper bound of the average code length is $\sum_{a \in \mathcal{A}} P(a) \lceil \log_2 \frac{1}{P(a)} \rceil$ which is smaller than $1 + h(\mathcal{A})$, thus the Huffman code is optimal within a term 1.

3.3.3   *Pattern matching lossless compression algorithm, Ziv–Lempel algorithm*

The drawback with Huffman's algorithm is that it compresses symbols separately and doesn't take advantage of possible pattern repetitions. For example, the binary file

    "abababababababababababababababababababab"

would not compress well via Huffman algorithm, since it would produce a very similar file:

    "((a,0),(b,1)),(01010101010101010101010101010101010101)".

The file could be further compressed in the form (ab,20) to indicate that the "ab" factor is repeated 20 times. If the number of repetitions were n, the length of the compressed Huffman code would be proportional to n, while the length of the pattern repetition code would be proportional to $\log n$, since the number of repetitions is $n/2$ and this could be indicated in the code, by encoding the number of repetitions in $\lfloor \log_2 n \rfloor$ bits.

In 1978, Lempel and Ziv invented a lossless compression algorithm known as *universal*, which takes advantage of repeating patterns in text [125]. The final algorithm, called the Lempel–Ziv–Welch algorithm (Figure 3.10), is the basis of most text compression applications such as zip, etc.

The algorithm is called "universal" because it has been proven to converge to the entropy limit in all known probabilistic models of text generation, as soon as the text length tends to infinity.

The compressed code of a text $C(X)$ is made up of tokens of the form:

$$C(X) = (T_1, T_2, \ldots, T_k).$$

A token consists of an index (smaller than $n = |X|$) and a symbol.

The decompression process takes place as follows (in fact, it's easier to examine the decompression process first, before the compression process). There is a dictionary, which is a list of words initialized with the empty word. When the decompression process reaches the i token, for $i \leq k$, the dictionary should contain i entries. Let Y_{i-1} be the part of the text decoded before the token T_i. Suppose the token is of the form $T_i = (m, y)$. The decoding algorithm updates

Fig. 3.10.   Lempel, Ziv and Welch.

the decoded part Y_{i-1} by adding the copy of the text stored in the dictionary to the input m, i.e. D_m, and adds the symbol y:

$$Y_i = Y_{i-1} + D_m + y.$$

A new dictionary entry D_i is created with $D_i = D_m + y$. We have $Y_k = X$.

The compression process proceeds as follows. The dictionary is created on the fly but will not be sent with the encoded text $C(X)$, since the dictionary can be recreated on the receiver's side by analyzing the token sequence.

The first symbol of X, x_1, is used to create the first token: $(0, x_1)$ and the segment D_1 is created by concatenating the segment 0 (the empty word) and the symbol x_1; the remaining part of the text is $X_2 = X_2^n$ (the complete text X with the first symbol omitted).

To create the token T_i and the dictionary entry D_i, we examine the remaining part X_i of the text and try to find the largest prefix of X_i which is a copy of a dictionary entry plus a symbol. Suppose this is the index entry m ($m < i$) plus the additional symbol y. The new token is $T_i = (m, y)$ and the new entry is $D_i = D_m + y$. Finally, $X_i = D_i + X_{i+1}$.

The compression of "ababababababababababababababababababababababab abab" will proceed as indicated in the following table.

Remaining text	dictionary	token	index
ababababababababababababababababababababababab	ϵ	-	0
babababababababababababababababababababababab	a	(0,a)	1
abababababababababababababababababababababab	b	(0,b)	2
ababababababababababababababababababababab	ab	(1,b)	3
babababababababababababababababababababab	aba	(3,a)	4
babababababababababababababababababababab	ba	(2,a)	5
ababababababababababababababababababab	bab	(5,b)	6
ababababababababababababababababab	abab	(3,b)	7
bababababababababababab	ababa	(7,a)	8
bababababababababab	baba	(6,a)	9
abababababab	babab	(9,b)	10
abab	ababab	(8,b)	11
	ababe	(7,ϵ)	12

The final compressed code is the list of the tokens:

(0,a), (0,b), (1,b), (3,a), (2,a), (5,b), (3,b), (7,a), (6,a), (9,b), (8,b), (7,ϵ).

We note that at each step, a dictionary entry is increased by one symbol. But since entries starting with "a" are prefixes of the sequence "abababab...", and entries starting with "b" are prefixes of the sequence "babababab...", and if for the first k tokens about half the entries start with "a" and half with "b", then the first k tokens cover about $k^2/4$ of the original text. We can see in the table the quadratic increase in the part of the text covered as a function of the token index. This means that for an original text composed of n repetitions of the "ab" factor, compression will take around $\sqrt{2n}$ tokens, leading to a compression ratio of the order of $\frac{\log_2 n}{\sqrt{n}}$ which tends toward zero when $n \to \infty$. This is because the length of a token in bits is limited by $\log_2 n$ (more precisely, by $\lceil \log_2 n \rceil + \log_2 |\mathcal{A}|$).

There is an alternative version of Lempel–Ziv's algorithm called the suffix tree version (invented in 1977) which considers two types of tokens: the symbol token (y) when a new, as yet unseen symbol is to be inserted, and the copy-paste token (n, m) which means that the segment of X between positions n and $n + m$ is to be copied and pasted at the end of the current decoded part. Decompression proceeds as follows (see Figure 3.11):

$Y_0 = \epsilon$, and recursively

- if the ith token is a symbol token (y): $Y_i = Y_{i-1} + y$,
- if the ith token is a copy paste token (n, m): $Y_i = Y_{i-1} + X_n^{n+m}$.

With this new version, the compressed code of "abababababab ababababababababababababababab" will be (a), (b), (1,37) since the "ab" factor is detected as being repeated. We note that the copy-and-paste area extends well beyond the already decoded part, but this is not a problem since the repetitive part can be extended indefinitely beyond the current decoded part. This leads to a compression factor of the order of $\frac{\log n}{n}$, more powerful than the basic Lempel–Ziv. In the compression phase, the algorithm searches for the largest prefix in

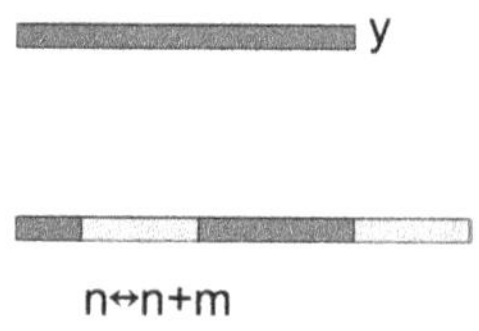

$n \leftrightarrow n+m$

Fig. 3.11. The alternative suffix tree version of Lempel–Ziv.

the remaining part of the text that has a copy in the previous part; this is done via the use of a suffix tree that we won't detail here.

An overlap between the decoded part and the part not yet decoded, as seen above, does not occur frequently in a reasonably random text and, in the present example, is only an artificial exception. A non-trivial analysis shows that when compressing a random text X^n of length n, the typical length of a sentence (the length of the input in the first version of the algorithm, the copied-and-pasted segment in its second version) is on average $\frac{\log_2 n}{h}$ [58], where h is the entropy rate, the limit (when it exists) of $\frac{h(X^n)}{n}$. More precisely, using the properties of renewal within text, the average number of tokens is asymptotically equal to n divided by the average sentence length, i.e. $h(X^n)/\log_2 n$. Since the length of a token is of order $\log_2 n$ for the basic Lempel–Ziv algorithm, and $\log_2 n + O(\log_2 n)$ for the suffix tree version. The term $O(\log n)$ comes from the encoding of the length of the segment to be copied and pasted, which must be of the order of the logarithm of its declared average length. This term in $\log \log n$ generally makes the suffix version less efficient than the classic Lempel–Ziv–Welch version (except in the highly unlikely case of pure repetitions).

3.4   Pattern Matching Predictor

The art of predicting the future has entertained mankind since the very beginning. But beyond the gift of prophecy and the art of crystal ball gazing (see Figure 3.12), there is today a vast science based on the very foundations of information theory. It follows from the previous section that pattern repetition detection for efficient text compression acts as a perfect transition to the predictor problem. The essence of the problem involves the existence of a sequence X^n of past events, spanning n elements, $X^n = (x_1, x_2, \ldots, x_n)$, and the game is to guess the next event x_{n+1}, or a finite set of future events. Let $\hat{x}_{n+1}$ be the prediction, and the difference between the prediction $\hat{x}_{n+1}$ and the actual realization is called the loss: $\text{loss}(x_{n+1}, \hat{x}_{n+1})$. The aim is to minimize the loss.

Before the discovery of global air circulation and pressure patterns, weather forecasting was not yet a science in the 19th century. If a local newspaper of the time commissioned a journalist to predict

Fig. 3.12.   The old times of predictions and prophecies.

the weather for each of the following days, with no other source of information than the history of the last n days. To do this, the journalist used an event predictor based on previous weather records in a given location. The report is made up of a small set of data (temperature, pressure, wind strength and direction, and other parameters, see Figure 3.13).

In his time, Claude Shannon invented the *mind reading machine* [106], a bluffing computer program whose stated aim was to guess the user's thoughts. The user would enter a random sequence of bits $(x_1, x_2, \ldots, x_n)$ and, on a separate terminal, the computer would try to guess the next bit x_{n+1}. The program can be adapted to a rock, paper, scissors type of quiz. Of course, a success rate of more than 50% would lead to the premature conclusion that the computer was exercising control over the user's mind, or that it possessed a prophetic ability of unknown origin. Of course, this is not the case. The secret lies in the fact that the human brain is not sufficiently adapted to imitate a random generator, and that the computer will easily pick up most of the repetitive bit patterns that

EXAMPLES.

ABERDEEN TO LONDON, 25th July, 1862, 8 A.M., received at 10.

South-west — very strong wind.*

06041   93453   94663   21072
60420   05628

CONVERSION IN REPORT.

1862 Friday, 25th July, 8 A.M. Aberdeen	B	E	D	W	F	X	C	I	H	R	S
29·39†	60	6	SW.	5	8	6	r	6	0·46†	S

Fig. 3.13. A weather record.
Courtesy: Wikipedia.

the user unintentionally introduces into his typing. The 19th-century weather forecaster did the same, detecting patterns in past weather records and exploiting them to make his predictions.

Prediction algorithms are related to text compression algorithms. In a way, text compression attempts to guess the next sequence of symbols in the text on the basis of its past sequence $(x_1, x_2, \ldots, x_n)$ and compress it by indicating only the difference between the actual sequence $(x_{n+1}, \ldots)$ and its prediction $(\hat{x}_{n+1}, \ldots)$. But the link is only apparent, as there are no in-depth results on this point. We, however, show a simple predictor directly inspired by Lempel–Ziv's compression algorithm and invented in [61].

A predictor is universal, as in the case of the pattern-matching predictor, when it tends to minimize the loss when the number n of past events tends toward infinity, and this over a very wide spectrum of sequence generation models. To work, the predictor needs the following three conditions (quite realistic, except perhaps the first one):

- Alphabet *smallness*: i.e. of size smaller than the logarithm of the event sequence length.
- Source *stationarity*: i.e. it does not change by time translation (not completely true for weather prediction).

- The source *amnesia*: events which are far apart in time tend to be independent.

The first condition can be difficult to meet in weather forecasting events, as the continuous spectrum of temperature values provides a potentially infinite alphabet. But this effect can be recovered through adequate quantization of temperature (and atmospheric pressure).

In the following, we describe the algorithm of the pattern-matching predictor. It takes the sequence of past events, written in the finite alphabet $\mathcal{A}$, $X^n = (x_1, \ldots, x_n)$. It looks for the largest integer k such that $X_{n-k}^n = (x_{n-k}, x_{n-k+1}, \ldots, x_n)$ has another copy inside X^n, say X_{m-k}^m for an integer $m < n$. In other words, it looks for the largest suffix of X^n that has a copy in X^n. At this point, we find ourselves in the same situation as with Lempel–Ziv's compression algorithm: Compressing X_1^n to position $n - k$ would find X_{m-k}^m as a new dictionary entry, since it is a copy of X_{n-k}^n. But from this stage onwards, the processes differ.

Let's call w^k the largest suffix of X^n that has a second copy in X^n: $w^k = X_{n-k}^n = X_{m-k}^m$ and is of size k. In most source models, k is of order $\log n$, in fact of order $\frac{\log n}{\log |\mathcal{A}|}$, more precisely $\frac{\log n}{h}$, where h is the entropy rate of the source of the sequential event (assumed invariant since the source is stationary). The precision of this last estimate establishes a clearer link with information theory. If the source were uniform, we'd have $h = \log |\mathcal{A}|$. A simple predictive scheme might be to take as prediction the symbol that comes just after X_{m-k}^m, $\hat{x}_{n+1} = x_{m+1}$ since this is the location in X^n that most closely resembles the final part of X^n. But the choice turns out to be far too limited and doesn't really minimize the loss. The idea is to have more candidates for prediction in order to select the best candidate likely to minimize the loss. The factor w^k has at least one copy inside X^n, but any suffix of w^k will have more copies. Assuming the algorithm has a parameter α between 0 and 1, the next step is to find all the copies of $w_{\lfloor \alpha k \rfloor}^k$ inside X^n, the fraction α of w^k (the fraction taken from the end). The character after each copy is called a "marker", and among all these markers, we select the symbol that appears most frequently, which will be $\hat{x}_{n+1}$. Figure 3.14 illustrates the process for $\alpha = \frac{1}{2}$.

With the procedure described above, the number of copies increases as α decreases, thus increasing marker diversity. They then

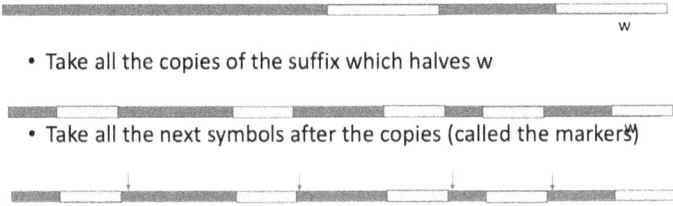

- Take all the copies of the suffix which halves w

- Take all the next symbols after the copies (called the markers)

Fig. 3.14.   The pattern-matching predictor.

give a more accurate representation of the distribution of the symbol that follows a sequence of past events similar to $w^k_{\lfloor \alpha k \rfloor}$. In most models, the number of markers is of the order of $n^{1-\alpha}$, ranging from $O(1)$ for $\alpha = 1$ (one or two copies only) to $O(n)$ for $\alpha = 0$. In this extreme case, $w^k_{\lfloor \alpha k \rfloor}$ will be limited to a single symbol, making as many markers; this symbol appears in X^n, so $O(n)$. The length of the copies will be of order $\alpha \frac{\log_2 n}{h}$ so that when $\alpha \to 0$ the diversity of markers increases but their importance decreases as the copies may not be long enough to significantly influence the symbol that follows them and the distribution will be less relevant. The analysis shows that the optimal value in many cases is $\alpha = \frac{1}{2}$, but the predictor remains universal whatever the value of α far from 0 and 1.

The algorithm description corresponds to the case where the loss is 0 when the predicted symbol is correct and 1 otherwise. For a more general loss function, the prediction will be the $\hat{x}_{n+1}$ value that minimizes the cumulative loss:

$$\hat{x}_{n+1} = \arg\min_x \sum_{y \in \mathcal{M}_n} \text{loss}(x, y),$$

where $\mathcal{M}_n$ is the set of markers.

The algorithm can be adapted to produce batch predictions, not only for tomorrow's weather but also for that of the whole week. It can also be adapted to accept copies that are not exactly identical to the suffix in X^n, to tolerate a certain degree of similarity.

A final word of advice though. The algorithm is easy to implement and you can try it out on your friends and family, but be careful with young children. They tend to amplify the algorithm's successes and could end up believing in your superpowers (only to be eventually super-deceived later).

3.5   Lossy Compression, Rate Distortion

In this section, we look at the other aspect of data compression: lossy compression. Information is compressed, but decompression does not return exactly the same information. This freedom makes it possible to escape the entropy limit and achieve much lower compression ratios. This property is not very interesting for text, as you don't want to risk losing important information, but it is particularly effective and interesting for music, images and videos. An image of several million pixels can be compressed into a few hundred kilobytes with no notable difference to the human eye. With just a few kilobits, the difference is visible, but the main characteristics of the image remain intact. This is why lossy compression algorithms such as JPEG [92] and MPEG [40] are the most widely used in information technology. Indeed, they enable the Internet to distribute video streams on a massive scale and television to be broadcast to cell phones, even with limited bandwidth. Figure 3.15 shows an example of various rendering of the same image under different compression rates (from left to right, 35 kbytes, 12 kbytes, 4 kbytes). Figure 3.15 shows an example of various rendering of the same image under different compression rates (from left to right, 35 kbytes, 12 kbytes, 4 kbytes).

The theoretical principles of lossy compression are as follows: Between the image to be compressed and the decompressed image $Y = D(C(X))$, there is mutual information that quantifies the amount of information shared between the original image and the reconstructed image Y. Mutual information is conventionally measured by $I(X, Y)$ and must contain the information transmitted by the compressed code $C(X)$. Therefore, in bits, we have

Fig. 3.15.   Left: Image compressed to 35 kbytes, center, to 12 kbytes, right, to 4 kbytes.

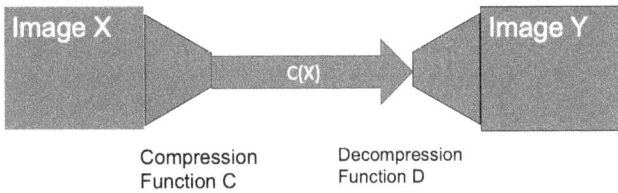

Fig. 3.16. The lossy compression channel: $|C(X)| \geq I(X,Y)$.

$|C(X)| \geq I(X,Y)$. In other words, the compressed code acts as a "channel" between the original image and the reconstructed image (see Figure 3.16).

It is essential to have a metric for measuring the distance between the original image X and the reconstructed image Y: $d(X,Y)$. This metric must be able to take into account the loss of quality perceived by the user. If $D > 0$, the rate of *distortion* $R(D)$ is the minimum mutual information between two images when the average distance between X and Y is less than D:

$$R(D) = \min_{E[d(X,Y)] \leq D} I(X,Y).$$

The reason we use the average distance between the two images is that mutual information is only defined when X and Y follow a probabilistic model. The rate of distortion gives a lower limit to the rate of compression, given the tolerance that would be imposed on the perceived loss of quality.

For a very trivial example, let's assume that X and Y are uniform bit sequences and that the distance between X and Y is the Hamming distance. Suppose that $Y = X \oplus \theta^n$ where θ^n is a Bernoulli bit sequence $X^n(p_0, p_1)$, as in the case of the Bernoulli channel (or the game of seven errors). We know that $E[d(X,Y)] = np_1$, so the condition $E[d(X,Y)] \leq D$ means that $p_1 \leq \frac{D}{n}$. Since $I(X,Y) = h(X) - h(\theta^n)$, hence $I(X,Y) = (1 + p_1 \log_2 p_1 + p_0 \log_2 p_0)n$. Note in Figure 3.17 that the fall in the function $I(X,Y)/n$ is abrupt when nD is small, as the slope of the function $\log(x)$ is infinite at $x = 0$. This is the main theoretical reason why lossy compression systems are so effective. In fact, when we take more realistic measurements on less trivial examples, the drop is even sharper, as shown in the figure (right) for four-bit coded colors, again with the simplistic Hamming distance.

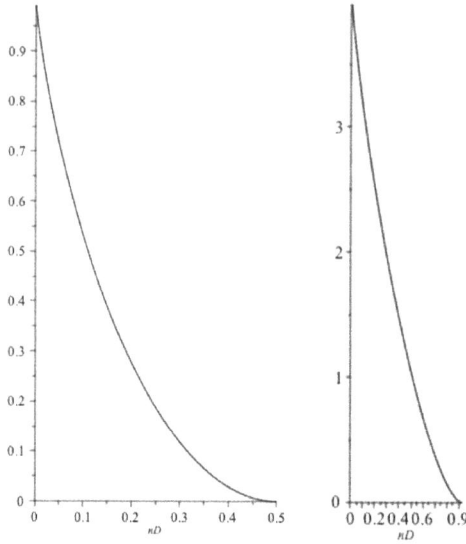

Fig. 3.17.   The distortion rate $R(D)/n$ versus D/n. Left: 1 bit color; right: 4 bit color.

3.6   Exercises

3.6.1   *Bernoulli channels superposition*

Let (p_0, p_1) and (p'_0, p'_1) be two probability distributions.

Exercise 11

Find (q_0, q_1) such that $X^n(p_0, p_1) \oplus X^n(p'_0, p'_1) = X^n(q_0, q_1)$ in distribution. *Corollary*: In seven errors games, prove that Y^n has same distribution as $X^n(\frac{1}{2}\frac{1}{2})$. See the Answer of Exercise 11 in Chapter 9.

Exercise 12

Prove positivity of mutual information $I(X, Y)$ in probabilistic setting. See the Answer of Exercise 12 in Chapter 9.

Exercise 13

Prove binomial expression for Bernoulli sources.

3.6.2 *Entropy of Markov chain*

Exercise 14

Find an expression of the entropy of a Markov chain of length k with transition matrix $\mathbf{P}$ and initial state ξ_0. See the Answer of Exercise 14 in Chapter 9.

Exercise 15

Prove with Stirling formula that $h^D(X^{n_0,n_1})$ and $h(X^n(n_0/n, n_1/n))$ are asymptotically equivalent. See the Answer of Exercise 15 in Chapter 9.

3.6.3 *Huffman code length*

We take the notation of Theorem 3.4. The proof is by induction. We first note that the theorem holds for $\mathcal{A}$ with only two elements.

Let $\mathcal{A}_k$ be an alphabet of size k for $k > 2$. To fix the idea: $\mathcal{A} = \{a_1, \ldots, a_{k-1}, a_k\}$ ranked in decreasing frequency. Applying Huffman algorithm merges the two last symbol a_{k-1} and a_k into a fake symbol $a_{k-1} + a_k$.

Exercise 16

Shows that the frequency of the merged symbol $a_{k-1} + a_k$ has a frequency $P(a_{k-1}+a_k)$ which is greater than $2P(a_k)$. See the Answer of Exercise 16 in Chapter 9.

Removing the two last symbols and replacing them by the merged symbol creates an alphabet $\mathcal{A}_{k-1}$ of size $k - 1$.

Exercise 17

Using the fact that the depth of the symbols a_{k-1} and a_k in the Huffman tree of $\mathcal{A}_k$ is one step below the merged symbol in $\mathcal{A}_{k-1}$ shows the theorem by recursion. See the Answer of Exercise 17 in Chapter 9.

Chapter 4

The Challenge of Information Networks, the Triumph of the Algorithms over the Complexity

Abstract

The road traffic has increased by a mere factor of 10^5 since the year 1900, which anyhow is enough to modify the commuting habits and the impact on the environment. Meanwhile, the traffic of information has grown by a factor of 10^{15}. The wireless telecommunications have taken a substantial share. Until WW2, the early progresses of data transmission were fully supported by advances in physics and has culminated with the invention of the transistor. We can call this period the "triumph over the matter". The transistor has been invented at the very same time as the information theory, and indeed when translated in circuits and in chips, it becomes very suitable for a "mathematization" via the information theory. The theory was so successful for improving telecommunication hardware that it made affordable point to point-to-point communication (telephone) and point to multipoint communication (TV broadcast), the two largest networks of these times. We can call this period the "triumph over the numbers". This triumph which culminated with the rise of large community TV networks and the invention of satellite communication finally reached its limits in the 1980s, when those networks met the wall of the complexity which blocks all new progresses before the birth of the Internet. Indeed, the initial goal of the information theory was to quantify the ultimate limits of transmission technology but did not actually tell much on how to reach them. Indeed, the closer we want to get to the limits, the more complex are the technologies and the algorithms needed. The challenge of Internet was a crucial step, in coping with the complexity of a universal network. The problem was to tackle the challenge of a communication network with multi-user and multiple servers permanently alternating in their roles. We show that classic approaches would inevitably lead to the collapse of the communication system and that an original solution brought by the early computer science solved the problem and gave birth to the Internet protocol and TCP-IP.

4.1   Building the Most Complex and Distributed System

As we saw in Chapter 1, the rise of the Internet was the latest step in telecommunications: the triumph over complexity. The science of complexity is not entirely within the scope of information theory. But information theory can help us find its limits. Methods for meeting the challenges of complexity are, in fact, an integral part of computer science. There's an old debate about whether computer science should be included in the broader field of applied mathematics, but it's not worth discussing here as long as everyone agrees on what we're talking about.

To cut a long story short, Claude Shannon introduced the foundations of information theory by treating a point-to-point telecommunication system: a single terminal, accessing data from a single server (see Figure 4.1) — one sender, one receiver.

The real challenge of the 1970s was to build a multi-user network, i.e. a plurality of terminals and servers operating on a plurality of networks. In this context, terminals and servers constantly exchange roles.

The Internet is not the first multi-user network. The television broadcasting network is an earlier example, but in this network, there is only one server and several terminals that never change roles. The telephone network is another early example; in this case, each device alternates the role of terminal and server, depending on the listen/speak mode. But this network is dedicated to a single application: the transport of the human voice over a dedicated, continuous physical circuit, unlike the Internet, which can carry any type

Fig. 4.1.   The point-to-point terminal to server network.

of digital data and information, and cross the boundaries between several distinct networks.

The main characteristic of the Internet is its ability to transport information across several distinct and heterogeneous networks. To do this, information is held in specific containers called "packets", which contain a finite number of bits inside a physical medium, a kind of electronic box. Packets are independent, in the sense that they contain the necessary information about the information's final destination, as well as its source. The physical box will change during the journey from source to destination. Indeed, it may start its journey as a WiFi packet, then be a multi-frequency modulation in Asymmetric Digital Subscriber Line (ADSL) in copper lines and then be included in a stream of photons in trans-oceanic optical fibers. The crucial devices that act as intermediaries between networks are *routers*, which are the small computers represented in the Internet cloud in Figure 4.2 (for some unknown reason, the Internet is often represented as a cloud in executive-level slides).

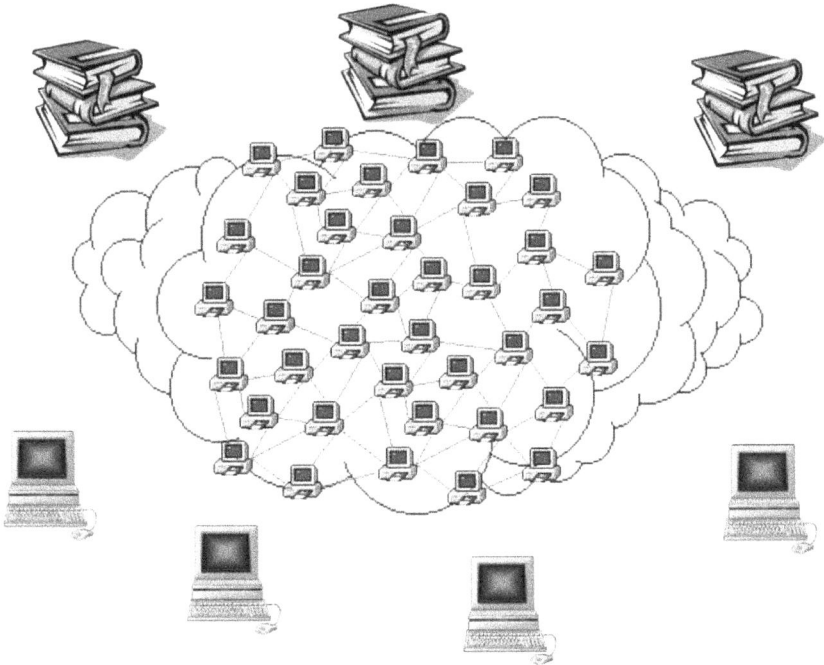

Fig. 4.2.　The multi-user Internet design.

The routers' task is to receive the packet and, using the information it contains, to select the network on which the packet will be routed. In doing so, the router reconstructs the packet's box according to the physical requirements of that next network. In fact, the Internet isn't a network, it's a set of networks. More precisely, it's a *protocol*, i.e. the set of rules that enable routers to work coherently to route packets to their destinations.

4.1.1 *Arpanet, Internet*

The origins of the Internet lie in a long period of "madness". In the 1960s, the term MAD (Mutual Assured Destruction [91]) was used to describe the doctrine of nuclear deterrence. The two main adversaries (at the time the USA and the USSR) had stored several thousand warheads in sufficiently protected silos (Figure 4.3) and in deep-sea submarines so that the first attacker would never have the capacity to destroy all his enemy's warheads to prevent automatic retaliation, thus ensuring their ultimate mutual destruction.

Fig. 4.3.   A nuclear missile silo, useless when off the telecommunication grid.

Scary as it was, the theory was effective and realistic: Missile silos were overprotected by several meters of steel, command centers were buried deep and sealed under mountains. But there was a weak point: the telecommunications network. How could a counterattack be ordered when the enemy had disabled the telecom links in the chain of command? This problem was obvious from the outset, and the military were quick to solve it. It was impossible to protect the tens of thousands of kilometers of cable linking the command centers to their strike centers, or to dig them very deep under them (see Figure 4.4). The idea was to allow orders and reports to be diverted via alternative routes, such as telecom networks dedicated to public use. Several million kilometers of cable with several thousand hubs become an impossible target even for a vastly equipped adversary.

In 1969, Arpanet was created [51]. Its initial aim was to implement a digital network redundant with the military command network. To amplify this redundancy, the army installed concentrators housing routers in administrative facilities, strategically important factories

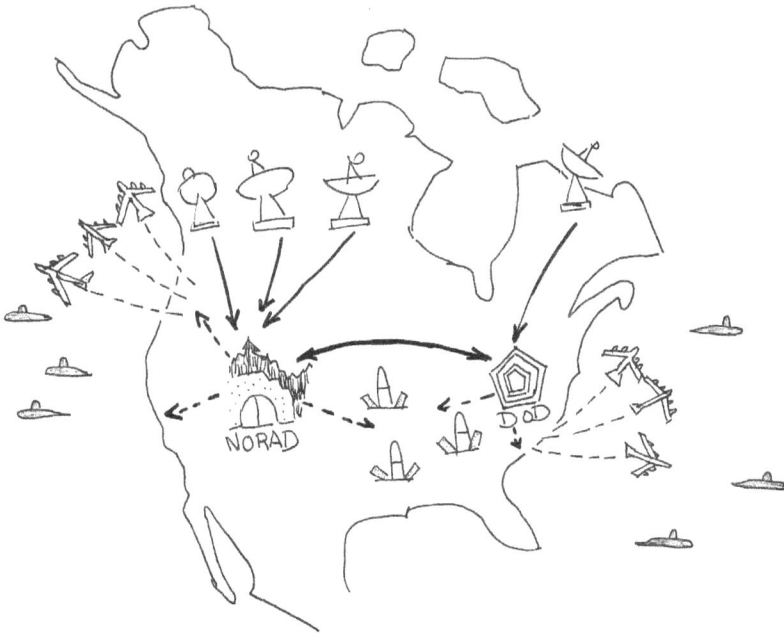

Fig. 4.4.   The early warning chain of command.

and major universities. Since missile firing orders didn't have to circulate every day, the military administration allowed faculties to use the new connection for their own research. To this end, Leonard Kleinrock and his team at UCLA built the first system to send the first recorded e-mail in 1969, the famous "LO" message (for "LOgin", but the system failed before the third symbol).

Arpanet became very popular and the digital network entered its golden age. Ten years later, the military decided to separate Arpanet from its civilian counterpart, as they had had time to build enough redundant internal lines for their own needs. The civilian part was christened *Internet*, as we know it today. The Internet Protocol standards body is the *Internet Architecture Board* (IAB) for supervision and the *Internet Engineering Task Force* (IETF) for its technical aspects [13], both created in 1986, the official birth date of the Internet, although the first original *Request For Comment* (RFC) dates back to 1969. Today, there are over 9,000 RFCs. The use of the term "Task Force", which originally referred to a wartime group of U.S. Navy ships, should prevent people from forgetting that these standard bodies are mostly American public institutions under US rules.

4.2   The Routing Internet Protocol

The aim of Arpanet was to divert the firing order if whole portions of the original network turned to be suddenly knocked out. For this purpose, the routers must collect enough information about the surrounding network topology, and this in an autonomous manner because a centralised network monitoring could be destroyed as well in a pre-emptive strike. The main features which allow this to happen are

- the addresses of internet hosts,
- the routing protocol.

4.2.1   *The organization of Internet*

Each terminal and each router has an Internet address, originally 4 bytes, but since it is outrageously low compared to the current development of Internet (only 4 billions addressable terminal), now we are with 16 bytes addresses with IPv6 (more than 3.4×10^{38} potentially addressable hosts).

The two first bytes are for domain and subdomain indication. For example, the domain 128 was for France, and in 128.92, 92 was for Inria Research Institute in France.

Each router contains a routing table which is a collection of IP addresses. Each IP address is associated with the transit interface identification, that is, the low-level network interface on which each packet toward a destination with this IP address must be routed through. The routing table entry also contains an estimate of the distance from the host to the destination. The distance can be arbitrary metric, but in most case, it is in hop count (i.e. the number of routers to get through on the path). Figure 4.5 shows a subdomain configuration with four routers/hosts: A, B, C and D. The low-level network interfaces are I1, I2 and I3. The routing table of host A will be

B	I1	1
C	I1	2
D	I1	3

In B, the routing table will be

A	I1	1
C	I2	1
D	I2	2

In C, it will be

A	I2	2
B	I2	1
D	I3	2

In D,

A	I3	3
B	I3	2
C	I3	1

The routing table lists all hosts inside the subdomain. It is like a sign post in a city, which indicates the different route toward the

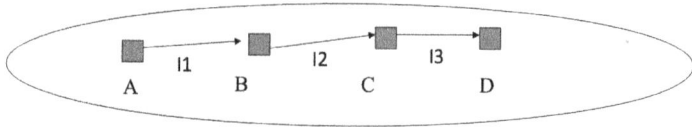

Fig. 4.5.   An Internet subdomain with four routers.

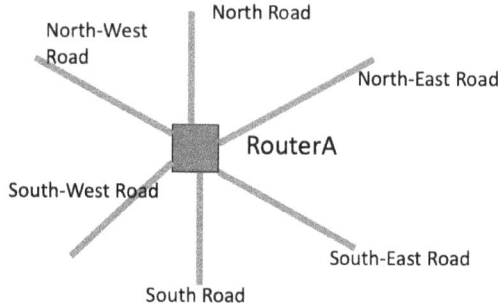

Fig. 4.6.   The routing table like a signpost set.

other cities, see Figure 4.6. In order to make the system scalable to worldwide Internet, the domain or subdomain beyond is only indicated by the address of the gateway. A gateway is a specific router standing at the boundary of two domains A and B and therefore has two addresses: the address in the domain A, with the A's prefix, and the address in domain B, with B's prefix. The routing table contains the two addresses of the gateway. A domain B can have several gateways toward different outside domains. The path length estimate will be the path length toward the gateway for all hosts beyond that gateway.

Since the gateway is supposed to be the unique port of entry from domain A to domain B, all cross traffic will be directed through the gateway. But since the routers in A do not "know" the hosts in B, the look-up in the routing table consists of searching the address with the longest common prefix with the destination address in B. The match will occur with the B address of the gateway, since the domain and subdomain are indicated in the address prefix.

Figure 4.7 illustrates the view of a router inside a domain (in blue) with a gateway (in green) toward an external domain (in orange). Every link in the host domain is a subnetwork which connects the

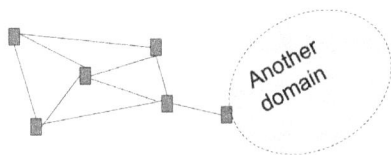

Fig. 4.7.   An internet domain with an external domain.

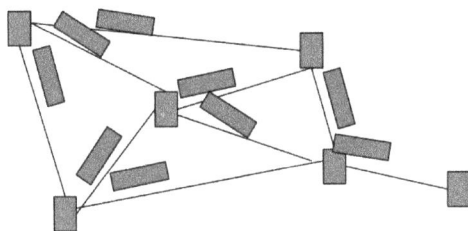

Fig. 4.8.   TC packets flowing inside a domain.

routers but is not indicated in the routing table. Indeed, each router is supposed to "know" only the subnetworks with which it has a physical interface (we see later how this early assumption was not satisfactory).

The key feature is the routing protocol, more precisely the operating protocol which will fill the routing table in each router in a distributed manner. To achieve this objective, there are two schools of routing protocols:

- the distance vector protocol,
- the link state protocol.

Both protocol types rely on control packet which we call *Topology Control* (TC) packets. In fact, those packets carry different denominations, depending on the type of protocol, but we will not go into that detail (see Figure 4.8).

If we forget about the inter-domain routing aspects we already discussed with the gateway stuff, the routing inside a domain is a kind of graph problem with a graph (V, E), where V is the set of host and E the set of links between them. The Internet graph is supposed to be undirected but we assume that the edge set contains directed edges, thus the links appear twice, in the two ways. This detail has some importance in the section about wireless Internet. On the network

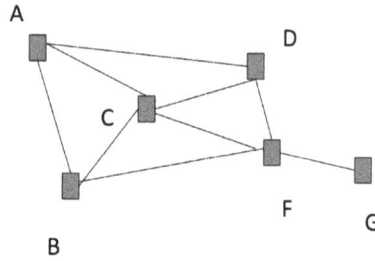

Fig. 4.9.   The routing inside a domain displayed as a graph.

displayed in Figure 4.9, we have $V = \{A, B, C, D, F, G\}$ and

$$E = \{AB, BA, AC, CA, AD, DA, BC, CB,$$
$$BF, FB, CD, DC, CF, FC, DF, FD, FG, GF\}.$$

4.2.2 *Routing internet protocol (RIP)*

The RIP protocol is of distance vector type [27]. Periodically (say every 20 minutes), each host copies its whole routing table in TC which is sent on each of its low-level network interfaces.

At the other end, the host receiving this TC from the neighboring routing updates its routing table the obvious way: For any neighbor host, the transit interface is the interface connecting to this host and the distance is 1. In some alternative implementation, the distance is 0 which can lead to some not so important deviation in the protocol functioning. For any distant router (beyond direct neighbor), it compares the received routing table from its direct neighbor and takes the one which advertises the smallest distance to this distant host. The routing table entry will be updated with the identification of the interface to this neighbor, the distance updated to the advertised distance plus one unit (representing the distance to the neighbor). If the distant router was absent in the routing table (it may happen when activating the host the first time), it is added as soon as a received routing table contains an entry with the previously unknown router.

Since the protocol operates in an asynchronous manner, each host will maintain a topological directory which will contain the last advertised routing table for each of its neighboring router. The protocol is just the distributed implementation of the Bellman–Ford algorithm.

A centralized version of the algorithm consists in manipulating matrices with tropical algebra [85]. We consider the subdomain graph given in Figure 4.9. The initial matrix I is as follows:

	A	B	C	D	F	G
A	0	∞	∞	∞	∞	∞
B	∞	0	∞	∞	∞	∞
C	∞	∞	0	∞	∞	∞
D	∞	∞	∞	0	∞	∞
F	∞	∞	∞	∞	0	∞
G	∞	∞	∞	∞	∞	0

Each line of the matrix corresponds to the (absence of) knowledge of the corresponding about the distance toward the other host. The host is only aware of itself (distance 0) and knows nothing about the other routers. The distance ∞ either means the distance is still unknown or the router is just still unknown.

After the first round of TC exchange, the distance matrix is now N:

	A	B	C	D	F	G
A	0	1	1	1	∞	∞
B	1	0	1	∞	1	∞
C	1	1	0	1	1	∞
D	1	∞	1	0	1	∞
F	∞	1	1	1	0	1
G	∞	∞	∞	∞	1	0

This matrix is the adjacency matrix, each "1" indicates when the routers are neighbors. In the second round, we have the matrix N_2 which indicates the pair of routers at one or two hops away:

	A	B	C	D	F	G
A	0	1	1	1	2	∞
B	1	0	1	2	1	2
C	1	1	0	1	1	2
D	1	2	1	0	1	2
F	2	1	1	1	0	1
G	∞	2	2	2	1	0

The third round provides the matrix N_3 which indicates the distance smaller or equal to 3:

	A	B	C	D	F	G
A	0	1	1	1	2	3
B	1	0	1	2	1	2
C	1	1	0	1	1	2
D	1	2	1	0	1	2
F	2	1	1	1	0	1
G	3	2	2	2	1	0

From this stage, the matrix is the routing table (per line) and won't change on further round. In fact, the transit interface ID is missing, which is an important omission, but it can be recovered in keeping an auxiliary matrix which keep track of the first relay in each path. We note the identity $N_2 = N * N$, $N_3 = N * N_2$, for all integer k: $N_k = N^{*k}$, where $*$ is the *tropical* multiplication defined as follow: For two matrices A and B with respective (i, j) coefficients a_{ij} and b_{ij}, the matrix $A * B$ is the matrix whose (i, j) coefficient is $\min_k\{a_{ik} + b_{kj}\}$. It is called tropical algebra because it is the algebra which replaces the addition operator with the "min" operator and the multiplication with the "+" operator. If you do the substitution, you indeed express a classic matrix multiplication. The initial matrix I indeed plays the role of the matrix identity in the tropical matrix multiplication.

The number of rounds for the protocol to converge is the diameter of the graph, i.e. the distance between the pair of routers which is the most remote from each other. With the general metric, this would be the largest number of hops in the minimal distance path between two routers. This delay can be improved, for example, by triggering TC emission every time there is a change in the routing table. Of course, this may cause much more intermediate TC transmissions out of the TC period.

The TC in RIP is particularly heavy, since they contain all the routing table of the host. By comparison, the TC is full like a truck but the truck travels only to the neighboring cities. Let's take an example. Figure 4.10 shows the operations. The two routers on the right send their "trucks" toward the router on the left. To stick with

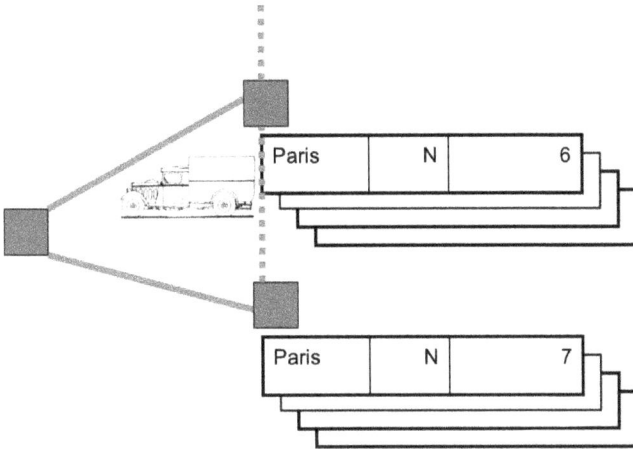

Fig. 4.10.   An example of RIP operation.

comparison with cities, assume that a router called "Paris" is at distance 6 from the top router and at distance 7 from the bottom router. The left router will receive the two entry and select the minimal one: the entry coming from its "North-East" exit road, the entry of Paris of the left router will indicate North-East exit as transit interface and distance 7 (one unit more than the distance received from North-East).

Now we address the question of the complexity of the protocol.

Theorem 4.1. *At each update period, RIP protocol transmit $|V| \times |E|$ IP addresses in TC, where $|V|$ and $|E|$ are respectively the number of vertices and the number of edges in the graph (V, E).*

Proof.   A TC packet contains $|V|$ IP addresses, plus one interface ID (can be omitted in the TC since it is the originator address), plus a distance integer (merely 4 bytes). The TC has to move toward all adjacent nodes of every host, thus every edges in the network must see $|V|$ IP addresses carried in TCs. The complexity of the RIP protocol consists in moving $|V|$ IP address in TC over $|E|$ links at each round. Thus, a $|V| \times |E|$ ID moved.   □

We see later how it compares with other protocols.

4.2.3 *The border gateway protocol (BGP)*

BGP is a link state protocol type [27]. It is also known under the name *Open Shortest Path First* (OSPF). It is an *a priori* light protocol, since at every TC period (20 mn as for RIP) each host issues a TC which only contains the entries of the routing table concerning the direct neighbors of the host. This should be significantly smaller than the whole routing table.

The difference is that at the other end the neighbor host will recopy the TC packet and retransmit it to its own neighbors and so forth; the copies of the TC will flood the entire subdomain. A sequence number added in the TC, associated with the IP address of the originator of the TC, will help avoid unnecessary duplication. This means that the TC will be transmitted $|E|$ times; this may significantly expand the quantity of information transmitted per round time compared to the RIP protocol. When the TC has been received by each router, it means that all the routers of the domain will "know" the list of neighbor hosts of the originator router. At the end of each TC period, each router in the domain "knows" the router graph of its subdomain. This information is stored in each router topological directory. Figure 4.11 shows an example of BGP operations. The router and the left issue a TC which contains the information about its direct neighbor routers; the latter retransmits the TC toward the rest of the subdomain. To illustrate that the payload of the TC is significantly reduced compared to the RIP truck, we have figured it with an icon representing a bike. In BGP, the TC

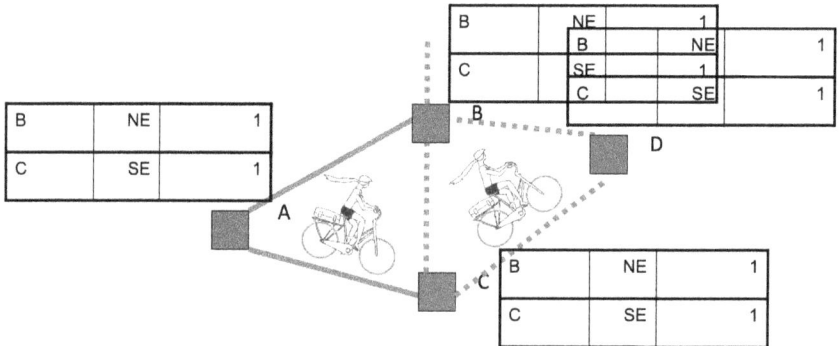

Fig. 4.11. An example of BGP operation.

is specifically called a *Link State Advertizement* packet, but we keep the TC denomination for consistency.

There is an apparent chicken-egg problem on how the host receives the information about the IP addresses of its direct neighbor. Logically, at its first round, in the ignorance of this information, the host could issue an empty TC containing only its own IP address, but this would be a waste of transmission since the TC should nevertheless be forwarded toward the whole subdomain. Instead, the protocol relies on a lower level packet, the "hello" packet which contains only the IP address of the originator and which is transmitted only to the direct neighbors and not beyond. This allows each router to catch the connect the information about the IP address of the direct neighbor with transit interface ID.

With the information stored in its topological directory, which is supposed to be the same in each host, the router computes the routing. More precisely, it computes the shortest path to any of the distant routers. The algorithm can be the Bellman–Ford algorithm, but the Dijkstra algorithm is more appropriate since it has lower complexity. The output is a shortest path tree. The root is the host; at each step, the tree is augmented of the link which creates the shortest path from the root to a new node. For the graph in Figure 4.9 for the host A, the tree evolves the following way: $\{AB\}$, $\{AB, AC\}$, $\{AB, AC, AD\}$, $\{AB, AC, AD, BF\}$ and $\{AB, AC, AD, BF, FG\}$. Figure 4.12 shows the five steps of the Dijkstra algorithm on the graph corresponding to the subdomain shown in Figure 4.9. Straight

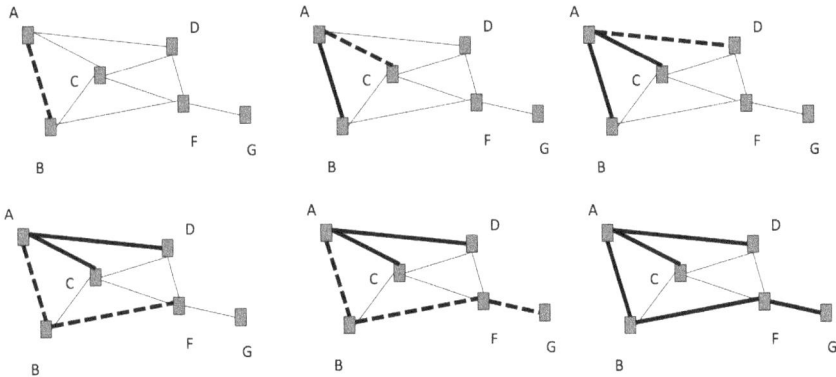

Fig. 4.12.   Shortest path tree building.

lines are the edges already added, and dashed are the paths currently augmented. The final host short path tree is displayed in the bottom right of the figure. All these computations are done from the host topological directory database and does not lead to telecommunication between the routers.

Coming again on the question of the complexity of BGP in terms of volume control traffic:

Theorem 4.2. *During each update period, BGP moves $|E| \times (|V| + |E|)$ IP addresses in hellos and LSAs (TCs).*

Proof.   Although the TC packets of BGP is objectively lighter than the RIP TC, the BGP protocol generates more ID exchanges. Indeed, each link in the subdomain must be advertised in TC in transit on all links. Therefore, this result in $|E|(|E| + |V|)$ IP addresses exchanged per TC period (including the IP address exchanged in the hello packets). □

4.2.4   *And the winner is the Arpanet accident*

A quick comparison of RIP and BGP does not bring conclusive remarks. RIP is significantly lighter than BGP ($|E| \times |V|$ IP addresses exchanged per update period compared to $|E| \times (|V| + |E|)$ for BGP). Anyhow, BGP shows a shorter convergence time than RIP: one update period compared to several with RIP, although RIP makes shorter by allowing reactive TC exchange but to the cost of more TC packets at the end of the day. How to make our own choice between the RIP truck, heavy but short distance, and the BGP bike, light but long distance like in *Tour de France* (see Figure 4.13)?

Fig. 4.13.   Heavy RIP truck (left) versus light BGP bike (right).

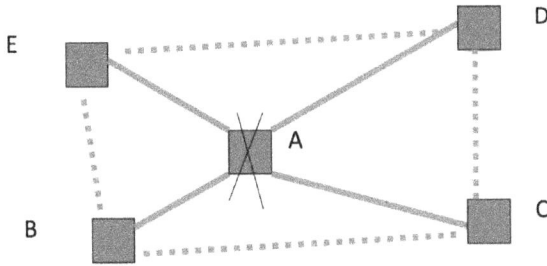

Fig. 4.14.   The count to infinity syndrome.

To tell the whole story, this was before the Arpanet accident on October 27 1980 [102]. Before that date, RIP was the sole routing protocol of Arpanet. That day, a virus was inserted on purpose for experiment in Arpanet. The virus knocked down two routers or at least the link between them and the loops created by this accidental removal caused overflow (like in a denial of service) in all other routers and caused the collapse of the entire Arpanet for four hour. Arpanet had to be manually rebooted in its entirety (fortunately less than 100 routers at that time).

To illustrate this event, let's us describe the "count to infinity" syndrome. To simplify our description, we assume the following sub-domain as illustrated in Figure 4.14 where the router A is suddenly knocked out.

Before the accident, the entry about router A routing table of B, C, D, E will be all identical to ┌───┬───┬───┐ A │ A │ 1 │, i.e. each router considers that the shortest path toward A is of length 1 through A itself (to simplify, we replace the transit interface ID with the name of router at the other end). After the knock out, all routers consider that their direct line to A is down and consequently remove the routing table received from A in their topological directory. At this moment, the router that the alternative route passes through any of their neighbors (which they still consider to be at distance 1 of A, according to their last routing table). Thus, the entries for router A in the routing tables now differ, but all with distance 2 (we assume that the router selects as next relay the smallest neighbor in lexicographical order):

- in B, │ A │ C │ 2 │,
- in C, │ A │ B │ 2 │,

- in D, | A | C | 2 |,
- in E, | A | B | 2 |.

We note that B and C are in a loop. Loops have a disastrous impact because assume A is a gateway, all packets directed to it will bounce forth and back on the neighboring router at the highest frequency permitted by the hardware and overflow the buffer.

On the next routing table roundup, the situation will remain unchanged with loops excepted that the distance has a new leg and is raised to 3:

- in B, | A | C | 3 |,
- in C, | A | B | 3 |,
- in D, | A | C | 3 |,
- in E, | A | B | 3 |.

The story could have had no end; following RIP, the distance continues to increase: 4, 5, 6, ... ∞. This is the "count to infinity". Many attempts were done to correct the flaw and detect loops but could not resist very little more sophisticated attacks. The network supposed to resist nuclear attacks was put down by a very harmless virus! The devil is in the details; indeed, RIP and BGP show similar performances in terms of time to convergence. But they completely differ in terms of *divergence time*, an abusively way to call the time to stable routing table after a link or a router breakage. This come from the fact that the operation of minimum distance vector in the building of RIP routing table does not perform symmetrically when a node is added versus when a node is removed. In the latter case, the RIP process will have to clear all the route toward the late before stabilizing. In BGP, the disappearance of a node or a link will just need few TCs to update the database about the neighbor routers of the late node and then each host recalculates its the shortest path tree.

In a climate close to pure panic, Arpanet engineering groups decided to switch from RIP to BGP (invented for this purpose). It was possible since at that time Arpanet was small enough, but it was like changing the engines of an aircraft in flight! From a design perspective, BGP is more complicated than RIP with the hello stuff, and the controlled flooding of LSA through the domains, the database synchronization, Dijkstra, etc. The specifications of RIP could stand in few pages; the full specification of BGP extends on several volumes of documents.

From the time BGP took the lead in Arpanet and Internet, the fate RIP was to remain as default routing protocols in some small stand alone networks. But the specification has been changed: The longest acceptable distance is 15, preventing the count to infinity.

4.2.5 *Why has the Internet become the universal digital telecommunications medium?*

The Internet was not the only candidate to become the universal telecommunication medium. The best-known (but definitely unsuccessful) challenger was called *Asynchronous Transfer Mode* (ATM [8]), which was favored for a time by the telecommunications industry. ATM was directly inspired by the telephone network, with the difference that information transfers were carried out in packets — in fact, in small packets because, at the time, the telephone was considered to be the main source of profit for operators. Larger packets would have exceeded voice sampling capacity, resulting in a considerable loss of bandwidth.

ATM was taking care of everything from the physical medium up to the routing protocol. But the successive ATM engineering committees comprehensibly failed to design a proper routing protocol. After each failure, the current committee dissolved and a new one was created. Its first job was to trash the work of the previous committees. It is true that the challenge was enormous. The operators were pressing hard to have routers capable of making contract with the user about the quality of services and giving priority to traffics accordingly. That way the operator would have been capable to charge more the user with quality of service. It would have been equivalent to having an electrical power provider capable of charging more electricity to a computer than the electricity to a lamp. Unfortunately, for the operator (and maybe fortunately for the user), the challenge was not attainable with the existing technology; the issue of the complexity was the deadlock for the scalability of the system. The problem was no longer a telecommunication problem but a computer science problem and Internet was the solution through some compromises.

Two technical key choices of Arpanet, the ancestor of the Internet, were, maybe involuntary at the time of their design, the cornerstones of the success:

(1) the blind forwarding,
(2) the end-to-end flow control.

In Arpanet, all packets are forwarded the same way, so there is no need to individualize the flows. Excepted for the TC packets generated inside the network, the only interaction of the router with any incoming packet is when its destination is read in order to determine its transit interface. This greatly increases the security of the telecommunication network; it is impossible that a packet generated outside Arpanet can be used for hacking the network. It is like in a bus, where the passengers cannot talk to the driver precisely for security reasons. There have been no known cases of Internet piracy over routers.

The overall concept over Internet packet delivery is the best effort routing with no distinction between packets. This led to the controversial concept of Internet neutrality, which has strong economic and political consequences around the question "who is going to make money in Internet" and we do not discuss that. But this concept, which was adopted in the early days of Arpanet and Internet, has an important technological impact on the design of the routers themselves.

Indeed, a router is not a computer with an usual architecture. In the early years of the Internet, a router was a classic PC with fast peripherals. The peripherals were the low-layer interface and a fast cache. In regular operations, a packet never penetrates inside the computer. The fast cache links the destination address to transit interface ID and the packet is directly forwarded to its transit interface without entering the router computer. When the packet address does not belong to the cache, then the packet is moved and temporarily stored inside the computer which contains the routing table. When the cache is updated, the stored packets are released to the fast peripherals.

The only operations made inside the computer are (i) the routing table look-ups and (ii) the topology database updates. The routing table updates do not occur so frequently, otherwise the routing table look-ups would be the main processing task of the processor. The

Root trie access

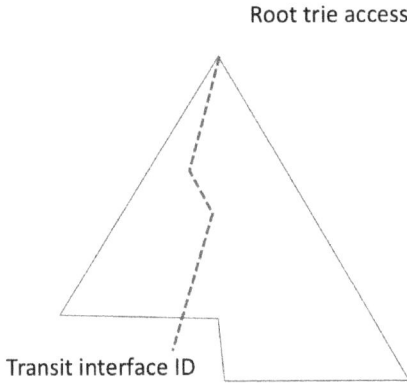

Transit interface ID

Fig. 4.15.   Routing table look-up as a trie search.

look-up consists of an improved search in the binary trie, or the prefix tree made of the host addresses stored in the routing table. The successive bits of the incoming addresses make a path from the root of the trie to the leaf of the tree which contains the ID of the transit interface (see Figure 4.15, the path to the leaf is dashed). The length of the path is the shortest prefix which distinguishes the incoming IP address with the other address stored in the trie. We know from the previous chapter about prediction and pattern-matching statistic that the average length of such a path is $\frac{\log_2 n}{h}$, where n is the total number of IP addresses stored in the trie and h is the entropy rate per bit of IP address. This proves that the information theory can still help the Internet.

In fact, the trie-based look-up is again too slow; the last versions are on compressed tries which give an average look-up cost of $\log \log n$. This illustrates the fact that the classic computer architecture would have been far too slow if all packets would generate routing table look-up. And this was the best that the existing technology could do at that time.

Therefore, it would have been hard to imagine that existing technologies could cope with the individualization and differentiated policies on packets as it was promoted by the ATM promoter. The remaining question is how to discipline all these flows of packet traffics in order to avoid remote congestion in the domains crossed. Indeed, the terminal is frequently on a high-speed local area network; if the user sends traffic at the speed permitted by its local connection, it would likely congest the remote networks, in particular, if the latter were mostly telephone lines or satellite links, as it was the case in the 1970s and 1980s. Here is the second early miracle design of Internet.

The second key is the *Transmission Control Protocol over Internet Protocol* (TCP/IP [26]). It is the protocol which enables the Internet traffic to cope with the congested areas of the Internet domains *without* connecting and interfering with the remote routers. TCP/IP is an end-to-end policy. The principle is that congested routers will start to discard packets because of buffer overflows. TCP/IP is a connection-oriented protocol between two hosts which respectively play the role of the server and the role of the terminal. The hosts can change their roles, but it will be another TCP/IP connection. The protocol TCP/IP opens connections from the server host to the distant terminal host. The server sends the packets of data (e.g. a file which needs to be fragmented into several IP packets) and the terminal acknowledges the data packets via specific acknowledgement packets. Missing such packets will indicate that most likely a buffer overflow occurred in the path to the terminal. The server resends the missing acknowledged packets.

The protocol works as follows. At the beginning of a connection, the server augments the throughput as long as it gets no packet loss. At packet loss (detected via missing acknowledgement), the server reduces the throughput. A packet can be correctly received but the acknowledgement will be lost; in both cases, this is an indication of congestion, and in some cases, some packets will be received twice (eliminated through their sequence number in the connection).

The throughput is controlled via the classic protocol of the delayed window of transmission. The window of transmission is the maximum number of packets authorized for transmission before having received their acknowledgement. If the round trip delay (packet delivery plus acknowledgement) remains the same, augmenting the transmission window would be equivalent to augmenting the actual throughput.

At the beginning phase of a connection, the TCP/IP protocol makes the transmission window augment one unit after each acknowledgment received. This would be equivalent to multiplying by two the next transmission window, when all packets of the previous transmission window have been duly acknowledged. This phase is called the *Slow Start* for strange reasons, which we discuss later. When a packet loss occurs, the transmission window is halved and the further increases in the transmission window will be linear, i.e. the window size increases by a fraction proportional to the inverse of the previous window size after each acknowledgment received by the server.

Thus, to summarize, in the "slow start" phase, the throughput doubles at each transmission window as long as there is no loss. At the first loss, the throughput is halved and then is augmenting linearly in the absence of loss or is again halved at each loss. The reason why the first phase is called "slow start" is because this phase was absent in the first TCP specifications, and the protocol was directly started on the *Linear Increase Geometric Decrease* mode which would take an infinite time when the capacity of networks drastically augmented leading to a slow start. Thus, the solution has the name of the problem it solves! Blindly doubling the throughput greatly accelerates the transmission process in reaching its cruise regime.

The slow start phase can reach a window maximal size of 1,048,576 with a maximum IP packet size of 65,536 bytes, thus an attainable throughput larger than 500 Gbit per round trip delay, which is enough dynamics for all existing network interfaces. And all this without making any interference between external user and core routers.

In conclusion, the early Internet benefited from luckily design choices which made it the natural winner in the competition to the universal digital telecommunication system.

4.3   The Wireless Internet, Topology Compression

Considerable progress in micro and nanotechnology has enabled the emergence of the wireless Internet. As Shannon's laws indicate, the challenge of high bandwidth could not be solved by increasing the energy dissipated during transmission but by extracting information from a low signal-to-noise ratio. This required more powerful and more complex chips, the generation of which took place in the late 1970s. Again, most of these new technologies would not have been possible without the considerable investments made by governments to counter the threat of other governments during the World

Wars and the Cold War. It's funny to think that the most unifying technologies in terms of communication between the peoples of the Earth were the result of one of the most difficult periods of confrontation in human history. However, the ability of the Internet, mobile networks and social networks to bring peace between peoples should be constantly reassessed. Nevertheless, the new generation of wireless equipment could change the wireless equipment of the early 20th century, which occupied an entire room, see Figure 4.16, to be finally concentrated in a tiny chip. What's more, the reduction in size and cost has enabled the technology to spread and be used on an unprecedented scale.

The wireless Internet consists in replacing the cloud of connected routers in Figure 4.2 by a cloud of wireless devices (see Figure 4.17). There is neither economical nor technical interest to replace the wired interface with a wireless interface, since the performance (range or throughput) of wireless interfaces is always inferior by several orders of magnitude to the performance of their wired counterparts. In fact, the most interesting applications are *Mobile Ad hoc NETworks*

Fig. 4.16.   Left: Early radio device.
Courtesy: Wikipedia. Right: A tiny WiFi interface; not to scale (it is actually smaller than any detail in the left picture).

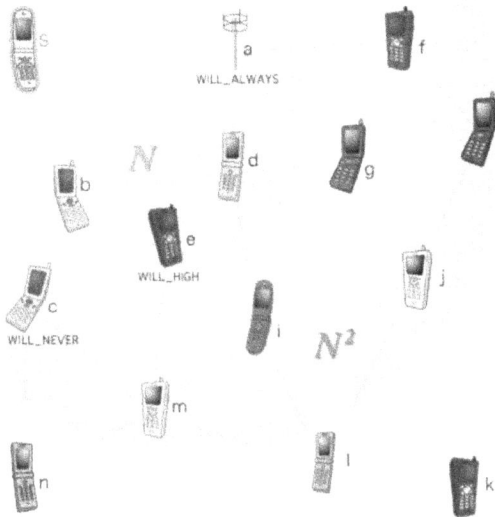

Fig. 4.17. Wireless Internet.
Courtesy: Wikipedia.

(MANET), which are dedicated to locally replace wired networks in the absence or destruction of infrastructures.

A MANET is a network made of wireless routers which are expected to move in some *ad hoc* mode, i.e. not necessarily planned movements, and in total absence of fixed infrastructure [76]. Most MANETs have military, or para-military applications. Imagine a city ravaged by an earthquake or by a hurricane, with most of the Internet routers knocked down. Rescue teams will progress in the devastated areas and communicate via MANETs. An invading force will first destroy the communication link of the enemy, therefore invading and defendant forces would need to rely on MANET for their own tactical communication. The invading force would need to bring their own telecommunication devices and not try to subscribe to the civilian network of the invaded country (some have tried this economy-saving communication method but with utterly failure due to easy eavesdropping).

The MANETs are limited by the performance of their wireless interface. If we take the performance of WiFi as an order of magnitude, the MANET's link should have the following characteristics:

- ten to few hundred meters of range inside buildings, up to one kilometer in free space,
- to few Mbit/s of throughput.

Furthermore, the speed of movement, between one to ten meters per second will make the link breakage or link apparition much more frequent than with the fixed Internet (supposedly much less mobile than a mobile networks). Indeed, a link failure in Internet is a *possible* event but not a *frequent* not even an *expected* event.

If we recall the purpose of Arpanet/Internet as the way to cope with the partial or total destruction of some network parts following a nuclear strike, nevertheless the time to recovery is far exceeding the TC period, 20 minutes, with many synchronisation phase under BGP protocol [27]. But 1 or 2 hours was an acceptable delay to order a counterstrike (and in fact a kind of good respite before Armageddon). But in the case of MANET, a link may experience a failure event within 10 or 100 seconds.

The main issue with MANETs is that their offered bandwidth may be too weak to support the control traffic needed for the routing management: too much movement on far too small bandwidth. Let's take the example of a stadium with 100,000 attendees (see Figure 4.18); remember that the support of communication in large-scale and peaceful events is also an application of MANETs. Assume that locally each user has a direct link to an average number of one thousand users. Indeed, for classic routers in fixed Internet, the number of neighboring routers is limited by the number of low-level interface, that is, the number of physical switches on the box. In radio, the link exists when two terminals are within radio range. All local links are gathered in a single interface (e.g. WiFi interface). Thus, in theory, the stadium network is with $|V| = 100,000$ and $|E| = 10^8$. If we apply the result of Theorem 4.2, we should get 10^{16} IP addresses exchanged during one update period with BGP. If the update period is of the order of 10 seconds, this would be completely unrealistic with WiFi like interfaces. To compare, 10^{16} is the number of meters traveled by a light beam during one year at 300,000 km/s.

In fact, early experiments of MANETs were done with an accelerated version BGP (by squeezing update periods) over WiFi interfaces to make them capable of tracking topology changes of walking users. They went into complete failures when the number

Fig. 4.18.   A large stadium event. Stade de France.
Courtesy: Wikipedia.

of mobile terminals exceeded 20, the WiFi network collapsing under the volume of topology control traffic.

4.3.1   *The topology compression*

A WiFi interface can move up to 1,000 packets per second. This would not pose any specific problem to any classic computer or portable phone. Contrary to the core of Internet which is moving Terabit per second, a wireless router does not need any specific hardware or software. The challenge is no longer a Computer Science problem, but it falls back in the garden of Information Theory. The main problem is that the routing protocol and the management of the wireless topology may give rise to too much information exchanges.

Indeed, there is no physical limitation in the number of neighbor routers except in the far unreachable maximum number of portable phones which can be packed in one hectare. Thus, the size of the graph can explode without warning. But in a dense topology (i.e. a topology with large neighborhood size), the graph comes with redundant path and route, which are not fully needed. It is not even needed for the balance of loads over the links because when the links share the same medium, two adjacent links will not be in capacity to carry

more bandwidth together than will do a single isolated link. Therefore, eliminating the knowledge about those redundant links may be
a good way to *compress* the topology without affecting the connectivity of the MANET.

4.3.2   *The multipoint relay selection of OLSR*

OLSR stands for *Optimized Link State Routing*; it is a simplified version of BGP specifically designed for MANET [29,59]. It is based on a
topology compression called *MultiPoint Relays* (MPR [98]). Invented
in Inria, it has become the standard of MANET routing (see the original logo in Figure 4.19).

Instead of advertising all its neighbor nodes (which would lead to
too many redundant path), an OLSR node advertizes a subset of its
neighborhood, called the node MPR set. The MPR set is expected
to be much smaller than the full neighbor set (see Figure 4.20).

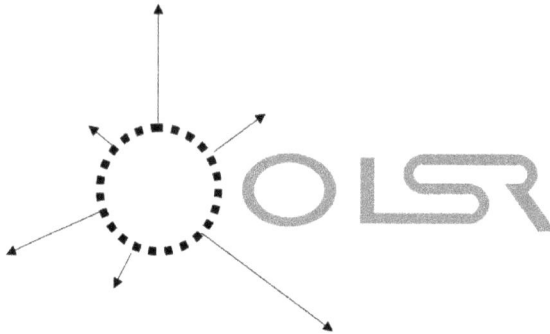

Fig. 4.19.   Optimized link state routing.

Fig. 4.20.   Left arrows: The full neighbor set. Right arrows: The MPR set.

Let E_r denote the set of MPR links, i.e. the edge from each host to its MPR set. We expect the size of E_r to be much smaller than the size of E when the graph is particularly dense. Of course, the choice of the multipoint relays cannot be done arbitrarily, since they will be the backbone of the construction of the routing table.

For a node A we denote $\mathcal{N}(A)$ the set of neighbor routers of node A and $\mathcal{N}_r(A)$ the MPR set of node A. We denote $\mathcal{N}^2(A)$ the two-hop neighbor set of A. With rigorous math, $\mathcal{N}^2(A) = \bigcup_{B \in \mathcal{N}(A)} \mathcal{N}(B) - \mathcal{N}(A) - \{A\}$.

The rule is that in all cases the set $\mathcal{N}_r(A)$ should cover the two-hop neighborhoods of node A: $\mathcal{N}^2(A) \subset \bigcup_{B \in \mathcal{N}_r(A)} \mathcal{N}(B)$ (see Figure 4.21). To keep the protocol distributed, each host is in charge of determining its own MPR set. The lazy choice is to do formally $\mathcal{N}_r(A) = \mathcal{N}(A)$, but this is not interesting since it does not "compress" the topology. The optimal MPR set $\mathcal{N}_r^*$ is when the MPR set has the minimal size. To make its MPR selection, each host A needs to "know" its neighborhood $\mathcal{N}(A)$ and its two-hop neighborhoods $\mathcal{N}^2(A)$; more precisely, it must know the neighborhood set of each of its neighbor routers. We see later that this "knowledge" is provided via hello packets.

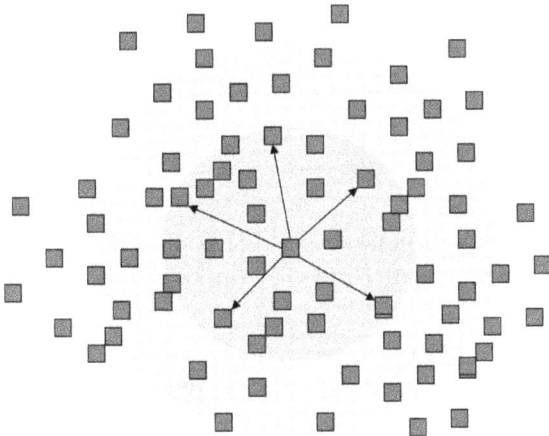

Fig. 4.21.   The neighbor set in pink, and the two-hop neighbor sets in pale pink. The arrows show a potential MPR set of the central node which is given.

We give two examples of MPR sets. If the graph (V, E) is a complete graph, then for all $A \in V$ the optimal MPR set is the empty set, since for all A, $\mathcal{N}^2(A) = \emptyset$. If the graph (V, E) is a tree, then for all $A \in V$ $\mathcal{N}_r(A)$ must be identical to $\mathcal{N}(A)$.

There is bad news and good news about the MPR selection process in general situation. The bad news is that finding the minimal MPR set $\mathcal{N}_r^*(A)$ is an NP-hard problem with respect to the neighborhood size. Without digging too much in complexity theory, it means that there are not much better strategy than checking all possible neighbor subsets. This is a consequence of the graph dominating set NP hardness. For more information about NP hardness, see Section 8.5.1. The good news is that the following greedy algorithm provides an MPR set which is no more than $\log |\mathcal{N}(A)| + 2$ larger than the minimal dominating set [1]:

$$\frac{|\mathcal{N}_r(A)|}{|\mathcal{N}_r^*(A)|} \leq \log |\mathcal{N}(A)| + 2.$$

In other words, the order of magnitude of $|\mathcal{N}_r(A)|$ and $|\mathcal{N}_r^*(A)|$ is similar compared to $|\mathcal{N}(A)|$. The following pseudocode describes the greedy algorithm:

procedure MPR(A)
  $\mathcal{N}_2 \leftarrow \mathcal{N}^2(A), \mathcal{N}_r(A) \leftarrow \emptyset$
  while $\mathcal{N}_2 \neq \emptyset$
    do $\begin{cases} \text{FIND}(\mathcal{N}(A)) \ y \ \text{which maximizes} \ |\mathcal{N}_2 \cap \mathcal{N}(y)| \\ \mathcal{N}_r(A) \leftarrow \mathcal{N}_r(A) \cup \{y\} \end{cases}$
  return $(\mathcal{N}_r(A))$

Note that the relation node B is MPR of node A is not necessarily symmetric. It means that the MPR link set is not necessarily made of symmetric links.

Let's denote τ_r the ratio $\frac{|E_r|}{|E|}$. Our aim is to give some evaluation of τ_r in some wireless network models. We investigate the following two classical models:

- the Erdos–Renyi graph model,
- the unit disk graph model.

4.3.3   The Erdos–Renyi graph model for indoor wireless network graph model

Inside a building the main obstacle to radio propagation is the presence of walls and furniture. The signal attenuation with distance plays a secondary role, since the equipments are not far enough to get significant signal attenuation because of distance (see a floor plan for the White House in Figure 4.22). In other words the existence, or absence of a link, between a pair of terminals is a random process, depending on the presence of a random obstacle in between.

A random Erdos–Renyi graph is a graph $G(V, E)$, where V is a fixed set of vertices and E a random set of edges created through a number $p \in [0, 1]$. Here we look to undirected random Erdos–Renyi graph. For each pair of vertices $\{u, v\} \subset V$, E contains the edges uv with probability p. To characterize such a graph, we don't need a formal set of vertices; it is sufficient to know the number n

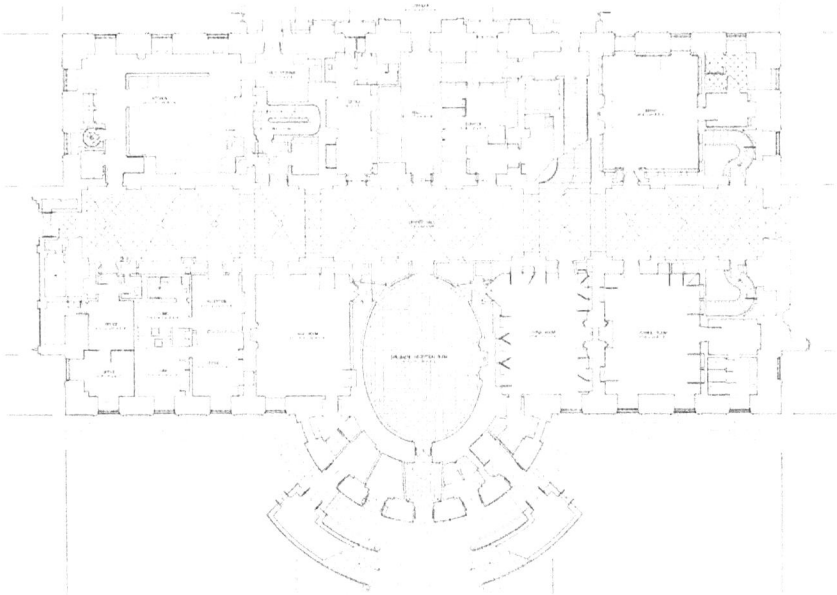

Fig. 4.22.   The White House floor plan with random obstacles with furniture and walls.
Courtesy: Wikipedia.

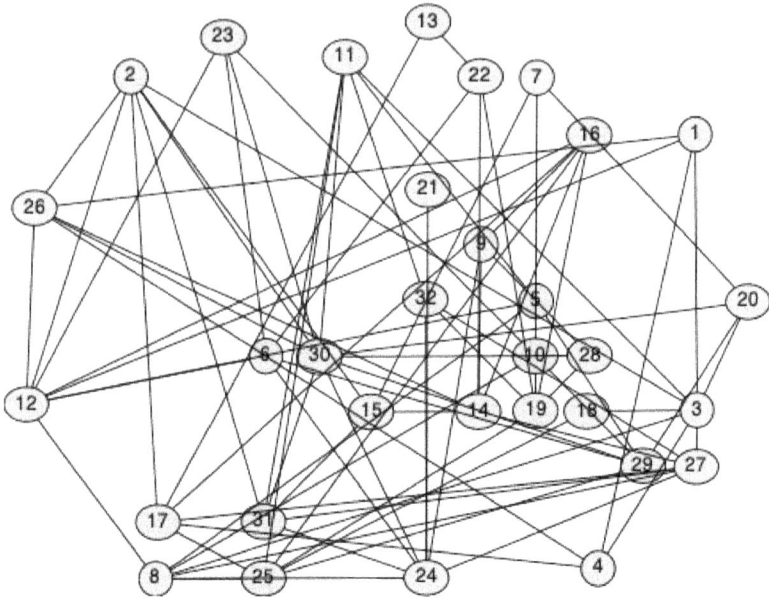

Fig. 4.23.   A random graph $G(32, 0.2)$.

of vertices and the probability p. We denote $G(n, p)$ such random graphs. Figure 4.23 gives an example of such a graph (the positions of the vertices are random in the plan).

The following theorem holds [60]:

Theorem 4.3. *Let $p \in [0, 1]$, when $n \to \infty$: $\tau_r = O(\frac{\log_{1-p} n}{n})$.*

Proof.   Since there are $\binom{n}{2}$ possible pairs of nodes, we have the average set size $|E| = 2p\binom{n}{2} \sim \frac{n^2}{2}$. Thus, the average density is pn which corresponds to a very dense network when $n \to \infty$. A very simple MPR selection consists in picking neighbors in a random order and stoping when the two-hop set is covered. After k MPR randomly selected, the probability that a given node is not covered is $(1 - p)^k$, and the probability that there exists a node which is not covered is smaller than $n(1 - p)^k$ which exponentially converges to zero when n increases if $k > \log_{1/(1-p)}(n)$.   □

4.3.4   *The unit disk graph model, outdoor wireless network graph model*

In outdoor situation, obstacles such as vegetation and relief play less role than the distance. The radio has a limited range determined by power and inter-symbol interferences. The interferences increase with distance to the point that they infer errors which are beyond the capability of standard error correction. In the unit disk graph models, terminals are random points and any pair is considered in the range of each other as soon as they are at a distance smaller than some distance threshold. Thus, such a graph is given by two-dimensional surfaces, a number of nodes and a threshold radius. The model is called the "unit disk graph" model; if we consider the radio range as the length unit, the unit disk around a node is supposed to contain all the neighbors of this node. Figure 4.24 shows an example of an unit disk graph on a 10×10 map.

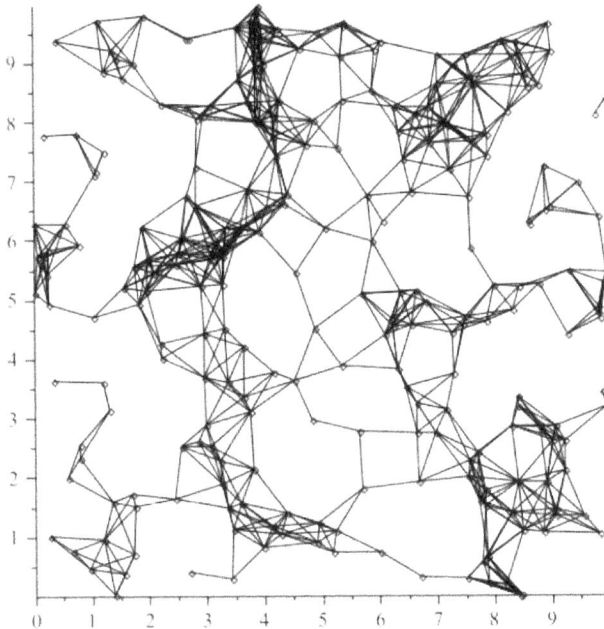

Fig. 4.24.   A unit disk graph on a 10×10 rectangular map.

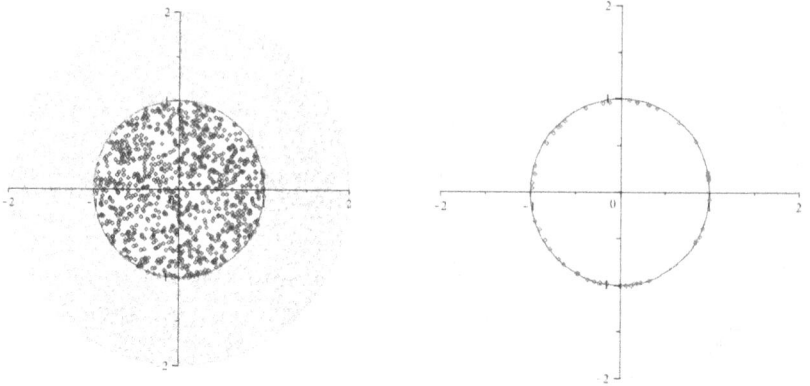

Fig. 4.25.   Unit disk graph. Left: A node with 1,000 neighbor, and 3,000 two-hop neighbor. Right: The MPR set.

We have the following theorem [1]:

Theorem 4.4. *In unit disk graph network model, we have $\tau_r = O(n^{-2/3})$, where n is the average neighbor size.*

The result of the theorem is related to the fact (but is not a direct consequence of it) that the convex hull of n random points inside a disk is of order $n^{1/3}$ (and not of order $n^{1/2}$ as one might wrongly think [84]). Figure 4.25 (left) shows a dense unit disk graph network, with 4,000 nodes: 1,000 blue nodes are in the unit disk; these nodes are neighbors of the central node; 3,000 green nodes are in the crown area of radius between 1 and 2. On the right, the figure displays 31 red nodes which are the MPR of the central node. The green nodes are the remaining nodes not covered by the MPR set because they finally don't belong to the two-hop neighborhood of the central node. Indeed, all the two-hop neighbor nodes belong to the crown area, but the reverse is not true, depending on the configuration of the nodes inside the unit disk.

4.3.5   *Computing OLSR routing tables*

The graph $G_r = (V, E_r)$ is a reduction of the full topology graph $G = (V, E)$, but computing a shortest path tree over it may not lead to optimal route. In fact, there is not even the guarantee that the reduced graph is connected. The rule of the game is to augment the reduced graph with the local topology, which is supposed to be acquired via local exchanges (hellos) independent of the TC traffic. Let $E(A)$ be the set of links connecting the node A to its neighbor nodes. The node A computes its shortest path tree over the graph $G(A) = (V, E(A) \cup E_r)$, more precisely on the set of reversed links of $G(A)$. We remind that the graph $G(A)$ is not necessarily undirected and the link may be not all symmetric.

Let $E' \subset E$, where E' is a *remote spanner* [118] of the graph (V, E) if for all $u \in V$ the graph $(V, E(u) \cup E')$ is reverse connected to u, that is, for all $v \in V$ there exists a path from v to u. A remote spanner is lossless if for all $(u, v) \in V^2$ the shortest path from v to u in $(V, E(u) \cup E')$ has the same length as the length of the corresponding shortest path in (V, E).

Theorem 4.5. *If the graph (V, E) is connected, then the MPR link set E_r is a lossless remote spanner.*

Proof. In an abusive use of language, we can say that the topology compression induced by the MPR link set is lossless. This property is interesting because it allows saving the performance of the network by reducing the number of retransmissions of each data packet while reducing the control traffic overhead of the TCs.

Let $A \in V$ and consider an other node $B \in V$. The proof is done by recursion on the distance k from A to B in the original graph.

If $k = 1$, then B is neighbor to A and $AB \in E(A)$, and the reverse link belongs to the reversed $G(A)$. Now assume that the hypothesis is true up to distance k and let us assume that B is at distance $k+1$. By hypothesis, there exists a path $B_1, \ldots, B_k$ in G which connects B to A. B_1 is a neighbor of B and B_2 belongs to the the two-hop neighborhood of B.

By property of the MPR set selection, there is a node G, neighbor of B which is also neighbor of B_2 (covers B_2), thus $BG \in E_r$. The node G is at distance k of A and by hypothesis there exists a path $G_1, \ldots, G_{k-1}$ in $G(A)$ which connects G to A. Since $BG \in E_r$, the path $G_1, \ldots, G_{k-1}, G$ belongs to $G(A)$ and connect sB to A with a length $k + 1$ (see Figure 4.26). □

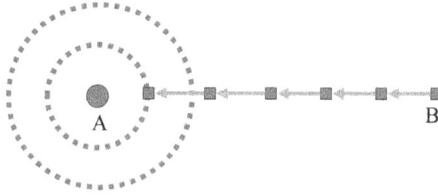

Fig. 4.26.   The proof of lossless MPR link remote spanner property.

4.3.6 *Flooding optimization*

With the current optimization with the topology compression, the complexity cost of OLSR in terms of volume of IP address exchanged per topology update period drops from the BGP TC exchange complexity of $|E|^2$ to $|E| \times |E_r|$, i.e. a factor τ_r of reduction. Even with τ_r of order $\frac{1}{|V|}$, this would not be sufficient to cope with the stadium network overhead of 10^{15} (which would only drop down to 10^{12}).

Another optimization is in the flooding process [83]. A flooding process consists of forwarding a packet toward all nodes in the network. The classic flooding operation makes the nodes retransmit the packet on all their adjacent links. An addition of a sequence number prevents multiple retransmissions which may loop forever. This is the flooding process of BGP, leading to a complexity cost of $|E|$ retransmissions for each flooded packet. In fact, one can drop down to $|E|/2$ by preventing the retransmission to neighbor which has already retransmitted the packet on the incoming link, but this does not change the order of magnitude. The following pseudocode describes the local flooding operations:

procedure UNICASTFLOODING(A)
  while true
    do
$\begin{cases} P \leftarrow \text{RECEPTION}() \\ \textbf{if } \text{TYPE}(P) = Flooding \\ \quad \textbf{then} \begin{cases} \textbf{if } P \notin AlreadyReceived \\ \quad \textbf{then} \begin{cases} \text{ADD}(P, AlreadyReceived) \\ B \leftarrow \text{LASTEMITTER}(P) \\ \textbf{for } C \in \mathcal{N}(A) - \{B\} \\ \quad \textbf{do } \text{SEND}(P, C) \end{cases} \end{cases} \end{cases}$

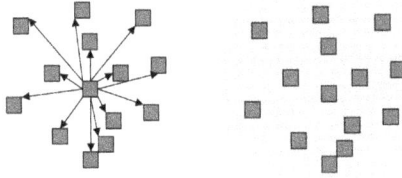

Fig. 4.27.   Left: Flooding via unicast. Right: Flooding via broadcast, a single transmission reaches all neighbors.

The first obvious flooding optimization is to take benefit of the broadcast nature of radio propagation: Instead of repeating the packet transmission as many time as there are adjacent links (we call this the flooding via unicast), the packet is retransmitted only once, to be received by all neighbor in one shot (we call this the flooding via broadcast, see Figure 4.27). That way the flooding complexity cost drops down to $|V|$ instead of $|E|$. And the OLSR complexity would be now downed to 10^9 instead of 10^{15}.

The pseudocode for the broadcast flooding is as follows:

procedure BROADCASTFLOODING(A)
  while true
    do $\begin{cases} P \leftarrow \text{RECEPTION}()\text{if } \text{TYPE}(P) = \text{Flooding} \\ \quad \textbf{then } \begin{cases} \textbf{if } P \notin \text{AlreadyReceived} \\ \quad \textbf{then } \{\text{ADD}(P, \text{AlreadyReceived})\text{BROADCAST}(P) \end{cases} \end{cases}$

The last OLSR flooding optimization is the *MPR flooding*. In this case, only the MPRs of a transmitter retransmit the packet. If the packet is received the first time by a non-MPR neighbor, the packet is not transmitted. The pseudocode is as follows:

procedure MPRFLOODING(A)
  while $TRUE$
    do $\begin{cases} P \leftarrow \text{RECEPTION}() \\ \textbf{if } \text{TYPE}(P) = Flooding \\ \quad \textbf{then } \begin{cases} \textbf{if } P \notin AlreadyReceived \\ \quad \textbf{then } \begin{cases} \text{ADD}(P, AlreadyReceived) \\ B \leftarrow \text{LASTEMITTER}(P) \\ \textbf{if } A \in \mathcal{N}_r(B) \\ \quad \textbf{then } \text{BROADCAST}(P) \end{cases} \end{cases} \end{cases}$

Note that when received a second time, the packet is not processed because of being already stored in the *AlreadyReceived* database. The gain is not easy to compute, unless making side assumptions, but it is around τ_r. Therefore, the gain in complexity performance from BGP to OLSR is of order $\frac{\tau_r}{|V|}$; with $\tau_r \approx \frac{1}{|V|}$ this would be a gain of 10^9 in the stadium situation. Anyhow, it is not clear from the above description that the packet will reach all the destinations. This is proved in the following theorem:

Theorem 4.6. *In case of lossless transmission, the MPR flooding reaches all nodes in the network.*

Proof.   By lossless transmission, we mean that when a node transmits a packet, all neighbor nodes receive it.

Note that we don't require that all nodes retransmit the packet, we only ask that all nodes receive the packet. Let us consider a random node A, not necessarily the originator of the packet to be flooded. Let k be the distance at which the packet has been (re)transmitted the closest to A during the whole MPR flooding process.

If $k = 0$, then A has retransmitted the packet (and thus has received it beforehand). If $k = 1$, then A has received the packet from a neighbor node but was eventually not eligible to retransmit it. We show that the case $k > 1$ is absurd and therefore node A has received the packet, as well have all the other nodes in the network. Let F be the first node to retransmit the packet at distance k of A.

Assume $k \geq 2$, and let $G, H, \ldots$ be one shortest path from F to A. G is at distance $k-1$ to A, and H is at distance $k-2$, which is possible since $k \geq 2$. If $k = 2$, then node H would be indeed A. Since H is in the two-hop neighborhood of F, then there exists an MPR node G' of F which covers H (see Figure 4.28). In the absence of transmission loss, the node G' would have received the packet from F. Since F is the first node to retransmit the packet at distance, the node G' would not have received the packet from another node (which would have been at distance at most k to A). Thus, G' must have retransmitted the packet since it satisfies the following two conditions:

- it is MPR of F,
- it has received the packet the first time from F.

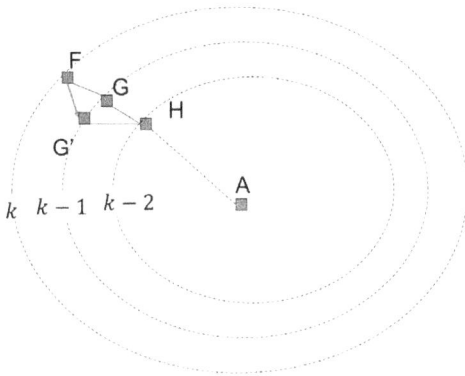

Fig. 4.28. MPR flooding proof.

If G retransmits the packet, then the packet has been retransmitted at distance $k-1$, which contradicts the hypothesis. □

4.3.7 *Hello packets*

As with BGP, OLSR has hello packets. But as with BGP, OLSR requires that the transmission links be symmetric. With wireless transmission, there is a risk of asymmetric links; this may occur when the transmission power or the sensibility parameters differ at both ends of the link. To cope with this problem, OLSR nodes periodically broadcast hello packets which contain a list of potential neighbor IP addresses:

- a list of "heard" nodes which have not been confirmed as symmetric neighbors,
- a list of the IP addresses of the symmetric neighbor nodes, i.e. neighbor nodes from which the host node has received a hello listing the IP address of the host node,
- the list of the IP address of selected MPR nodes of the host.

We note that the hello exchange allows the host node to build its two-hop database by reporting the symmetric neighbor (including MPR nodes) in the hello received from its neighbor nodes. Thus, the MPR selection can be performed locally.

The complexity cost of the hello process is therefore $|E|$ IP address moved per update period, since each link is advertised in the hellos of the node at one end of the link. Finally, the complete OLSR complexity is $|E| \times (1 + \tau_r^2 \frac{|E|}{|V|})$ IP addresses moved per update period. One should note that the main complexity component is likely the hello complexity. But the latter can be greatly reduced by transmitting partial hellos, containing only the IP address of the originator. A full hello is transmitted only when either a new neighbor appears or when one symmetric neighbor is lost (in this case, one must add a list of "lost" neighbors in the full hello). In this case, the OLSR complexity, in stationary situation, would be $|V| + \tau_r^2 \frac{|E|^2}{|V|}$ IP addresses moved per update period.

4.4   Exercises

4.4.1   *RIP count to infinity*

We imagine a network made of a chain of k routers: $A_1, A_2, \ldots A_k$. We assume that time is slotted with a time slot of one unit. At the beginning of each slot, all routers send their TC to their neighbor. At the end of the slot, each router computes the length of its shortest path to any other router.

At time $t = 0$, a new router A_0 connects to router A_1.

Exercise 18

Assuming that at time $t = 0$ all routers show an infinite distance to router A_0 in their routing table shows the evolution of this distance with respect to t and the router index. See the Answer of Exercise 18 in Chapter 9.

At some time T' in the future of time k, the link between router A_0 and router A_1 disconnects.

Exercise 19

Show the evolution of the distances to A_0 in router A_i for all $1 \leq i \leq k$. Comment the count to infinity. See the Answer of Exercise 19 in Chapter 9.

4.4.2   *MPR selection in OLSR*

The greedy selection of MPR is very efficient but in theory can be away from optimal by a mere factor $\log n$. But, in most case, it gives the optimal MPR set.

Exercise 20

Find a simple counterexample where the greedy MPR selection does not give an optimal MPR set. See the Answer of Exercise 20 in Chapter 9.

4.4.3   *Remote spanner and topology compression*

We consider wireless network with a set V of vertices and an edge set E. A remote spanner is a set of links $\mathcal{S}$ such that for all vertices $v \in V$ the set of links $\mathcal{S} \cup E(v)$ (the set of neighbor links to v) makes a connected set which covers all the vertices in V.

Exercise 21

Let $G(V, E)$ be a complete graph (i.e. a graph where for all pairs (u, v) of distinct vertices $(u, v) \in E$. What is the minimal size of a remote spanner of $G(V, E)$? See the Answer of Exercise 21 in Chapter 9.

Exercise 22

Show that $|\mathcal{S}| \geq |V| - 1 - \frac{|E|}{|V|}$. See the Answer of Exercise 22 in Chapter 9.

    We note that with a complete graph we have $|E| = (|V| - 1)|V|$, thus the lower bound on $|\mathcal{S}|$ is exactly 0 which corresponds to the bound found at the first question.

    In the general case, we get the minimum topology compression rate $\frac{|\mathcal{S}|}{|E|}$ which tends to be equivalent to $\frac{|V|}{|E|}$ which cannot be smaller than $\frac{1}{|V|}$ (since in every case $|E| \leq |V|^2$) and is very close to the topology compression τ_r obtained with the MPRs.

Chapter 5

The Performance Paradoxes of Wireless Networks Caused by Physics

Abstract

Wireless networks are the most interesting incarnation of telecommunication technologies. Wireless networks are those that are closest to the nature of the physics of space and time. The medium of information is the electromagnetic field, which interacts strongly with the environment and can undergo rapid variations, given the mobility of users and the variability of their environment. In the wired world of the Internet, a transmission link is stable. It can fail, but this event is so infrequent that the line restoration protocol is cumbersome and cannot be invoked too frequently. In a mobile wireless network, mobile users may not be within the range of each other, so connectivity will be ensured by internal routing between moving users. In this context, the appearance and disappearance of a link are natural and frequent events, so the network must constantly monitor its topology, and the resulting performance is highly dependent on mobility as we seen in the previous chapter. We present three theoretical consequences of this situation. The *space capacity paradox* shows that increasing user density can increase total transport capacity, i.e. the amount of information that can be transported in the network per unit of time. This unexpected result, which occurs without modifying the information medium, seems highly counterintuitive. Total capacity scales as $n^{1-1/D}$, where n is the total number of users and D the dimension of the integrated space (1, 2 or 3). Note that each user's share decreases in $n^{-1/D}$, which means that if the cake increases in volume, individual shares decrease.

The *time capacity paradox* shows a similar phenomenon but linked to time. In simple terms, networks are based on internal routing transmit packets as soon as they are received. By introducing time delay, transport capacity is increased, as the geometry of the network can evolve and become more favorable to the transmission of information. Under certain (sometimes unrealistic) conditions of mobility, with infinite tolerance to delays (a packet can take months to be delivered), total capacity can be increased to the order of n, i.e. individual slices of the cake no longer decrease.

A final paradox *fractal capacity* takes advantage of the geometry of the network, and an unexpected gain in capacity when the geometry of the substrate space is fractal.

5.1   Wired versus Wireless Networking

5.1.1   *The "brave old world" of wired networking*

The advantage of cable networks is that they respond to a "reasonable" universe. In a 100 kbps physical cable, no matter how many users are connected, each user can't get more than its share, and when we add up all the shares of all the users connected simultaneously, we again get 100 kbps. Or perhaps less, if we subtract the overheads wasted in ensuring fair access to the resource (see Figure 5.1). This is the *isochoric* conservation of volume (from the Greek *iso* "same", *koros* "volume").

Similarly, if the physical properties of the cable do not change over time, there is no point in waiting for more bandwidth. The sum of the parts of the cake will always be 100 kbps at 10:00 a.m. as it is at 11:00 a.m. This is the conservation of volume over time (see Figure 5.2).

All these conservation properties underline how a reasonable universe should be where there would be no question of free meals.

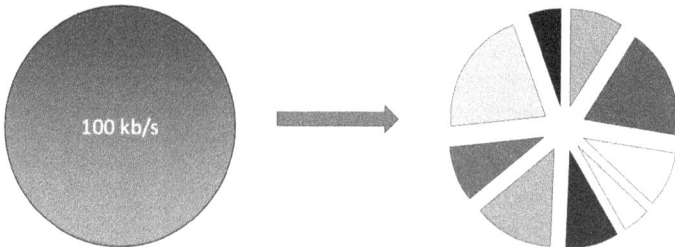

Fig. 5.1.   The splitted resource still adds up to 100 kbps.

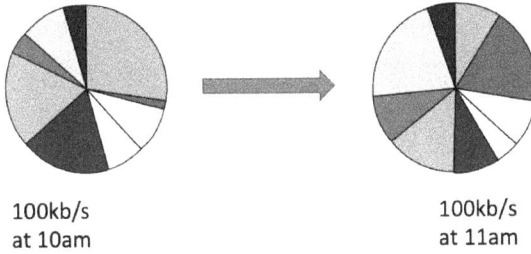

100kb/s
at 10am

100kb/s
at 11am

Fig. 5.2.   The splitted resources still add up to 100 kbps independent of time.

However, we're going to show that wireless networks break these rules, i.e.

- space creates wireless capacity,
- time creates wireless capacity.

This is mainly due to the specific interactions of wireless transmissions with the physical properties of our environment. In the following subsection, we describe these interactions.

5.1.2   *The physics of wireless networks*

Compared with conventional wired communications, wireless networks are much more linked to the physical properties of our environment. This is because radio signals have to pass through spaces and encounter material obstacles, just like any other electromagnetic wave, such as visible light. Among the most important physical effects that wireless transmissions have to combat are:

- the signal attenuation and degradation by distance,
- the lack of isolation of wireless links.

The average attenuation of the signal is a function of distance to a given power. If $S(\mathbf{z}_R, \mathbf{z}_T)$ is the energy intensity received at a location $\mathbf{z}_R$ from a signal transmitted from a location $\mathbf{z}_T$, the mean value of $S(\mathbf{z}_T, \mathbf{z}_R)$ decreases in $\|\mathbf{z}_T - \mathbf{z}_R\|^{-\alpha}$, where α is the attenuation coefficient (also called the path loss exponent). In an isotropic vacuum, $\alpha = 2$, but when radio propagates in a non-vacuum medium, we have in general $\alpha > 2$ [2,4,7]. In the theoretical case of a radio wave propagating on a flat, absorbing surface, we have $\alpha = 4$ as we get when radio propagates over the sea. Not only does the medium attenuates the signal, but scattering obstacles create a multitude of alternative paths for the propagation of the signal, which can lead to constructive or destructive interference. Locally, this is equivalent to random attenuation or amplification of the signal. The received signal is therefore of the form $S(\mathbf{z}_T, \mathbf{z}_R) = \mathbf{F}\|\mathbf{z}_T - \mathbf{z}_R\|^{-\alpha}$, where $\mathbf{F}$ is a random factor of mean 1 and the effect is an increase in the entropy of the communication system. Rayleigh fading, where $\mathbf{F}$ is the exponential distribution of intensity 1, is the most popular and best analyzed fading distribution [110,121].

When simultaneous transmitters use the same frequency band and the same modulation scheme, their signals interfere. Without a Faraday cage, there is no way to isolate radio links. In general, a simultaneous yet distant secondary transmission is considered as a noise by the receiver of the primary transmission and consequently will degrade the signal-to-noise ratio parameter, and in turn will compromise the ability to extract information from the signal, in conformity with Shannon's law. If the density of transmitters and their traffic increase in the vicinity, this will naturally reduce the area around each transmitter where information can be safely received.

Wireless networks entail another intersection with the physical world: Wireless terminals are often mobile (in general, this is the basic reason why they are wireless), and mobility affects the ability to send information. We saw in the previous chapter on mobile *ad hoc* networks (MANET) that managing a mobile topology entails a non-negligible traffic overload that can considerably hamper network operation. In general, the trajectories and speeds of nodes are unpredictable and introduce another source of entropy into the system.

5.2   Space Adds Capacity

As we saw in the previous section, distance hinders transmission. As a result, frequency reuse, especially when the two pairs (transmitter and receiver) are far enough apart, is now possible. The area in which a transmission can be successfully received depends on a number of factors:

- the density and proximity of simultaneous transmitters,
- the configuration of obstacles around the transmitters,

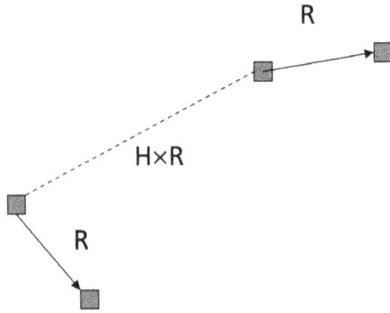

Fig. 5.3. Minimal distance between transmitter in spatial reuse.

- the attenuation coefficients,
- the radio access protocol.

The spatial reuse condition is illustrated in Figure 5.3. To achieve the signal-to-noise ratio required for the requested data rate, a minimum distance must be maintained between the receiver and the secondary transmitters. If R is the required radio range, the distance must be greater than HR with $H^{-\alpha} = K_0$, where K_0 is the signal-to-noise ratio required for correct reception of information from the main transmitter. In other words, around a transmitter, there must be an area of interdiction $\pi H^2 R^2$, and consequently, the maximum density of simultaneous transmitters will be of the order of $\frac{\beta}{R^2}$, where β sums up all the previous constraints in terms of geometry, power and signal-to-noise ratio, e.g. $\beta = 0.2$ (we see later that this is a realistic figure, see Figure 5.18). This implies that the cumulative throughput of the entire network within the zone is $\frac{\beta}{R^2}|\mathcal{A}|C_0$, where $|\mathcal{A}|$ is the area of the $\mathcal{A}$ zone occupied by the network and C_0 is the modulation rate of each source (in bits per second, assumed identical for all transmitters). Note that C_0 is at most equal to $\Phi \log_2(1 + K_0)$ if Φ is the transmission frequency bandwidth.

The area covered, i.e. the area actually receiving error-free packets, is a percentage of the total area, with the value $\pi\beta$. This analysis is simplified because it does not cumulate signals from simultaneous transmissions. It is also simplified because it assumes that the coverage area is circular. In Section 5.4, we show a finer-grained analysis with cumulated signals and coverage areas more complicated than circles.

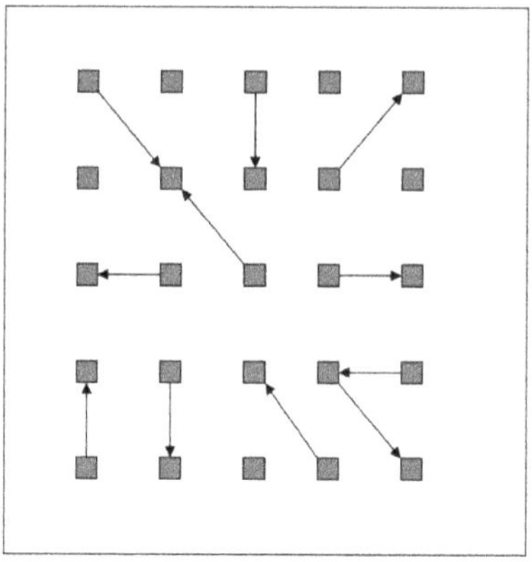

Fig. 5.4. A network with closest neighbor communication.

If the cumulative demand for communication traffic on the network in the area is ρ, this implies that the radio range R of each transmission satisfies the inequality $\rho \leq \frac{|\mathcal{A}|\beta}{R^2} C_0$, where $|\mathcal{A}|$ is the physical area occupied by the network. In other words, the greater the demand, the smaller the range.

Figure 5.4 illustrates a fictitious case. Imagine a grid square of $\sqrt{n} \times \sqrt{n}$ with n nodes. To reach at least one of their nearest neighbors, you need at least $R = 1$. Cumulative throughput is therefore less than or equal to $\rho_n = \beta n C_0$.

The consequence is that spatial reuse effectively increases network capacity and makes it linear with the number of nodes. But this is a kind of cheating. Indeed, in this model, a node can only communicate with other nodes if they are its nearest neighbors. This is not how a network works, as it would limit it to close-neighbor interactions. In a practical network, the author of a piece of information must be able to deliver it to any other node, not necessarily the nearest one.

We now look at the scenario in which each node chooses a random destination in the network map. Figure 5.5 shows an example where node S_1 chooses node D_1 as its destination and S_2 chooses D_2. If we maintain the overall throughput at ρ_n, the communication process must use a chain of relays between the nearest nodes. Let R_n be the average number of relays required. In a grid network such as the one we've used for illustration, the average number of relays will be of the order of $\sqrt{n}$. So, if we need R_n relays, i.e. to retransmit R_n

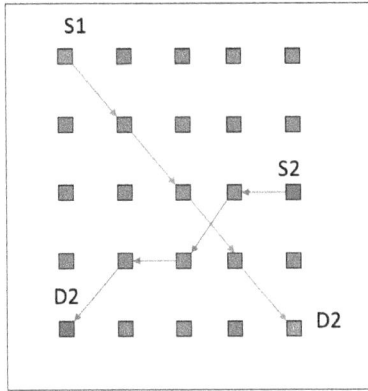

Fig. 5.5. A network with routing.

times the same packet en route to its destination, this means that the transport capacity T_n, i.e. the average number of unique packets created by each source node en route to its destination per unit time, satisfies

$$\rho_n \geq R_n T_n.$$

In fact, this inequality is only correct when all nodes have the same information generation rate T_n/n to be transmitted. Gupta and Kumar [47] also stated the general result in a more general framework, see Figure 5.6, than our toy model, and their work has become a major reference in the field of wireless networks. The inequality $\rho_n \leq \beta n$ and the estimate R_n of order $\sqrt{n}$ lead to the fact that

$$T_n = O(\beta \sqrt{n} C_0).$$

In fact, it has been shown that this estimate is in fact the order of the maximum capacity of the wireless network. Note that this order increases with n and tends toward infinity when $n \to \infty$. This capacity creation is due to the spatial reuse of frequencies. In Figure 5.7, we assume a nominal capacity of 100 kbps. Taking a network with $n = 1{,}000{,}000$ (in a large city, for example) and $\beta = 0.2$, we obtain a network capable of moving 20 Mbps of data. We note that the share per node is $T_n/n = O(\beta C_0/\sqrt{n})$ which actually decreases with n but not as radically as $O(C_0/n)$ with a wired network. So there's a free lunch of volume $1/\sqrt{n}$.

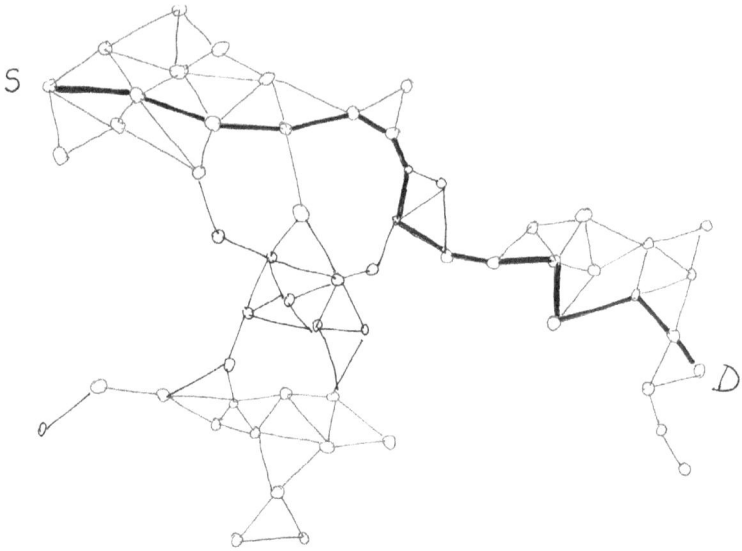

Fig. 5.6.   The space capacity: A realistic situations beyond the grid model.

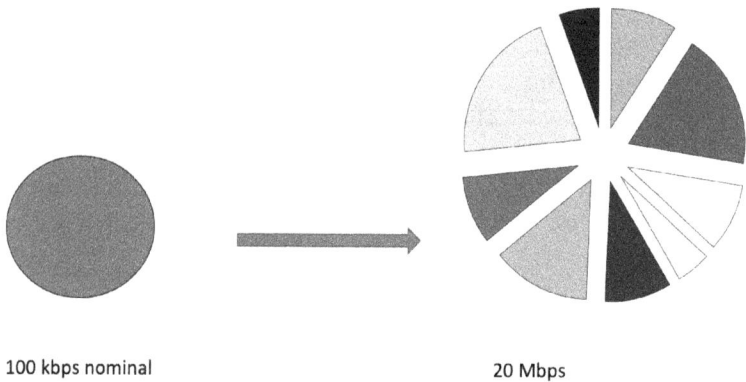

100 kbps nominal   20 Mbps

Fig. 5.7.   Space creates capacity: The capacity increases with $n = 1{,}000{,}000$.

The effect depends on the Euclidean dimension of the space in which the network map is inscribed. If the network is mainly linear, e.g. a line of vehicles on a road, then R_n will be of order n, and $T_n = O(\beta C_0)$, there is no gain as n increases. If the network map is embedded in a cube, for example, a fleet of drones in the sky, then R_n

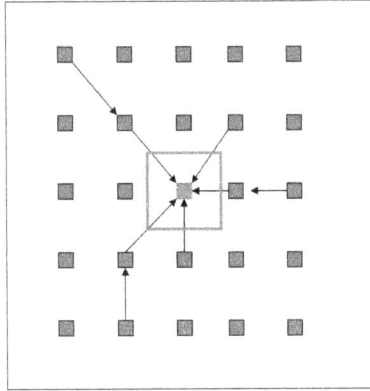

Fig. 5.8. The space capacity: The counterexample with the central sink.

will be of order $n^{1/3}$ and, consequently, T_n will be of order $\beta C_0 n^{2/3}$. If the map is embedded in a Euclidean space of dimension D, then

$$T_n \sim \beta C_0 n^{1-1/D}.$$

However, it's hard to imagine an Euclidean space of dimension 4 or more. For $D = 4$, we could add the dimension of frequencies, but the distances between frequencies are not quite of the same type as Euclidean distances. We could also add the dimension of time, but this alternative is very specific, and the following section deals more specifically with this dimension.

But this result is *fragile* and relies heavily on the assumption of uniform distribution of source–destination pairs. In the case where all destinations are concentrated on a single central node (see Figure 5.8), the magic no longer works. The system's capacity is once again limited to the local capacity, i.e. the capacity of the area around the central node, and now $T_n \le C_0$. Unfortunately, this type of traffic configuration is to be expected when a central node plays the role of a subdomain gateway.

5.3 Time Adds Capacity

In this section, we look at the fourth dimension: time. Time can have a singular effect on the communication process when nodes are

mobile and, in particular, when there is significant relative movement between them. Indeed, the source node may move closer to
its destination; in this case, the number of hops the packets must
cover decreases and the wireless capacity increases. Conversely, as
the source moves further away from its destination, the number of
hops increases and the capacity decreases. If the group moves in uniform translation or in uniform rotation, propagation conditions may
not change (but fading parameters will nevertheless change more
rapidly than in a network made of stable nodes).

In the previous sections, we assumed that the router treated packets as "hot potatoes", also less trivially known as the *store and for-*
ward process. In short, when a packet is received, the host consults
its routing table and immediately forwards the packet to the transit
interface, or stores it if the host is the final destination. Or it rejects
the packet if the destination is not in the routing table. We're now
going to look at a new store, hold and forward process, which allows
the host to hold the packet in its memory until favorable conditions
are met.

Our basic model is that mobile nodes move independent of each
other, walking at random. To fix the idea, we assume that the nodes
move on a finite map (for example, a square area). We assume that
the random walks are *uniform ergodic*, meaning that all areas of
the map are visited equally. Ergodicity is also a property of Markov
chains. The uniform ergodic random walk is a kind of generalization
of the Markov chain that has the uniform distribution as its stationary distribution. We also assume that the nodes' movements are not
affected by their communication needs and that the nodes have no
knowledge of any possible planning of their movements.

If each flying node has its buffer full of packets waiting to be
transmitted, and all possible destinations in the network are represented in the queue, then the *bees* protocol will consist of the flying
node transmitting a portion of the packets tagged for this destination to its nearest neighbor at any given time. At the same time, the
node will receive packets from this neighbor. More rigorously, the
node sends packets whose destinations are one hop away in the routing table (meaning that there is control traffic likely to detect close
neighbors). In this way, packets are physically delivered in a single
hop $R_n = 1$, leading to a transport capacity of $\beta n C_0$. This protocol is

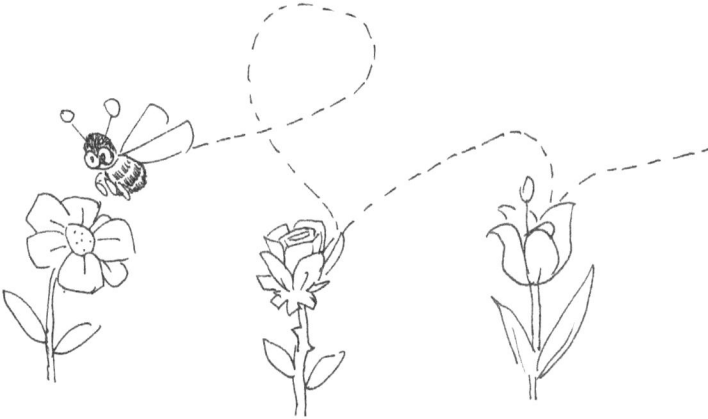

Fig. 5.9.   The "bees" protocol.

reminiscent of the way bees distribute pollen to all the flowers they visit (see Figure 5.9).

But it's not a reasonable assumption that every flying node maintains simultaneous, permanent conversations with every other node in the network. In general, a node has few favorite destinations. To be more realistic, we assume that each node has only one or a few favorite destinations and, for simplicity, that these destinations are evenly distributed throughout the network. The "postman" protocol is described in the following:

This protocol, originally invented by Grossglauser and Tse [44], operates in two phases, which alternate rapidly with time:

- the upload phases,
- the download phases.

During an upload phase, the node forwards the packets it originated to its nearest neighbors (i.e. nodes located one hop away in the routing table). They may not be able to transmit all their pending packets, as they may not have the necessary time and bandwidth before change; if this is the case, the node will continue to transmit packets during the next upload phase but perhaps to another neighbor. In addition, nodes store packets received during an upload phase in a *transit* queue. We can also call the queue that stores the authentic packet the packet generator queue.

During a download phase, nodes forward their packets in the transit queue to their destination, if this is one hop away. Of course, packets in both queues whose destinations are one hop apart are forwarded directly to their destination, regardless of the phase.

Note that in this protocol, each node acts as a postman: In the upload phase, it collects packets from its neighbors and stores them in its transit queue, and in the download phase, it distributes packets to their destination when it is nearby (see Figure 5.10).

Even when each node generates packets to a single preferred destination, the transit queue will be populated with packets whose destinations are evenly distributed across the entire network. When the network is very active, both queues are full, meaning that every node has something to transmit in the upload and download phases. Neglecting the case where nodes get closer to the destination of their original packets (which happens $1/n$ of the time), all packets eventually reach their destination in two hops, so $R_n = 2$ ($R_n = 1$ for original packets when the destination is one hop away). Thus,

$$T_n = \beta/2nC_0.$$

This result is interesting because it shows a free lunch sign: The share per node T_n/n does not decrease with n! Figure 5.11 illustrates this property. Still with $n = 1,000,000$, we start from the actual gain obtained with the routing protocol, which has enabled us to go from a nominal throughput $C_0 = 100$ kbps to a throughput of 20 Mbps. With the postman protocol, we're up to 10 Gbps!

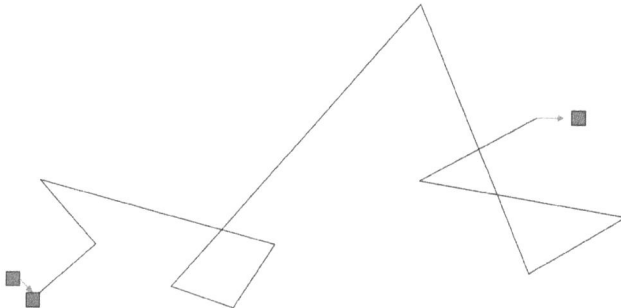

Fig. 5.10.   The "postman" protocol.

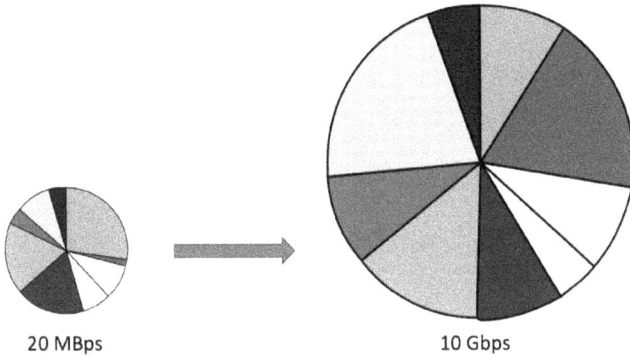

20 MBps   10 Gbps

Fig. 5.11.   The time growth of capacity with the "postman" protocol.

But the result is also fragile, as it depends heavily on the ergodic hypothesis. It's highly unrealistic to assume that an inhabitant of Paris is likely to meet every other inhabitant on his or her walks. Nevertheless, in theory, the result can be adapted to the nesting hypothesis: Each node gravitates around its favorite location (say, its home address) up to a certain distance (say, one kilometer). By refining the above protocol, there would be a finite number of nests to pass through before reaching the destination. If there are k nests to cross, the capacity will be β/kC_0. In protocol terms, this would mean building routing tables on the nests, which can be complicated unless you use geo-routing, as we see later.

The result seems to be equally impractical. Let's consider a mobile network laid out in a square of $L \times L$ such that the speed v of the mobile nodes if they traveled straight ahead would give an average time to cross the network diagonally equal to T. If we consider n nodes (for simplicity, we assume they are arranged as shown in Figure 5.12 (left)), it's clear that during the diagonal traverse the moving node will visit $\sqrt{n}$ other nodes, since the distance to the nearest neighbor is of the order of $L/\sqrt{n}$. But to visit an arbitrarily chosen node, a mobile node will have to visit an average of n nodes, which will take an average $T\sqrt{n}$ (see Figure 5.12 (right)). If we take a city with 1,000,000 mobile nodes moving at a speed such that the traversal time is around one hour, the average packet delivery time with the Postman protocol will be 1,000 hours, or around one month! Clearly too long! At the other end of the scale, the hot

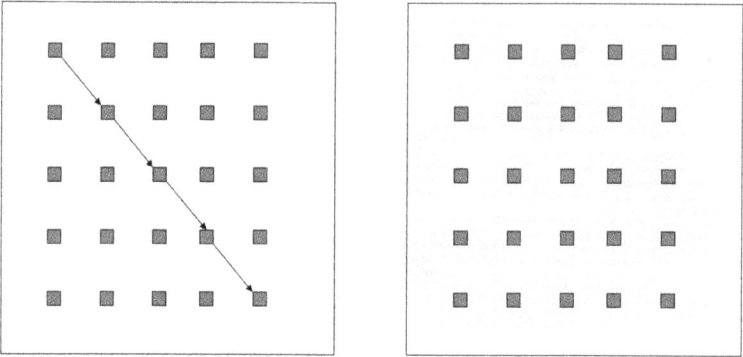

Fig. 5.12.   Left: The hot potatoes packet trajectory. Right: The postman packet trajectory.

potatoes protocol would only take a second, if the store-and-forward operation takes just 1 ms on each router.

The geo-routing [108] protocol would be an interesting compromise. But it assumes that (i) all destinations are fixed with known GPS coordinates, (ii) each mobile node has its own live GPS coordinates, and (iii) the trajectory of mobile nodes is made up of more or less straight lines (which is the case along streets in a city). The protocol consists of storing the packet in the first mobile node heading toward the destination at more or less a certain angle, and changing router when it loses the angle, for example, by changing direction. Under these conditions, the packet path resembles a logarithmic staircase with $\log n$ relays (see Figure 5.13) and therefore a transport capacity of order $\beta \frac{n}{\log n} C_0$. But the delivery time remains of the order of the crossing time T. The performance of our different protocols is summarized in the following table:

protocol	capacity	delay
hot potatoes	20 Mbps	1 second
postman	10 Gbps	1 month
geo-routing	2 Gbps	1 hour

In fact, we can go even further. Mobility and time can do more than increase telecommunication capacity, they can actually create it from scratch in a situation where the network is permanently disconnected. In a delay-tolerant network (DTN), the network can be

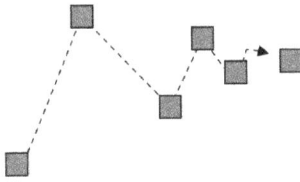

Fig. 5.13.   The logarithmic staircase of the geo-routing protocol.

permanently disconnected simply because radio ranges are too limited, as illustrated in Figure 5.14. Thanks to the possibility of storing packets in transit in the mobile nodes' buffer memory, as is the case with the Postman protocol, the information will eventually reach its destination, but as we saw earlier, the delivery delay can be significant. Even with the most optimistic logistics, there aren't many applications that can accept a delivery delay of several months, and the information may lose its value. We can imagine a maximum waiting time of T for each packet. With the source at position $\mathbf{z}$, the destination at position $\mathbf{z}'$, the packet can arrive at the destination at time $T(\mathbf{z}')$ at the earliest, with a certain small but non-zero probability, due to the physical speed of the mobile nodes and the randomness of their trajectories, so when $T < T(\mathbf{z}')$, the capacity between position z and position $\mathbf{z}'$ is zero. When $T > T(\mathbf{z}')$, the probability of on-time packet delivery increases and, consequently, capacity increases (see Figure 5.15). A more thorough analysis shows that in the Poisson shot model with random walk, the average quantity $T(\mathbf{z}')$ varies proportionally to the distance $\|\mathbf{z} - \mathbf{z}'\|$, the ratio calling to an information propagation speed in the map [77]. This information propagation speed is linear with the average mobile node speeds. It also has the property to exponentially increase when the density of mobile nodes increases, provided the coverage area of radio transmissions is fixed (i.e. the density of actually transmitting mobile nodes remains constant).

There is an interesting consequence of this result in dimension 1 when cars travel on highway lanes. This case is interesting because there are only two directions of movements: westward or eastward. If all cars move toward the same direction (e.g. eastward) at the same constant speed, then the information propagation cannot be greater than the actual motion speed of the cars. Indeed, the information propagation wave will remain locked at the last car before an interval

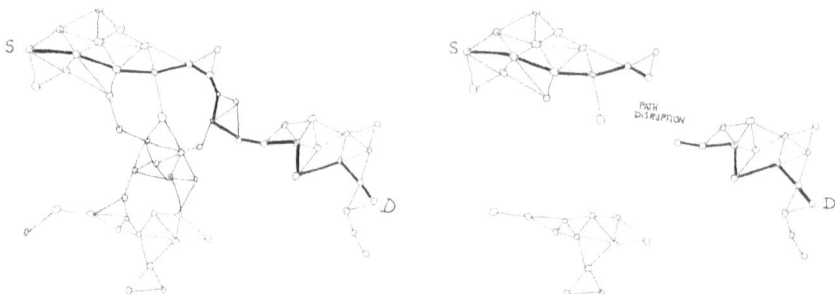

Fig. 5.14.   The delay tolerant networks: permanently connected (left) and permanently disconnected (right).

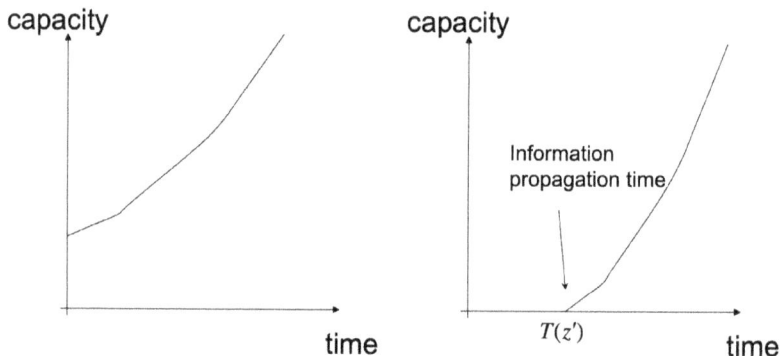

Fig. 5.15.   Capacity increases with time (left) and creates capacity in disconnected mobile network (right).

void of cars larger than the actual radio range. But if some fraction of cars are moving in the opposite direction (imagine a two lanes highway), then the gap will sooner or later be temporarily filled by eastward cars and let the information jump over the gap. In this case, it is the paradoxical effect that letting some cars to move in the opposite direction actually increases the information propagation speed [10].

5.4   Geometry Can Add Capacity

5.4.1   *The Poisson shot model*

A Poisson distribution of points in two-dimensional space gives a realistic model of the location of nodes in an area on the plane [9].

Such a distribution is defined by a single parameter: the point density (also called intensity). We consider an infinite but countable set of nodes to be distributed over the infinite plane $\mathbb{R}^2$. We denote by $\mathcal{I}$ the set of their positions. The Poisson shot model has the following properties:

- For any bounded measurable domain $\mathcal{A}$, the number of points contained in $\mathcal{A} \cap \mathcal{I}$ is finite and follows a Poisson distribution: $P(|\mathcal{A} \cap \mathcal{I}| = k) = \frac{\lambda |\mathcal{A}|}{k!} e^{-\lambda |\mathcal{A}|}$.
- For any pair of measurable areas $\mathcal{A}$ and $\mathcal{B}$ having a non-measurable intersection, the sets of points $\mathcal{A} \cap \mathcal{I}$ and $\mathcal{B} \cap \mathcal{I}$ are finite and independent (both in size and coordinates).

We note that a Poisson shot of density λ remains a Poisson shot of density λ after any translation, rotation or symmetry. We also note that for any $a > 0$, a Poisson shot of density $a\lambda$ is rigorously equivalent to the Poisson shot of density λ shrunk by a factor $1/\sqrt{a}$.

The consequence of these two properties is that, given a point at an arbitrary position z, the positions of all other points also constitute a Poisson shot of intensity λ. This is the conditional Poisson property.

From now on, we assume that the set of simultaneous transmitters is a Poisson shot of density $\lambda > 0$. This type of assumption is a two-dimensional generalization of the theory of random accesses in a cable. Assuming that all nodes transmit with the same nominal power P_0 and experience the same attenuation factor α, let $\mathbf{S}(\mathbf{z})$ be the total energy received at a given point $\mathbf{z}$ from all transmitters. For simplicity's sake, we omit other sources of energy, in particular the internal noise generated by the receiver's circuits. The quantity $\mathbf{S}(\mathbf{z})$ is a random variable. Figure 5.16 shows an example of a map of the variable $\mathbf{S}(\mathbf{z})$ as a function of $\mathbf{z}$ in a given Poisson shot. Note that $\mathbf{S}(\mathbf{z})$ diverges toward infinity whenever $\mathbf{z}$ lies on a point of $\mathcal{I}$. We can fully characterize the distribution of this random variable by its Stieltjes–Laplace transform: $w(\theta) = E[\exp(-\theta \mathbf{S}(\mathbf{z}))]$. Surprisingly, the function $\mathbf{S}(\mathbf{z})$ has a closed form:

$$w(\theta) = \exp\left(-\lambda \pi \Gamma(1 - \gamma) P_0^\gamma \theta^\gamma\right),$$

with $\gamma = \frac{2}{\alpha}$, supposedly strictly smaller than 1, e.g. $\gamma = 1/2$ with $\alpha = 4$. The fading is the factor which makes the signal from an emitter to differ from the expected average $P_0 r^{-\alpha}$ when received at

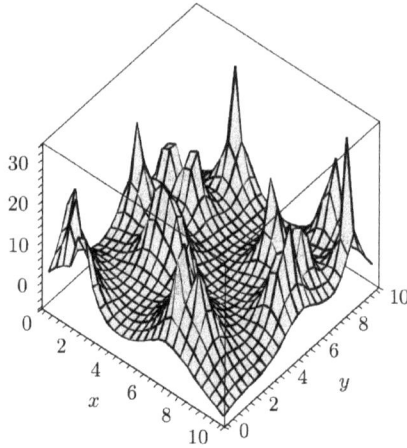

Fig. 5.16.   Landscape example of $S(\mathbf{z})$ as function of $\mathbf{z}$ in the plan.

distance r. With a fading random variable $\mathbf{F}$ such that $E(\mathbf{F}) = 1$, the signal at distance r is $P_0 \mathbf{F} r^{-\alpha}$. We get

$$w(\theta) = \exp\left(-\lambda \pi \Gamma (1 - \gamma) P_0^\gamma E[\mathbf{F}^\gamma] \theta^\gamma\right).$$

If the nominal power of the node is also a random variable, we need to replace P_0^γ in the previous formula with $E[P_0^\gamma]$. This formula will not play an important role in the following, but it is interesting because it has a closed form. When a node transmits a packet, the area in which the packet can be received correctly takes on complex shapes and is highly dependent on the signal-to-noise ratio K_0 required by the technique of modulation [120], as shown in Figure 5.17 extracted from the $\mathbf{S}(\mathbf{z})$ landscape presented in Figure 5.16. This oyster-plate like image shows the correct reception zones for different values of K_0.

The average area of correct reception $\sigma(\lambda)$ for a given fixed transmitter has also a closed form:

$$\sigma(\lambda) = \frac{1}{\lambda} \frac{\sin(\pi \gamma)}{\pi \gamma} K_0^{-\gamma}.$$

We assume that the other transmitters take up random positions according to a Poisson distribution, in line with the conditional Poisson property. The factor $1/\lambda$ is simply explained: As we increase the density of points, we also reduce the area of correct reception by

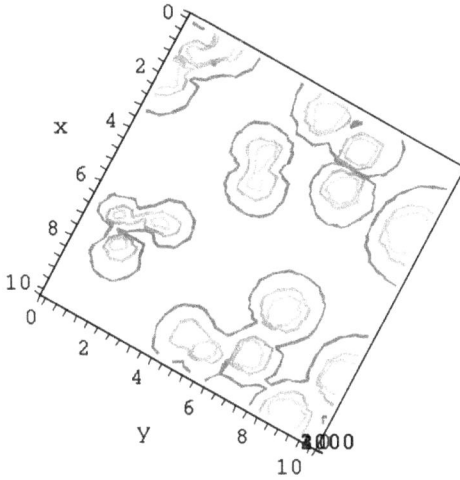

Fig. 5.17. The correct reception areas from external to internal, for $K_0 = 1$, 2, 10.

the same factor. Note that the formula is independent of the fading parameter, which seems surprising but not so much, for subtle reasons. Let's assume that all nodes transmit without random fading, with the exception of the given fixed transmitter. If all points receive the central node with $\mathbf{F}$ fading, then correct reception will be as with the formula above but with an apparent required SNR of $\frac{K_0}{\mathbf{F}}$. Averaging all $\mathbf{F}$ values, we obtain

$$\frac{1}{\lambda} \frac{\sin(\pi\gamma)}{\pi\gamma} K_0^{-\gamma} E[\mathbf{F}^\gamma].$$

Now we need to take into account the effect of fading from other nodes. Since we're calculating an average, it doesn't change the problem to consider that fading is just a variation in the node's transmission power (i.e. it transmits with a power $\mathbf{F}P_0$). Low-fading nodes will be as if they were a factor $\mathbf{F}^{-\gamma}$ away, or a factor $\mathbf{F}^\gamma$ diluted, as will high-fading nodes. Averaging all the fading values again, the resulting signal distribution will be equivalent to a Poisson shot without fading, with an intensity $\lambda E[\mathbf{F}^\gamma]$. In this case, the factor $\frac{1}{\lambda}$ becomes $\frac{1}{E[\mathbf{F}^\gamma]}$, the factors $E[\mathbf{F}^\gamma]$ vanish and we return to the original formula.

The correct reception zone also has the expression $\sigma(\lambda) = \frac{\beta\pi}{\lambda}$, where β is the reuse parameter described in the previous section and for which we currently give a closed-form expression in the Poisson firing model (see Figure 5.18).

The formula assumes that all nodes transmit with the same nominal power. If the nominal power is different, say P_1 which is now a random variable, it would be equivalent to consider a fading $\mathbf{F}\frac{P_1}{P_0}$. We know that the average reception area will remain unchanged, but if we condition it by a nominal power P_1, the correct average reception area of a node transmitting with power P_1 will now be [62]

$$\sigma(\lambda) = \frac{1}{\lambda} \frac{P_1^\gamma}{E[P_1^\gamma]} \frac{\sin(\pi\gamma)}{\pi\gamma} K_0^{-\gamma}.$$

But most people are surprised by the appearance of the trigonometric function "sine" in the expression. Since $\gamma < 1$, the evaluation is always positive, but we note that when $\gamma \to 1$, then $\sigma(\lambda)$ tends toward zero. This is normal because, when $\alpha = 2$, an easy calculation shows that the signal received at any point is infinite, so the SNR of a signal from an arbitrary transmitter would be zero, leading to a reception area of size 0. This is particularly apparent in the remark that $w(\theta) \to 0$ as soon as $\mathrm{Re}(\theta) > 0$ when $\gamma \to 1$ (because $\Gamma(1 - \gamma) \to \infty$). Less obviously, when $\gamma \to 0$ (i.e. when $\alpha \to \infty$), we have $\sigma(\lambda) \to \frac{1}{\lambda}$. This means that, whatever the required signal-to-noise ratio, the union of the correct reception zones of each transmitter covers the whole plane. In fact, when α is large, between the signals received by a receiver at an arbitrary position, the largest becomes massively dominant compared to the other signals. The packet it carries will therefore be received correctly, even if the required signal-to-noise ratio is high.

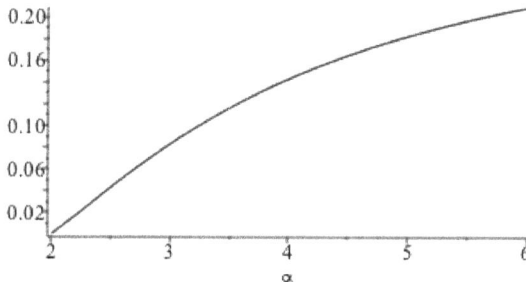

Fig. 5.18.   The plot of parameter β in Poisson shot model for $K_0 = 2$.

When fading is Rayleigh-like, the probability $p(r, \lambda, K)$ that a packet is correctly received at a distance r when the density of simultaneous transmitters is λ and the required SNR is K gives a closed formula:

$$p(r, \lambda, K) = \exp\left(-\pi\lambda \frac{\pi\gamma}{\sin(\pi\gamma)} K^\gamma r^2\right).$$

Rayleigh fading is the only known model of fading for which the quantity $p(r, \lambda, K)$ corresponds to a closed formula. In general, we have $p(r, \lambda, K) \sim \exp(-Cr^2)$ for a certain constant C. In all cases, the homothety and symmetry conditions give the identity: $p(r, \lambda, K) = p(r\sqrt{\lambda}K^{1/\alpha}, 1, 1)$. Let's suppose that time is divided into slots and that at each slot, the set of nodes is authorized to transmit changes but remains a Poisson process of intensity λ (which can be done if, at each slot, the nodes randomly draw the authorization to transmit with a fixed probability). In this situation, $\frac{1}{p(r,\lambda,K)}$ is the number of times a packet must be retransmitted before it is corrected and received by an intended receiver at a distance r from the sender. Clearly, direct transmission to a destination at a distance R would be too costly in terms of packet retransmissions (increasing as an exponential of the square of R).

If we consider the possibility of routing through relays between two nodes at a distance R, we have to choose between a large number or a small number of relays (see Figure 5.19). If we consider a number of hops k, the number of packet retransmissions on the path is equal to $\frac{k}{p(R/k,\lambda,K)} = \frac{R}{p(R/k,\lambda,K)R/k}$, the term $\frac{R}{k}$ assumes equidistance between relays more or less assured if the density of potential relays is large enough to allow it. If we forget the constraint that k must

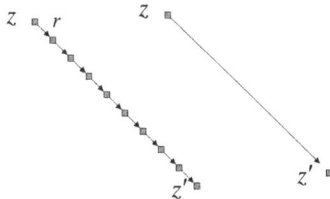

Fig. 5.19.   The compromise over routing in Poisson shot model.

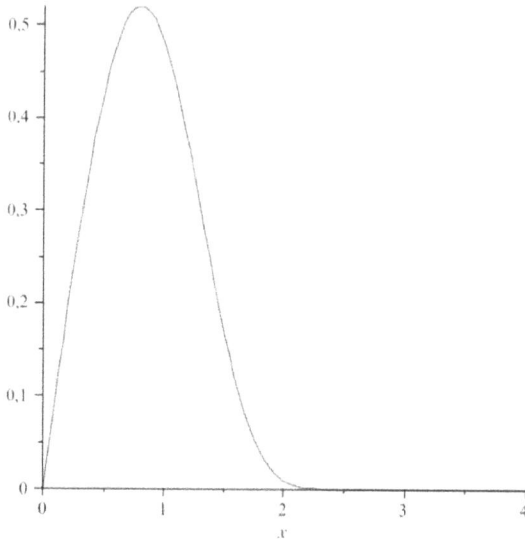

Fig. 5.20. The quantity $rp(r, 1, 1)$ versus distance r.

be an integer, the optimal value is obtained by [64]

$$\frac{R}{\max_r rp(r, \lambda, K)}$$

and $k = \frac{R}{\max_r rp(r, \lambda, K)}$. Figure 5.20 illustrates the function to be maximized. It turns out that for the Rayleigh fading $\arg\max_r rp(r, 1, 1) = \sqrt{\frac{\sin(\pi\gamma)}{2\pi^2\gamma}}$, thus $\arg\max_r rp(r, \lambda, K) = \sqrt{\frac{\sin(\pi\gamma)}{2\pi^2\gamma}} \frac{1}{\sqrt{\lambda}K^{1/\alpha}}$.

5.4.2   *The Shannon capacity of wireless networks*

When Shannon established that the theoretical capacity of a wireless link is equal to $\Phi \log_2(1 + K)$, where Φ is the physical width of the transmission frequency band (expressed in Hz) and K is the signal-to-noise ratio at the receiving end, he actually limited himself to a single, isolated transmitter-receiver pair.

Since radio links cannot be isolated by Faraday cages, we focus on the multi-user aspect of the problem. Contrary to the single user mode, the multi-user mode is still a research problem. Indeed, it might look as elusive as the n-body problem in celestial mechanics compared to the 2-body problem solved by Kepler and Newton [25]. But we can try to solve it under simplifying hypotheses, for example, by considering that the signal of the other transmission forms a gaussian noise. As in the previous section, we consider a potentially infinite population of transmitters described by the set of their locations $\mathcal{I} = \{\mathbf{z}_1, \mathbf{z}_2, \ldots\}$. Let's denote $C(\mathcal{I}, \mathbf{z})$ the sum of the Shannon capacities of all simultaneous transmitters, assuming they have an unique receiver at location $\mathbf{z}$ (see Figure 5.21). We consider that for each transmitter in $\mathcal{I}$, the signal from the other transmitters contributes to the noise. More precisely, if $\mathbf{z}' \in \mathcal{I}$, we denote $S(\mathbf{z}', \mathbf{z})$ the signal received from the transmitter at position $\mathbf{z}'$ by the receiver at position $\mathbf{z}$, and $\mathbf{S}(\mathbf{z})$ is the sum of the signals received from all transmitters in $\mathcal{I}$: $\mathbf{S}(\mathbf{z}) = N(z) + \sum_{\mathbf{z}' \in \mathcal{I}} S(\mathbf{z}', \mathbf{z})$. The noise that opposes the signal $S(\mathbf{z}', \mathbf{z})$ at location $\mathbf{z}$ is $\mathbf{S}(\mathbf{z}) - S(\mathbf{z}', \mathbf{z})$, where $\mathbf{S}(\mathbf{z})$ can include the internal noise $N(\mathbf{z})$ generated by the receiving device. Thus, assuming that all transmitters operate on the same bandwidth Φ,

$$C(\mathcal{I}, \mathbf{z}) = \Phi \sum_{\mathbf{z}' \in \mathcal{I}} \log_2 \left(1 + \frac{S(\mathbf{z}', \mathbf{z})}{\mathbf{S}(\mathbf{z}) - S(\mathbf{z}', \mathbf{z})} \right). \tag{5.1}$$

Surprisingly, the mean value of this very complicated expression actually has a very simple mean, which we call the *wireless Shannon capacity*.

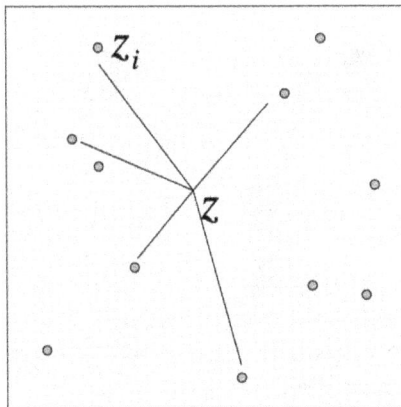

Fig. 5.21.   Multi-input wireless network.

Theorem 5.1. *With no internal noise* $(N(\mathbf{z}) = 0)$, *we have*

$$C(\lambda) = E[C(\mathcal{I}, \mathbf{z})] = \Phi\frac{\alpha}{2\log 2},$$

whatever the fading and nominal power distribution. The quantity Φ *is the nominal bandwidth of the receiver and* α *is the attenuation factor.*

We note that the parameter z is omitted from the mean value, as the Poisson shot distribution is a translation invariant. The parameter *ambda* does not appear in the formula, which is no surprise, as our problem implies that the capacity is invariant when we change the scale of the transmitter distribution by an arbitrary factor. This very simple formula can be proved by expressing $w(\theta) = E[e^{-\theta \mathbf{S}(\mathbf{z})}]$, but this approach is rather complicated. Instead, we use a proof via the energy-differentiated field theorem, which we see later.

We note that when $\alpha \to \infty$, then $C(\lambda) \to \infty$; this is consistent with our earlier observation that as the attenuation factor increases the signal from the nearest neighbor becomes preponderant with an SNR tending toward infinity, allowing this neighbor to transmit at the highest rate. But when $\alpha \to 2$, $C(\lambda)$ doesn't tend toward 0 but toward $\frac{\Phi}{\log 2}$, even though, as we saw earlier, the SNR of each reception tends toward zero (and thus leads to an individual capacity tending toward 0). This apparent paradox is easily explained by the fact that the infinite sum of all individual capacities does not tend toward 0. Indeed, in formula (5.1), we have $\mathbf{S}(z) \to \infty$ when $\alpha \to \infty$, hence the expression

$$\log_2\left(1 + \frac{S(\mathbf{z}', \mathbf{z})}{\mathbf{S}(\mathbf{z}) - S(\mathbf{z}', \mathbf{z})}\right) = \frac{1}{\log 2}\left(\frac{S(\mathbf{z}', \mathbf{z})}{\mathbf{S}(\mathbf{z})} + O\left(\frac{S(\mathbf{z}', \mathbf{z})^2}{\mathbf{S}(\mathbf{z})^2}\right)\right).$$

Since $\mathbf{S}(\mathbf{z}) = \sum_{\mathbf{z}' \in \mathcal{I}} S(\mathbf{z}', \mathbf{z})$, summing the above leads to

$$C(\mathcal{I}, \mathbf{z}) = \frac{\Phi}{\log 2}\left(1 + O\left(\frac{1}{\mathbf{S}(\mathbf{z})}\right)\right)$$

which tends well to $\frac{\Phi}{\log 2}$ when $\alpha \to 2$ (i.e. when $\mathbf{S}(\mathbf{z}) \to \infty$). Note that the property will hold even when $\alpha < 2$ (it would be possible, for example, in the unrealistic case where the propagation space is sandwiched between two reflecting surfaces).

The following theorem is the corner stone of our analysis and of the subsequent analyses.

Theorem 5.2 (Energy-differentiated field theorem). *For* $\mathbf{z}$ *in the propagation space, we have the identity*

$$C(\lambda) = \Phi \frac{\partial}{\partial \log \lambda} E[\log_2 \mathbf{S}(\mathbf{z})].$$

The theorem is valid even when the internal noise $N(\mathbf{z})$ is non-zero and also when the Poisson distribution is non-uniform. The theorem does not depend on the unit used for energy evaluation (Joule, erg, or Mev); the change of unit just introduces a constant factor that disappears after the derivation. Similarly, it does not depend on the density unit. The theorem does not depend on the dimension of the propagation space, which will prove to be useful for more complicated geometries. We draw attention to the fact that in $\mathbb{R}^2$ with the Poisson model, the mean absolute $E[\mathbf{S}(\mathbf{z})] = \infty$ is mainly due to the contributions of the nearest emitters (in $r^{-\alpha}$ when $\alpha > 2$, whereas the mean number of emitters at a distance of r or less is $\pi \lambda r^2$). When $\alpha < 2$, the infinitness of $E[\mathbf{S}(\mathbf{z})]$ is due to the far emitters. But the expectation of $\log \mathbf{S}(z)$ remains finite. However, we postpone the outline of the theorem's proof until the end of the section. Before that, we complete the proof of Theorem 5.3. The method does not require us to calculate the exact value of the quantity $E[\log_2 \mathbf{S}(\mathbf{z})]$ but to take advantage of the scaling properties of this quantity.

To make the proof more complete, we introduce the parameter λ to define $\mathbf{S}(\lambda, \mathbf{z}) = \mathbf{S}(\mathbf{z})$. Let a be a positive real number; we consider the transformation of the set of transmitters in $\mathcal{I}$ via a homothety of factor a and center $\mathbf{z}$. In so doing, each signal is increased by a factor $a^{-\alpha}$. At the same time, the density decreases by a factor a^{-2}. In other words, $\mathbf{S}(a^{-2}\lambda, \mathbf{z}) \equiv a^{\alpha}\mathbf{S}(\lambda, \mathbf{z})$ in the distribution. Thus, $E[\log_2 \mathbf{S}(a^{-2}\lambda, \mathbf{z})] = -\alpha \frac{\log a}{\log 2} + E[\log_2 \mathbf{S}(\lambda, \mathbf{z})]$. The identity is true because we assume that the intensity $\mathbf{S}(\mathbf{z})$ includes random noise insensitive to the above scaling properties.

By choosing $a = \sqrt{\lambda}$, we get the identity

$$E[\log_2 \mathbf{S}(\lambda, \mathbf{z})] = \frac{\alpha \log \lambda}{2 \log 2} + E[\log_2 \mathbf{S}(1, \mathbf{z})].$$

Clearly, the derivative with respect to λ satisfies the identity

$$\lambda \frac{\partial}{\partial \lambda} E[\log_2 \mathbf{S}(\lambda, \mathbf{z})] = \frac{\alpha}{2 \log 2}$$

since $E[\log_2 \mathbf{S}(1, \mathbf{z})]$ is a constant with respect to the variable λ.

The proof of the differentiated energy field is developed as an exercise. In short, the conditional property of the Poisson distribution leads to the claimed identity (we focus on the capacity received at the origin). For simplicity's sake, let $\Phi = 1$ and $\mathbf{z} = 0$:

$$C(\lambda) = \iint E\left[\log_2\left(1 + \frac{S(\mathbf{z}, 0)}{\mathbf{S}(\lambda, 0)}\right)\right] \lambda d\mathbf{z}^2$$

$$= \iint \left(E\left[\log_2\left(\mathbf{S}(\lambda, 0) + S(\mathbf{z}, 0)\right)\right] - E\left[\log_2\left(\mathbf{S}(\lambda, 0)\right)\right]\right) \lambda d\mathbf{z}^2.$$

In the second term on the right, we formally recognize the derivative $\lambda \frac{\partial}{\partial_\lambda} E[\log_2 \mathbf{S}(\lambda, 0)]$.

The exercise shows that the energy-differentiated field theorem remains valid even when the Poisson distribution of emitter positions is non-uniform and also when the internal noise is non-zero. The same applies when the Euclidean dimension of the Poisson distribution support is extended to all arbitrary non-zero integers D. Figure 5.22 shows a Poisson shot in dimension 3.

Theorem 5.3. *With no internal noise $(N(\mathbf{z}) = 0)$, in a Poisson shot in dimension D we have*

$$\text{when } \alpha > D: \ C(\lambda) = E[C(\mathcal{I}, \mathbf{z})] = \Phi \frac{\alpha}{D \log 2},$$

$$\text{when } \alpha \leq D: \ C(\lambda) = E[C(\mathcal{I}, \mathbf{z})] = \frac{\Phi}{\log 2},$$

whatever the fading and nominal power distribution. The quantity Φ is the bandwidth and α is the attenuation factor.

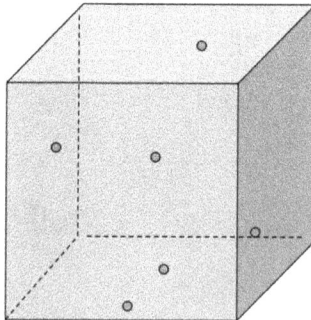

Fig. 5.22.   Multi-input wireless network in dimension 3.

Note that the above formula summarizes the influence of the main ingredients of wireless networks: The attenuation factor α sums up the impact of the physics of radio propagation, the factor D sums up the impact of geometry, and the factor $\log 2$ is a shortcut to the implication of information theory.

Proof.   The proof of the theorem is as simple as when $D = 2$. Indeed, a homothety of center $\mathbf{z}$ by a scalar $a > 0$ makes $\mathbf{S}(a^{-D}, \lambda, \mathbf{z}) \equiv a^{\alpha}\mathbf{S}(\lambda, \mathbf{z})$. Thus,

$$E[\log_2 \mathbf{S}(\lambda, \mathbf{z})] = \frac{\alpha}{D}\frac{\log \lambda}{\log 2} + E[\log_2 \mathbf{S}(1, \mathbf{z})]. \qquad \square$$

5.5   Fractal Geometries

This section is based on the analysis that the Poisson uniform shot model is a powerful modelling tool but that it is far from realistic when considering large-scale structures. Natural and man-made structures, such as forests and cities, are not static and permanent. In general, they obey a largely unplanned, long-term historical development that leads to fortuitous groupings and splits. This leads to self-similarities in structure, meaning that many small structures have similarities and, in the course of their development, eventually merge to form larger structures that resemble the smaller ones, but with a difference in scale. When similarities in fact lead to identical structures *modulo* a rotation, a translation and a scaling, we speak of a *fractal* structure. Fractal structures were invented by Mandelbrot [73]:

> Clouds are not spheres, mountains are not cones, coastlines are not circles, and bark is not smooth, nor does lightning travel in a straight line.

The most emblematic example in nature is the fern and the snowflake (see Figure 5.23). Of course, not all substructures can be absolutely identical to the overall structure, since repetition down to the atomic level would require sculpting inside atoms! But the model has enough nested levels to have consequences for the internal properties of the structure.

Fig. 5.23. The self-similar structure of fern (left), of the snowflake (center), and of the river ramifications (right).
Courtesy: Wikipedia and Paul Bourke.

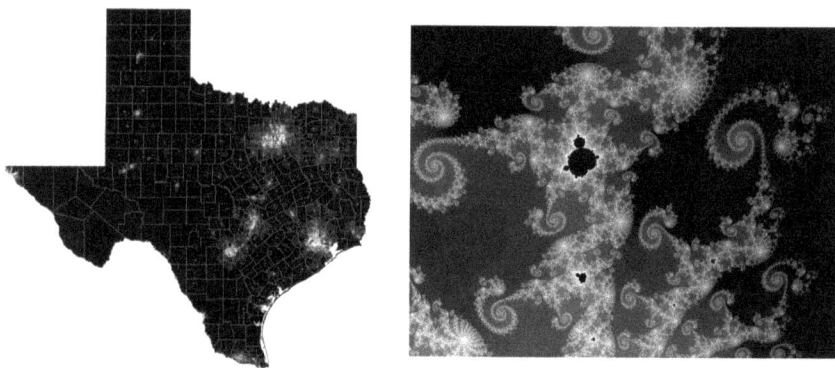

Fig. 5.24. Left: The self-similar structure of Texas population density (U-Haul Growth States, 2022). Right: The Mandelbrot object.

In Figure 5.24, we show the self-similar structure of population density in the state of Texas. We also show the mathematical object known as the Mandelbrot set. This object is obtained by iterating a simple quadratic function on the complex plane. The Mandelbrot set is not a pure fractal object, it is only asymptotically at points of attraction.

The fractal is also reflected in man-made structures, particularly in cities. Figure 5.25 shows the current map of Paris, which has evolved from a decentralized network of small suburbs to an urban continuum absorbing all these suburbs and making them the centers of new districts. On the right-hand side of this figure, we show a fractal simulation of a city map. Indeed, fractals are geometric objects that can be simulated by recursive algorithms. A famous example of a fractal object is the Sierpinski triangle shown in Figure 5.26 [107].

Fig. 5.25. Left: The self-similar structure of Paris. Right: A fractal simulation of a city map. http://chrisbence.com/geometrics/fractal_city/.

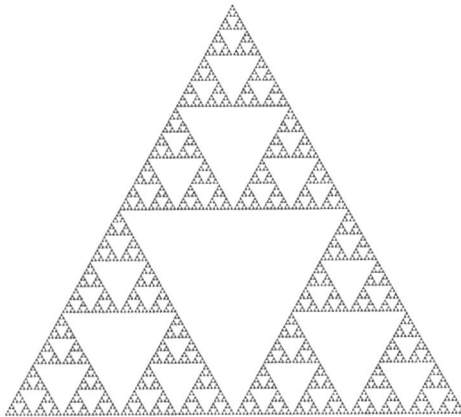

Fig. 5.26.   The Sierpinski triangle.

Figure 5.27 illustrates the construction of the triangle. The initial phase of the algorithm is a whole triangle. The second phase divides the triangle into four parts, marking parallel lines at mid-length. The central part is removed and the algorithm is applied to each of the three remaining parts, again and again. The resulting object is a fractal, mostly vacuum-filled object, i.e. more rigorously of zero Lebesgue measure.

We note that at each step, we create three identical subparts by dividing the object at mid-length. If we do the same with a complete square, we obtain four identical subparts (see Figure 5.28).

Fig. 5.27.   The construction of the Sierpinski triangle.

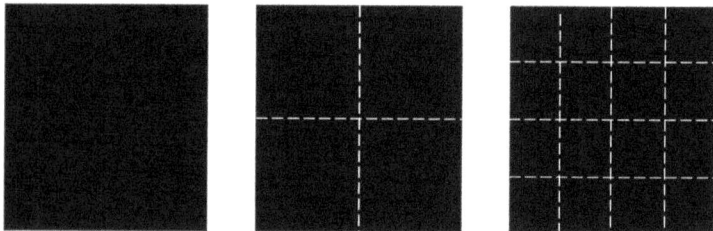

Fig. 5.28.   Splitting a square.

The reason for this is that reducing a square by a factor $\frac{1}{2}$ yields four identical parts, and dividing the lengths by $\frac{1}{4}$ yields sixteen parts. In dimension D, this would be 2^D and 4^D respectively because

$$\left(\frac{1}{2}\right)^D = \frac{1}{2^D},$$

$$\left(\frac{1}{4}\right)^D = \frac{1}{4^D}.$$

Therefore, if we try to define a *dimension* to the Sierpinski triangle, it would be the quantity d_F such that

$$\left(\frac{1}{2}\right)^{d_F} = \frac{1}{3}.$$

This leads to $d_F = \frac{\log 3}{\log 2} = 1.584962501 \cdots$ which is neither an integer nor a rational. The quantity d_F is the *fractal* dimension of the object. Each fractal object has a dimension which is smaller in value than the Euclidean dimension of the space containing the object *per se*. However, the fractal dimension is mainly a descriptive parameter, as no classical vector operation can be performed in a fractal object, since it is not a Euclidean space. Other fractal objects have dimensions listed in the following table:

Koch snowflake	$\frac{\log 4}{\log 3} \approx 1.26186$
Snow crystal	2.07–2.13
Fern	1.7576
River	1.6844

The first fractal object listed in the table above, the Koch snowflake [71], is a purely mathematical object that is an outline resembling a snowflake; its fractal dimension is therefore given precisely. The other objects are natural objects, whose fractal dimension is given by measurements and therefore includes inaccuracies. The snow crystal is a three-dimensional object [28], so its fractal dimension is greater than 2 (but less than 3). The fractal dimension of a fern depends on its type [san pedro] and the dimension of a river flow depends on its slope, etc. [127].

A fractal is a good way of describing a network deployed within a city. It takes into account the similarities between large neighborhoods and its smaller components: a neighborhood plan (see Figure 5.29), a block of buildings, an apartment in a building, a room in an apartment, a dresser in a room, a drawer in the dresser, etc (although I don't have enough imagination to go lower than the drawer level).

For example, we can take the fractal Swiss flag shown in Figure 5.30. The construction of a fractal Swiss flag is similar to the construction of a Sierpinski triangle and gives a model for a fractal city map. We take a unitary square and engrave a central cross inside it, like two intersecting avenues. In doing so, we divide the square into four smaller identical squares. Let $a < 1/2$ be the side length of these subsquares, the parameter a being the key parameter of the fractal Swiss flag. The width of the central avenue is $1 - 2a$. The fractal dimension of the fractal Swiss flag is equal to

$$a^{d_F} = \frac{1}{4},$$

so $d_F = -\frac{\log 4}{\log a}$. We note that when $a \to 0$, $d_F \to 0$ and when $a \to \frac{1}{2}$, $d_F \to 2$ (the Swiss flag tends to be a complete 2D square). Of course, we can imagine many different ways of constructing such

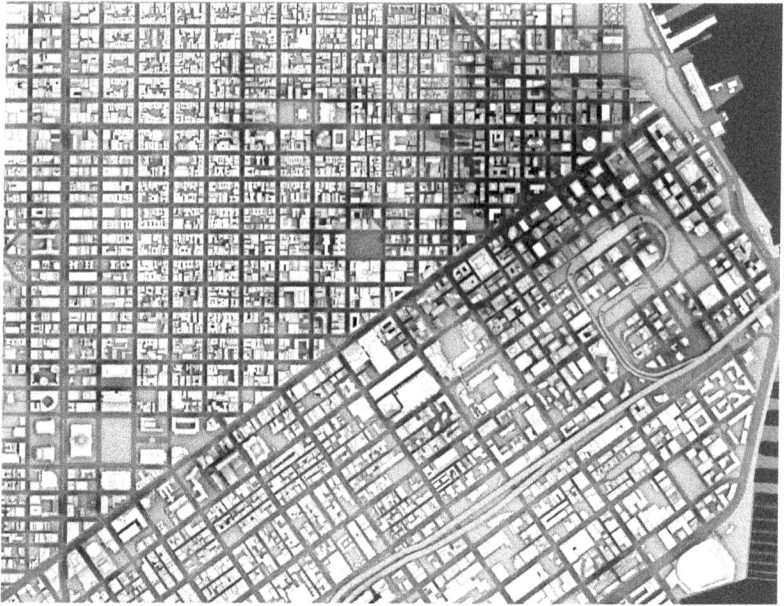

Fig. 5.29.   A San Francisco district.
Courtesy: Mark Schwettmann.

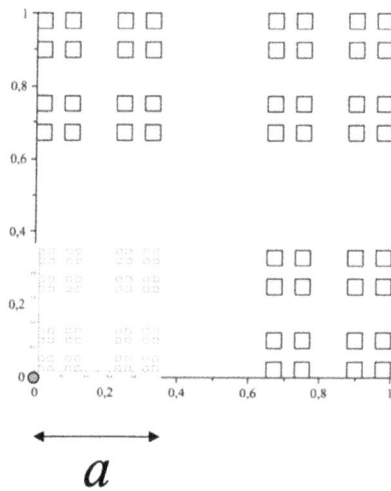

a

Fig. 5.30.   The Swiss flag fractal.

a fractal map, for example, a square with four central crosses, or nine central crosses, and so on.

The key to this discussion lies in the definition of a Poisson process having the fractal set as support. As we see in the exercises, a Poisson shot can have any arbitrary measure as its support. Since the fractal set has a Lebesgue measure of zero, the support will be defined by a combination of Dirac distributions. However, it is simpler to use the construction of the fractal set via a recursive method. To place a point in the Swiss fractal set, select one of the four quadrants at random and repeat the process indefinitely within the quadrant. In fact, uncertainty over position decreases in a^k, k being the number of steps, and in practice, iterations can be stopped after just a few steps. Figure 5.31 shows different examples of fractal Poisson shots for $n = 100$, 400 and $1,000$. This property is not limited to the fractal Swiss flag; in fact, all fractal sets can support a Poisson shot. Figure 5.32 shows a Poisson shot on the Sierpinski triangle.

For the following theorem, we assume that all our fractal maps are extended to infinity. In this case, the density λ is the intensity over the square of unity. As the fractal is extended toward infinity, the density naturally tends toward zero, as the void spaces between the components tends to get larger and larger.

Theorem 5.4. *With no internal noise ($N(\mathbf{z}) = 0$), in a Poisson shot in a fractal set of dimension d_F with $\mathbf{z}$ arbitrary chosen in the*

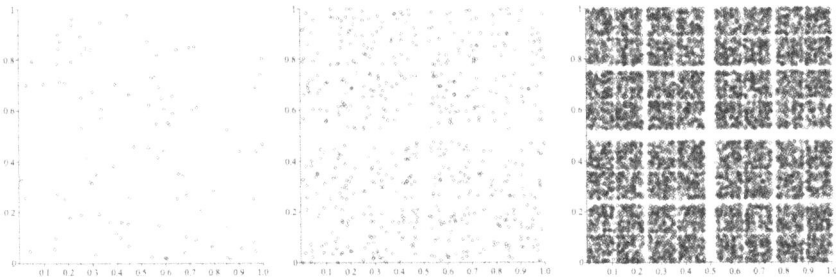

Fig. 5.31. The fractal Swiss flag with Poisson shot: 100 nodes (left), 400 nodes (center), and 4,000 nodes (right).

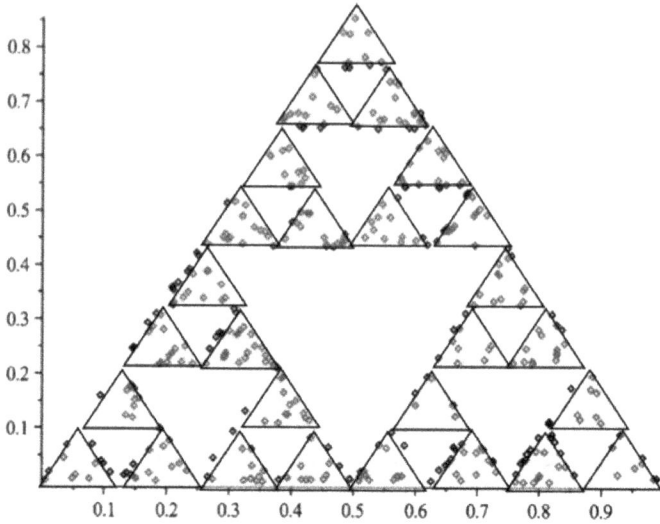

Fig. 5.32.   A Poisson shot on the Sierpinski triangle.

fractal set, we have

$$\text{when } \alpha > d_F : \ C(\lambda) = E[C(\mathcal{I}, \mathbf{z})] = \Phi \frac{\alpha}{d_F \log 2} \left(1 + Q(\log \lambda)\right),$$

$$\text{when } \alpha \leq d_F : \ C(\lambda) = E[C(\mathcal{I}, \mathbf{z})] = \frac{\Phi}{\log 2},$$

where $Q(.)$ is a periodic function of small amplitude and mean 0. The result holds whatever the fading and nominal power distribution. The quantity Φ is the bandwidth and α is the attenuation factor.

The wireless networking industry struggles to squeeze out the last few centimetres at the limit of Shannon's law to gain a few fractional bits of capacity. Here, we can gain much more simply by bending the geometry to our advantage. Switching from a uniform 2D Poisson process to a Poisson process on the Sierpinski triangle already delivers a gain of 26%.

Proof.   To prove the theorem, we use the energy-differentiated field theorem. We take the example of the fractal Swiss flag with associated parameter a. To begin with, we fix z at the origin (in the bottom left-hand corner of the map). By homotheticating the

fractal set with ratio a, we multiply the density *lambda* by 4, i.e.
$S(\lambda/4, 0) \equiv a^{-\alpha} S(\lambda, 0)$. Therefore,

$$E[\log_2 \mathbf{S}(4\lambda, 0)] = -\alpha \log_2 a + E[\log_2 \mathbf{S}(\lambda, 0)].$$

And it appears that the function $f(x) = E[\log_2 \mathbf{S}(e^x, 0)] + \alpha \frac{\log_2 a}{\log 4} x$
is a periodic function of period $\log 4$. Indeed,

$$f(x + \log 4) = E[\log_2 \mathbf{S}(4e^x, 0)] + \alpha \frac{\log_2 a}{\log 4}(x + \log 4)$$

$$= -\alpha \log_2 a + E[\log_2 \mathbf{S}(e^x, 0)] + \alpha \frac{\log_2 a}{\log 4}(x + \log 4)$$

$$= E[\log_2 \mathbf{S}(e^x, 0)] + \alpha \frac{\log_2 a}{\log 4} x = f(x)$$

So $E[\log_2 \mathbf{S}(\lambda, 0)] = -\alpha \frac{\log_2 a}{\log 4} \log \lambda + f(\log \lambda)$. We've almost arrived at
our result, but at this stage the function $f()$ isn't necessarily mean 0.
By virtue of the energy-differentiated field, we have $C(\lambda, 0) = \Phi \lambda \frac{\partial}{\partial \lambda} E[\log_2 \mathbf{S}(\lambda, 0)]$, thus

$$C(\lambda, 0) = -\Phi \frac{\alpha \log a}{\log 4 \log 2} + f'(\log \lambda),$$

where $f'()$ is the first derivative of the function $f()$, which is also
periodic and of mean 0 since, by virtue of the derivation, the constant
term of its Fourier transform disappears in the operation. Noting that
$-\frac{\log a}{\log 4} = \frac{1}{d_F}$ completes the proof for $\mathbf{z} = 0$. To generalize to the mean
on the location $\mathbf{z}$ of the receiver chosen at random in the fractal
set, simply note that it always verifies $S(\lambda/4, a\mathbf{z}) \equiv a^{-\alpha} S(\lambda, \mathbf{z})$ and
conclude that in the fractal set $a\mathbf{z} \equiv \mathbf{z}$ in probability. □

The result of the theorem is interesting beyond pure curiosity
because it shows that the Shannon capacity of a wireless network is
much greater in a Poisson fractal process in $\mathbb{R}^D$ than in a uniform
Poisson process in $\mathbb{R}^D$. This is because $d_F < D$. Of course, this
is the mean value. One could imagine that the periodic function
could possibly affect this conclusion. Fortunately, or unfortunately
depending on whether you're at the top of the hill or at the bottom
of the valley, the amplitude of the oscillations of the periodic function
is very small (of the order of 10^{-3} see Figure 5.33).

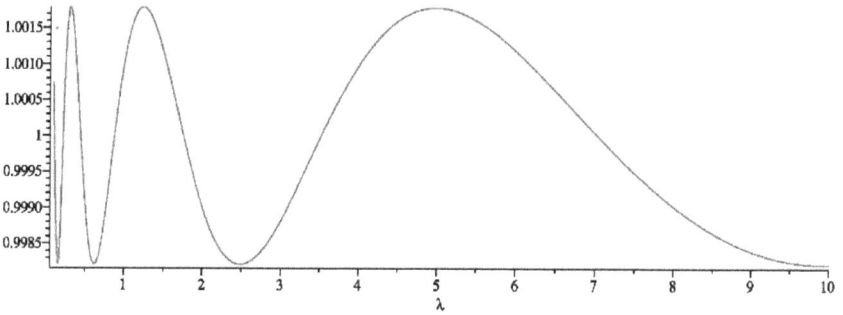

Fig. 5.33.   Oscillation of the fractal capacity with respect to intensity λ.

For the second part of the theorem, i.e. when $\alpha < d_F$, we follow the same steps as for uniform Poisson firing in dimension 2. Given a transmitter density of 1, let $f(R, \mathbf{z})$ be the average number of simultaneous transmitters located at most one distance R from a receiver located at location $\mathbf{z}$. As λ varies, the average number of transmitters at a distance R from $\mathbf{z}$ is $\lambda f(R, \mathbf{z})$. In the fractal Swiss flag, when $\mathbf{z} = 0$, we have the identity $f(aR, 0) = \frac{1}{4}f(R, 0)$ or, in other words, $f(R, 0) = R^{d_F} P(\log R)$ where $P()$ is a periodic function. When $\mathbf{z} \neq 0$ we have $f(R, \mathbf{z}) \leq f(R, 0)$ but this is a peculiarity of the Swiss fractal flag, in a general fractal set we have $f(R, \mathbf{z}) \geq R^{d_F} P_1(\log R)$ for a non-zero periodic function.

The average contribution of emitters beyond the distance R is given by the integral $\int_R^\infty \frac{\partial}{\partial r} f(r, \mathbf{z}) P_0 r^{-\alpha} dr$ which is formally $f(R, \mathbf{z}) P_0 R^{-\alpha} + \alpha P_0 \int_R^\infty f(r, 0) r^{-\alpha-1} dr$ and is finite when $\alpha > d_F$ but tends to infinity when $\alpha \to d_F$. In the latter case, we have $C(\mathcal{I}, z) = \frac{\Phi}{\log 2}$ when $\alpha \leq d_F$ as in the Euclidean case. The most interesting aspect is that the average energy received at the $\mathbf{z}$ point, $\mathbf{S}(\mathbf{z})$, remains finite when $\alpha > d_F$ and therefore α can be much smaller than the Euclidean dimension D.

Here there is a light on a classic paradox, although unexpected in this discussion. Let's place us in a uniform Poisson process in dimension 3 ($D = 3$), we know that the received energy $\mathbf{S}(\mathbf{z})$ is already infinite when $\alpha = D$. But in vacuum, the attenuation factor is 2, so the amount of energy received by an arbitrary observer should be infinite.

If we substitute the network map with a map of the entire universe, and the emitters with the stars shining in the sky, then $\mathbf{S}(\mathbf{z})$

would be equivalent to the amount of light received by the observer at location z (i.e. on Earth). If the location of stars in the universe were the result of a uniform Poisson shot in dimension 3 with non-zero λ intensity, then in theory the amount of light received by an observer at night would be infinite. This phenomenon is known as the "dark night sky" paradox [48]. This paradox has intrigued astronomers since the 16th century: Thomas Digges in 1576, Johannes Kepler in 1610, Jean Philippe Loys de Cheseaux in the 18th century, but most notably by Heinrich Olbers in 1823. Of course, the conditions are not completely identical with the Poisson shot model, as the stars are not points without measure and have a non-zero diameter. This means that all the stars behind them are masked, so we wouldn't have infinite energy, but the night sky dome would in fact have the brightness of a star's surface, i.e. several thousand Celsius degrees.

In order to reconcile theory and experience of the dark night, astronomers have come up with a number of different explanations. Here are three of them

(1) The universe is finite;
(2) The universe is expanding;
(3) The universe is organised under self-similar structures.

The hypothesis (1) remains open, as we know from the big bang theory that the portion of the universe visible to a given observer is finite. In any case, the night sky does not have a uniform appearance, even after cooling from the initial explosion. The trace of this event is the fossil background radiation at 2.7 degrees Kelvin. The assumption (2) is not sufficient, as a large majority of visible deep-space objects are not very red-shifted. If the assumption were valid, the night sky would be a bright white milk, with deeper portions of reddish milk and large portions in the infrared. In fact, fossil radiation is the main form of radiation that undergoes a significant red shift, making it difficult to detect. Only hypothesis (3) remains, but most likely the real cause is a combination of all three hypotheses, but we put forward the self-similar nature of the universe as the main explanation for the dark night paradox. Indeed, stars are grouped into galaxies, dwarf galaxies and giant galaxies, which are grouped into clusters, and these clusters are grouped into super-clusters with ever larger void zones between them, and with extremely tenuous

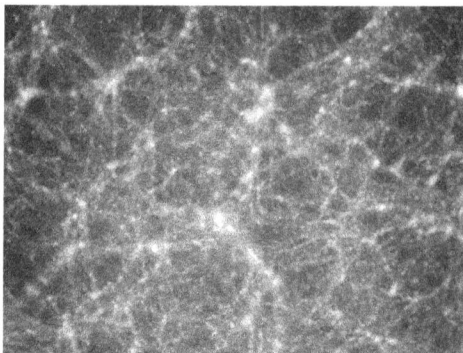

Fig. 5.34. Dark night sky paradox: Self-similar repartition of galaxies and dark matter in the universe viewed at large scale.
Courtesy: Paul Bourke and Alexander Knebe, Swinburne University of Technology.

filaments of dark matter connecting them. Thus, the overall density of matter in the universe is probably very low (even including the elusive dark matter) and if the universe were a fractal object, it would have a fractal dimension slightly greater than 1, 1.4 for [112] (see Figure 5.34).

All previous analyses were based on the assumption that geometry would have no impact on wave propagation. This is certainly true for stars in space (neglecting the slight effect of radio beams bent by gravity), but it's not necessarily true for the propagation of radio waves between buildings in a city. Every time a radio wave hits a wall, it undergoes both refraction and absorption. This gives rise to some interesting theoretical developments. In the following section, we address this issue with hyperfractal geometries.

5.6 Hyperfractals

In the previous section, we discussed a realistic model in which the layout of buildings in a city represented a fractal set. This satisfactorily models the distribution of users inside buildings. There is a dual model in which users are located in streets. In this case, a *hyperfractal* distribution makes sense and seems to best reflect a very branching street network with highly variable user density. When we fly over a

Fig. 5.35.   Aerial view of a city during night (left).
Courtesy: Wikipedia; The corresponding hyperfractal model (right).

Fig. 5.36.   Distribution of street traffic density: Minneapolis (left) and Seattle (right) [95].

city at night, the most striking aspect is the lines of cars lit up on streets with different densities (see Figure 5.35). In every city, high-density streets are very close to very low-density streets, and these alternating densities are reflected at every scale. Figure 5.36 shows traffic density in two American cities: Minneapolis and Seattle.

An example of a map pattern is the grid shown in Figure 5.37. This is a simplified situation where streets run north-south and east-west, a configuration that is typical of many modern cities. However, this right-angle pattern is not mandatory in the hyperfractal model. Here, we take a binary geometric arrangement of the streets. This means that in this example, "vertical" streets have abscissas that

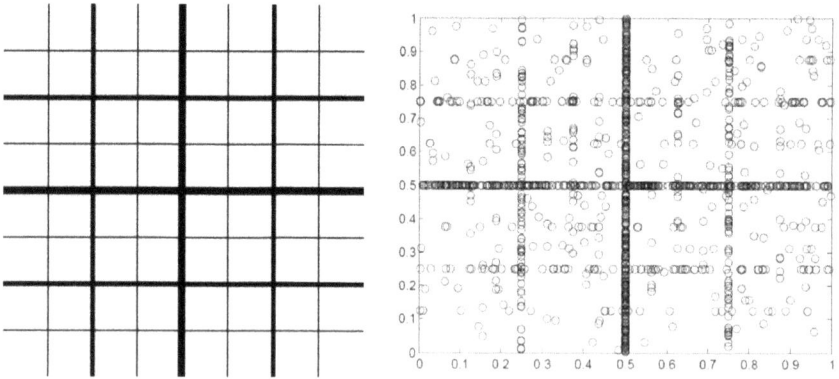

Fig. 5.37. Left: A grid network for hyperfractal city model, the line width is proportional to mobile nodes traffic density. Right: A Poisson shot on the hyperfractal grid.

are multiples of inverse powers of 2, while "horizontal" streets have ordinates that are multiples of inverse powers of 2. If we want to recursively build this map, we start with the central cross, then add a central cross in each quadrant, and so on. We define the depth of a street as the index of the street creation step during recursive construction. There are 2 streets of depth 0 (the central cross), and there are 2.2^H streets of depth H. In the simplest model, the density of users on a street depends only on the depth of the street H: λ_H and this quantity decreases geometrically with a certain rate p:

$$\lambda_H = \frac{q}{2}(p/2)^H$$

with $q = 1 - p$. We note that the set of streets whose depth is greater than a given integer H presents a dense network, as expected for the hyperfractal model. Of course, this is a purely theoretical model: We don't expect the street network in a city to grow to infinite depth. In fact, in the model, statistics mean that streets are empty beyond a certain maximum level H_n of the order of $\log n$ when n is the total population moving through the streets.

Mobile nodes can also be placed recursively. A node to be placed will have a probability p of being uniformly placed on the central cross. Otherwise, with the complementary probability $q = 1 - p$, it is directed to one of the quadrants, then with probability p, it is placed

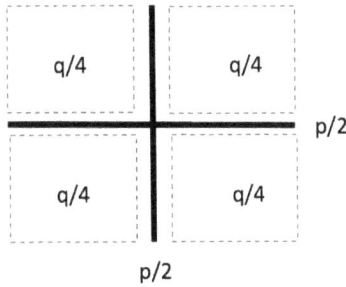

Fig. 5.38. The recursive placement in the hyperfractal map.

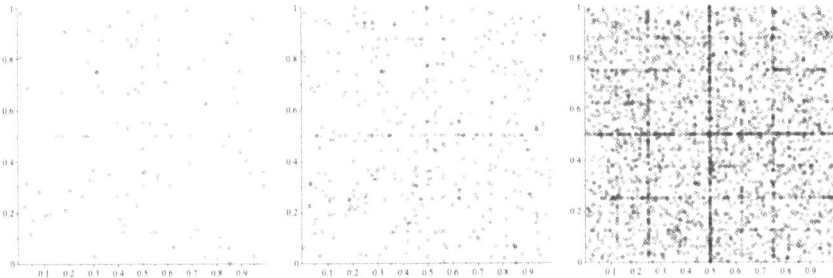

Fig. 5.39. The progressive placement of mobile nodes on the hyperfractal map: 100 nodes (left), 400 nodes (center), and 1,000 nodes (right).

on the local central cross of that quadrant, otherwise it goes to one of the quadrant's subquadrants, etc. See Figures 5.38 and 5.39.

Recursive placement, as described above, naturally calls for fractal qualification. To summarize, the operation consists in cutting the map into four quadrants and moving a fraction q of the volume of the nodes, i.e. $q/4$ for each quadrant, the remaining part being assigned to the central cross, which has a Lebesgue measure of zero. This is equivalent to the fact that dividing the unit of length by 2 implies a reduction in volume by a ratio $q/4$, so the fractal dimension d_F would be satisfied [93]:

$$\left(\frac{1}{2}\right)^{d_F} = \frac{q}{4}.$$

So $d_F = \log_2(4/q)$. Although the above formula doesn't make it obvious, it still implies that $d_F > 2$. For example, $p = 0.5$ leads to $d_F = 3$.

This seems surprising, since the fractal dimension of a set is always smaller than the Euclidean dimension of the space surrounding it. This apparent paradox stems from the fact that the hyperfractal map is not a set but a *distribution*, and a distribution (or measure since it is positive) is agnostic of Euclidean dimension. Indeed, for any integer D, $\mathbb{R}^D$ is in a Lebesgue measurable bijection with $\mathbb{R}$, i.e. a measure on $\mathbb{R}^2$ is also a measure on $\mathbb{R}^{100}$. To see this for yourself, consider a number $x \in [0,1]$ written in binary $x = 0.10011101001101001$ and translate it as the coordinates of a point $f(x) = (y_1, y_2, y_3) \in [0,1]^3$, where y_1 consists of the first, fourth, seventh, etc. bits of x, y_2 consists of the second, fifth, eighth, etc. bits, y_3 consists of the third, sixth, ninth, etc, bits of x. The function is continuous at all points, with the notable exception of the countable set made up of multiples of inverses of powers of 2. Treating these numbers separately since each admits two infinite binary sequences, we obtain a measurable bijection between $[0,1]$ and $[0,1]^3$, see Figure 5.40.

Above all, it's important to understand that the hyperfractal model is not a variant of the Manhattan grid model. Indeed, the grid-like appearance that appears in the above description when only the densest streets are represented is merely a convenience of presentation. In fact, the essence of the model lies not in the general shape of the city plan but in the local organization of the street network according to its density. It's not necessary for a city plan to have the shape of a grid with a binary organization of street levels to correspond to a hyperfractal model. The two overriding conditions are that street densities decay with polynomial decay and that low-density streets alternate with high-density streets at any scale.

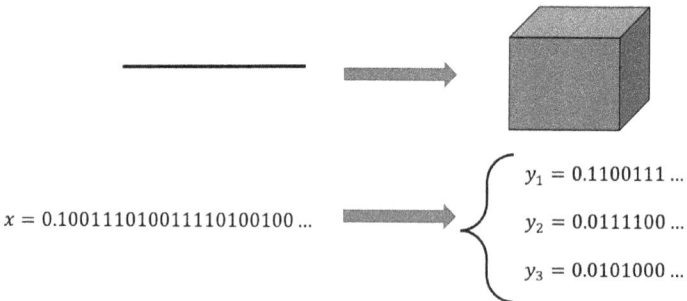

$$x = 0.100111010011110100100\ldots \quad \begin{cases} y_1 = 0.1100111\ldots \\ y_2 = 0.0111100\ldots \\ y_3 = 0.0101000\ldots \end{cases}$$

Fig. 5.40. $[0,1]$ and $[0,1]^3$ in Lebesgue measurable bijection.

Of course, this would imply an infinite number of streets, which is obviously an unrealistic assumption, but this problem is blunted by the fact that very low-density streets in theory won't appear because they won't bear any users in practice. Again, the general shape of the city map doesn't have to be a square; it can be any shape, and streets can take on arbitrary orientations.

The following procedure calculates the hyperfractal dimension of various real cities. The difficulty lies in the fact that roads rarely have an explicit hierarchy of levels, since the data we have on cities generally concern the length of road segments and the average density of road traffic on each. To get around this problem, we rank road segments in descending order of traffic density. If S is a segment, we denote $\eta(S)$ its density and $\mathcal{L}(S)$ the cumulative length of the segment ranked before S (i.e. of greater density than $\eta(S)$). For $\xi > 0$, we denote $\mu(\xi) = \eta(\mathcal{L}^{-1}(\xi))$. Formally, $\mathcal{L}^{-1}(\xi)$ is the road segment S with the smallest density such that $\mathcal{L}(S) \le \xi$. The hyperfractal dimension will appear in the asymptotic estimate of $\mu(\xi)$ of order ξ^{1-d_F} when $\xi \to \infty$ thanks to the following property.

The following table shows the hyperfractal dimensions of real cities calculated using this method [97]:

City	d_F
Adelaide	2.8
Minneapolis	2.9
Nyon	2.3
Seattle	2.3

5.6.1   *Urban Canyon effect and fixed relays*

The hyperfractal nature of maps has no specific effect on radio telecommunication performance, unless we take into account the "canyon effect", which generally accounts for the impact of urban buildings on the propagation of radio waves. The low penetration capacity of building walls leads to what is known as the canyon effect, which essentially reveals that radio signals emitted by the mobile user propagate mainly along the axis of the streets, with little or no penetration of buildings. In this section, we consider a hop-by-hop routing strategy between mobile nodes in the classic store-and-forward mode.

We assume that the speed of user movement will be considerably lower than the speed of packet switching between nodes.

In addition to mobile nodes capable of relaying as in a MANET, the network will also use fixed relays placed on existing infrastructures [93]. These fixed relays will use the same technology as the mobile nodes and will be placed at street intersections to extend radio coverage. Indeed, without fixed relays at certain street intersections, packets could not be transmitted outside their street of origin due to the low penetration of millimeter-wave radio waves in buildings. Indeed, cars would not spend enough time at the street intersection when they are in line with the two streets to allow efficient routing of packets off the vehicle's axis of progress. Fixed relays can be located, for example, at traffic lights at major intersections, according to their respective traffic density. This calls for a hyperfractal distribution, see Figure 5.41.

We set a specific $0 < p_r < 1$ and $q_r = 1 - p_r$ parameter for relay placement. For each coordinate, we have a recursive placement. Abscissa and ordinate will be independent, so let's concentrate on selecting the abscissa. With probability p_r the abscissa is exactly

Fig. 5.41.   Map of traffic light locations in Adelaide [95].

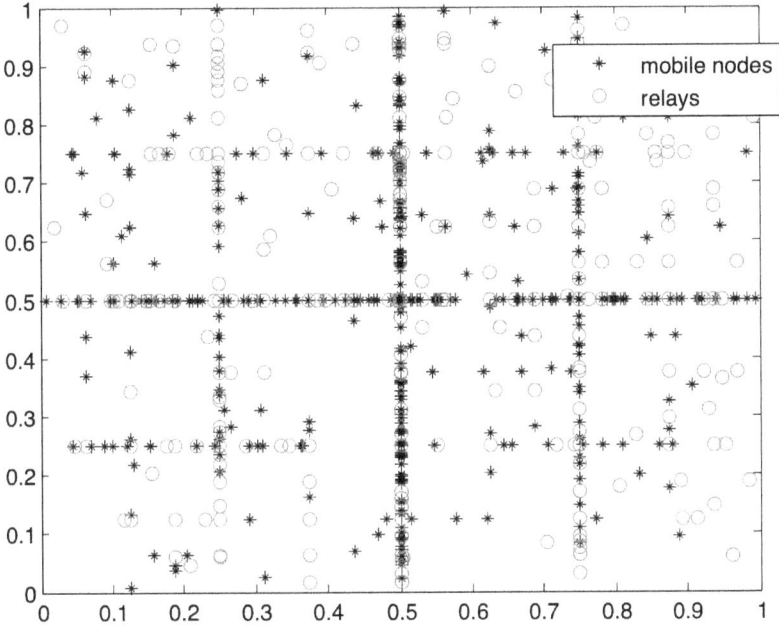

Fig. 5.42. Hyperfractal map with mobile nodes and fixed relays.

on the main vertical route (i.e. at level 0). Otherwise, it will be in the left or right half-plane, the process repeats itself, and the level increments, until it is placed on a vertical road. Let V be the level of the road reached. We do the same with the ordinate, so let H be the level obtained. The point obtained is exactly at the intersection of a North-South street of level V and a West–East street of level H. See Figure 5.42. The probability of a relay being located at a given intersection of two streets with respective levels H and V is

$$p(H,V) = p_r^2 \left(\frac{q_r}{2}\right)^{H+V}.$$

The projected set of relay positions on a coordinate axis is a hyperfractal set of dimension $\frac{(2/q_r)}{\log 2}$. The combination of abscissa and ordinate doubles this dimension:

$$d_r = 2\frac{(2/q_r)}{\log 2} \geq 2.$$

The estimation process for assessing the hyperfractal dimension of the relay set is similar to the determination of the hyperfractal dimension of the mobile node locations, except that for the relay set, both intersecting street densities must be taken into account. For Adelaide, we find that the location of the traffic lights has a hyperfractal dimension $d_r = 3.5$.

As with the hyperfractal distribution of mobile nodes, the distribution of fixed relays tends toward the uniform Poisson distribution when $d_r \to 2$. The following formula shows the transport capacity of the hyperfractal map as a function of the hyperfractal dimension of mobile nodes and relays. In the limit where d_r tends toward 2 (the distribution tends toward a uniform Poisson distribution), the contribution of the relays means that the transport capacity tends to be equivalent to the transport capacity of a uniform Poisson map of dimension $D = 1 + d_F$ [93,95]. In other words, in terms of wireless capacity, a city is at least of apparent dimension 3! The reason for this gain in dimension is that the routing path between two mobile nodes can be optimized by taking shortcuts through less populated streets where the distances between relays are larger:

$$T_n = O\left(n^{\frac{2}{(1+1/d_F)d_r}}\right) \to O\left(n^{1-\frac{1}{1+d_F}}\right).$$

This result holds when the number of relays ρ is of order n. In the general case, we have

$$T_n = O\left(\rho^{\frac{2}{(1+1/d_F)d_r}}\right).$$

The condition ρ of order n means that the fixed relays may be in similar number than the mobile nodes. In general, we state ρ of order n^θ with $\theta \le 1$. When $\theta < 1$, the wireless dimension of the city cannot be larger than $\frac{1}{1-\theta}$.

5.7   Exercises

To simplify, in the following we assume the device bandwidth as one unit: $\Phi = 1$ and we denote $\mathbf{S}(\lambda)$ the random variable $S(\mathcal{I}, 0)$ when $\mathcal{I}$ follows a Poisson shot of density λ.

5.7.1 *Laplace transform of signal level in Poisson shot model*

5.7.1.1 *Fading $F \equiv 1$*

We cut the plane in a partition $A_1, A_2, \ldots, A_k, \ldots$. We denote $\mathbf{S}(\lambda, A)$ the sum of all the contributions from the emitters located in A and $w(\theta, A) = E[e^{-\theta \mathbf{S}(\lambda, A)}]$ its Laplace transform.

Exercise 23

Give an expression of $w(\theta)$ involving $w(\theta, A_k)$. See the Answer of Exercise 23 in Chapter 9.

Exercise 24

Assume that A is a small $dx \times dy$ square at position $\mathbf{z} = (x, y)$. Show that

$$w(\theta, A) = 1 + \lambda dx dy (e^{-\theta \|z\|^{-\alpha}} - 1) + O(dx^2 dy^2), \quad (5.2)$$

$$\log w(\theta, A) = -(1 - e^{-\theta \|z\|^{-\alpha}}) \lambda dx dy. \quad (5.3)$$

See the Answer of Exercise 24 in Chapter 9.

Exercise 25

Thus by making dx and $dy \to 0$ show that

$$\log w(\theta) = -\lambda \pi \int_0^\infty 1 - e^{-\theta \|z\|^{-\alpha}}) r dr.$$

See the Answer of Exercise 25 in Chapter 9.

5.7.1.2 *Case F general*

In this case, we have to take into account all possible values of F. We denote $P_F(u)$ the probability density of F on real u. We have

$$w(\theta, A) = 1 + \int_0^\infty \lambda dx dy (e^{-u\theta \|z\|^{-\alpha}} - 1) P_F(u) du + O(dx^2 dy^2),$$
$$(5.6)$$

$$\log w(\theta, A) = -\int_0^\infty (1 - e^{-u\theta \|z\|^{-\alpha}}) \lambda P_F(u) du dx dy, \quad (5.7)$$

and finally $\log w(\theta) = -\lambda C \int_0^\infty (u\theta)^\gamma du = -\lambda C_F \theta^\gamma$ with $C_F = \pi \Gamma(1-\gamma) E[F^\gamma]$.

Exercise 26

Compute C_F for Rayleigh fading:
   See the Answer of Exercise 26 in Chapter 9.

5.7.2   *Probability of correct reception*

Exercise 27

Compute the probability $p(r, \lambda, K)$ to be received at distance r when the emitter density is λ and when the SINR required threshold is K assuming a relay fading F. See the Answer of Exercise 27 in Chapter 9.

5.7.3   *The field differentiation theorem*

Exercise 28

Compute the average throughput $T(\lambda, \mathbf{z})$ of an emitter at a location $\mathbf{z}$. See the Answer of Exercise 28 in Chapter 9.

Exercise 29

What is $C(\lambda)$ expressed with $T(\mathbf{z})$ function? See the Answer of Exercise 29 in Chapter 9.
   Let μ be an arbitrary density measure on the plan. We denote $\mathbf{S}(\mu)$ the signal received from A Poisson shot following μ.
   We denote $L(\mu) = E[\log \mathbf{S}(\mu)]$ and $T(\mu, \mathbf{z})$ the throughput of an emitter at position $\mathbf{z}$.

Exercise 30

Show that $L(\mu + \delta_{\mathbf{z}} dt) - L(\mu) = T(\mu, \mathbf{z}) dt$ at first order when dt is small. See the Answer of Exercise 30 in Chapter 9.
   Thus,

$$\frac{d}{dt} L(\mu + t\delta_{\mathbf{z}})|_{t=0} = T(\mu, \mathbf{z})$$

or by abusive notations

$$\frac{\partial}{\partial \delta_{\mathbf{z}}} L(\mu) = T(\mu, \mathbf{z}).$$

Extending the notation we have

$$\frac{\partial}{\partial \mu} L(\mu) = \frac{d}{dt} L(\mu + t\mu)|_{t=0} = \iint T(\mu, \mathbf{z})\mu(\mathbf{z})dxdy.$$

Thus with U indicating the uniform distribution

$$\lambda \frac{d}{dt} L(\lambda U) = C(\lambda).$$

Chapter 6

The Limit of Artificial Intelligence Imposed by Information Theory

Abstract

In this chapter, we try to draw the moving frontier between AI, information, and computer science. To simplify, Shannon laid the foundation of information theory by voluntarily skipping the algorithmic aspects while Turing formalized the transformation of information by means of algorithms and computations. Meanwhile, AI has made revolutionary (and sometimes scary) changes to our lives. There are plenty of questions which haven't yet received very few satisfying answers: How can modern information and computer sciences help shape the future of artificial intelligence? While AI is identified as a challenge to human intelligence? Can AI be the starting point of the next age of computer science or just a kind of last resort when classic algorithms fail? Are all problems learnable by AI? Can learnability be learnable? Can the classic algorithms still provide some ersatz of artificial intelligence? These are audacious questions that must be pondered. We have no ambition to provide conclusive answers to these probing questions. However, they will serve as starting points of the discussion about the contact points between AI and information theory.

6.1 Artificial Intelligence versus Information Theory

Artificial intelligence has become an important and fundamental tool with an exceedingly abundant spectrum of applications. However, we are still waiting for a proper full theory explaining its performance. Consequently, it extends in front of us like a *terra incognita*.

Fig. 6.1.　Last interview of Stephen Hawking.

In one of his last interviews (see Figure 6.1), the late astrophysicist Stephen Hawking manifested his concerns about the long-term perspective of artificial intelligence:

> If AI would take off on its own and redesign itself at an ever increasing rate, Humans, who are limited by slow biological evolution, couldn't compete and would be superseded.

Our point is to confront some information theoretic points to these concerns.

6.1.1　*Shannon point of view*

Shannon worked on some of the earliest concepts of artificial intelligence (see Figure 6.2), in particular with his famous mouse-in-the-maze experiment. A mechanical mouse placed in a labyrinth learned to exit by randomly deciding its direction at each new fork in the path. By simply back tracking on the forks in the path, the mouse (and indeed the computer behind it) was able to find and remember its way out.

Shannon comment about Hawking worries could have been that a system cannot naturally evolve into something more complicated. A simple coffee grinder cannot become a racing car engine (see Figure 6.3) without the introduction of a good bunch of external

Fig. 6.2. Claude Shannon experimenting machine learning.

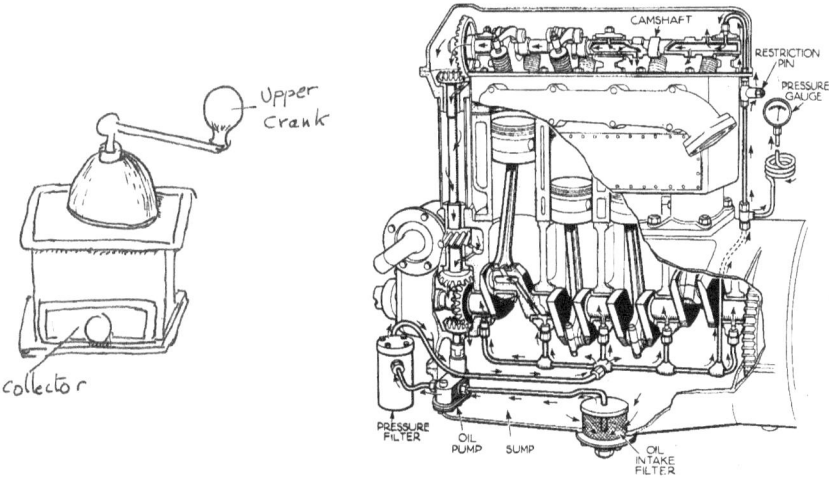

Fig. 6.3. A simple system cannot evolve into a more complicated version of itself. *Courtesy*: Wikipedia.

complexity. Isolated within a natural evolution, some parts of the machine would get damaged so that some states of the system become unreachable, the complexity of the resulting system would hence be smaller than the original one. In other words, the future system Z' can only be a function $f(Z)$ of the original system, and its complexity (i.e. its deterministic entropy) can only decrease:

$$h(f(Z)) \leq h(Z).$$

The only way to make the complexity increasing is to introduce some *alea* or, in other words, to borrow some entropy from the outside word. With some abuse of notation, one could have the inequality enabling complexity increase:

$$h(f(Z)) \leq h(Z) + h(f).$$

The best example of the increase in complexity of a system subject to external influences can be found in the evolutionary history of life on Earth. Life is based on nucleotide chains located in cell nuclei (DNA), for which the main source of mutations is cosmic radiation. A powerful X-ray photon can easily eject a base from the chain [86]. Other sources of mutation are viruses, which are able to insert new material into the DNA chain during reproductive duplication. All mutations are random, and sometimes randomness can be traced back to the quantum level (see Figure 6.4), where a tiny event can have vast consequences.

Charles Darwin (see Figure 6.5) was the first proponent of the theory of evolution [23]. In his time, the theory of genetics around DNA had not yet been discovered, and the main idea behind the evolution of life in Darwin's theory revolved around the pressure of the environment. This pressure would lead to the best species being

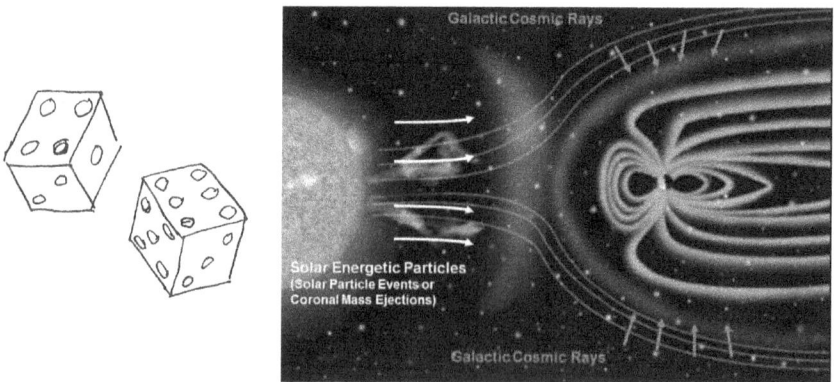

Fig. 6.4.   Introducing *alea* in system evolution (left) and random mutations are main causes in life evolution (right).
Courtesy: Wikipedia.

Fig. 6.5.   Charles Darwin.

selected according to their ability to survive. The difficulty with this theory lay in the fact that, since the environment itself could also be modified by the evolution of competing species, the definition of the "best" adaptation could become elusive. For example, the best adaptation might mean the highest rate of reproduction, as can be seen in the world of viruses or bacteria. In the end, we can hardly come up with a sufficiently autonomous definition of the "best" species, apart from the tautological loop whereby the best species covers the species that has actually survived (see Figure 6.6).

Beyond the problem of trying to better qualify the mechanism of evolution, we can try to quantify the quantity of information that has been created over the entire history of the evolution of life over the last 3.5 billion years. Biomass is the weight of all living species currently on Earth. Surprisingly, it hasn't changed much over these billions of years. Unsurprisingly, it consists mainly of bacteria. It is estimated that up to $5 \cdot 10^{30}$ bacteria live on Earth at any one time (mainly in the oceans). On average, there is one mutation (a change of base) every twenty minutes for every million bacteria. By functional mutation, we mean a mutation that leads to the survival of a new variant, since most mutations are not viable. All the calculations made, with a base equivalent to 2 bits of information (the alphabet of nucleotide bases is of size 4, symbolically noted $\{A, G, C, T\}$), lead to $7.2 \cdot 10^{38}$ bits of information created since the beginning of life.

Another way of obtaining this figure is to consider a single liter of seawater. We all remember the classic elementary school experiment of taking a close look at the ecosystem of a small liter of seawater:

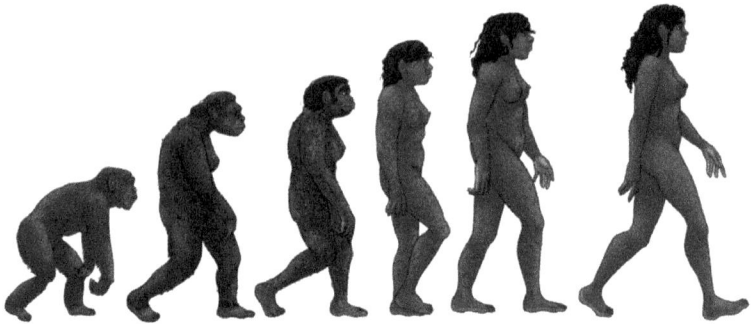

Fig. 6.6.   Human evolution (top), extinct species (bottom).
Courtesy: Wikipedia

You can discover a micro-world populated by tiny shrimps, worms, plankton and algae, not to mention bacteria invisible to the naked eye. This whole world is in constant activity, with species feeding on each other, socializing and reproducing. As a rule of thumb, we can imagine that around 100 kbytes of new genetic information is created every day. If we consider that the oceans contain around 1 billion cubic kilometers of water, each made of 1,000 billion liters of water, and multiply this figure by the 1,000 billion days corresponding to the 3.5 billion years of life on Earth, we get 10^{38} kbytes of information. Just because this amount of information can be quantified in numbers doesn't mean we can easily grasp its full extent. However, if we could imagine a technology where one bit would be stored per atom of rock,

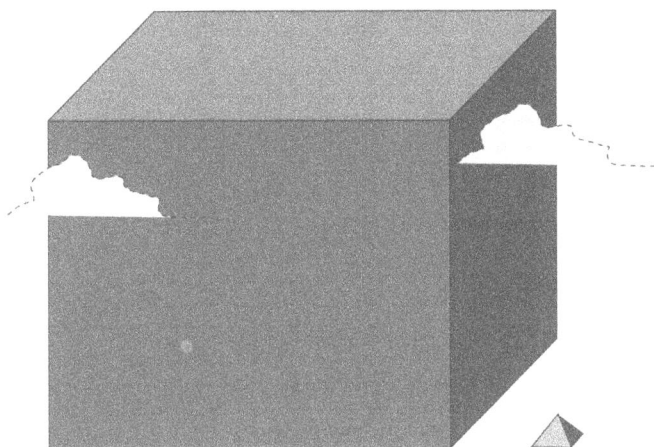

Fig. 6.7. The cube of life (compared to the Great Pyramid on the right).

we'd need a storage facility the size of 4,782 Great Pyramids or a 2.3 km square cube of limestone (Figure 6.7). On the other hand, if we wished to store the total quantity of information created to date by mankind (around 10^{18} bytes), we would need, with the same technology, a storage space of a few grains of sand (10^{-5} grams). What's more, this amount of information consists mainly of videos, images and audio files, which are raw data, not executable code. If we limit this quantity to actually executable codes, we need to put into perspective that the total quantity of genuine executable codes written by computer engineers represents less than 100 gbytes, if we don't count the numerous copies disseminated in individual computers, this quantity would be comparable to a nanodrop of air exhaled by a single bacterium.

However vast the user space for life may be, the systematic extinction of species can lead to a significant reduction in the complexity of life, even without a reduction in biomass. For example, human genetic information is written on 3 billion bases, crossed by 20,000 genes. To simplify, each gene in a baby's genetic information has the same probability of coming from the mother or the father, due to crossover during meiosis (this number may be higher due to permitted overlap). Thus, the amount of information needed to define a baby's genetic heritage is of the order of 20 kbits, if we take the convention that bit 0 would indicate that the gene comes from the mother and bit 1 that it comes from the father. 20 kbits is less than the information content

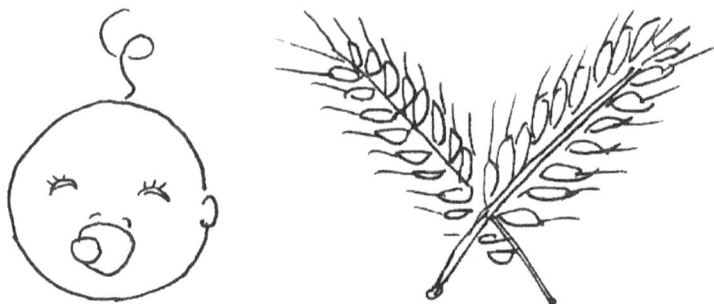

Fig. 6.8. A baby adds little to life complexity (left) and a specie extinction removes a lot (right).

of 5 tweets (in general, a birth generates many more tweets). On the other hand, the extinction of an entire living species removes much more complexity. *Paris Japonica* is the plant with the largest DNA sequence at 150 billion bases, while the DNA sequence of common wheat is 15 billion bases. Plant DNA is generally larger than that of animals, because plants can only move with difficulty when climatic conditions change, so their genetic information must contain all the options needed to adapt to these changes. In any case, this adaptation doesn't work when changes are too rapid, and this is why we see many species extinctions (even among animals). Losing a species is equivalent to losing the genetic information of several million babies (see Figure 6.8). Biomass complexity and diversity are diminishing at a tragic rate.

Of course, the evolution to self-optimising AI may not need to start from the initial null state of life on Earth. Limiting working codes to be written in a single language and running them under a single operating system would allow pruning several orders of magnitudes. But this would still result in a significantly large user space capable of allowing all the working codes to contend and evolve in parallel. Furthermore, there would be need to trigger the evolution of the codes toward interesting codes because having the most reproducible codes might not result in the most capable codes for driving a car. Bending the evolution toward a final goal is extremely difficult if every intermediate step does not significantly lean toward this goal. For example, the deep neural network capable to recognize individual human face is in fact initialized with the weight of the network trained to detect cats in pictures.

The other obstacle is code certification. We need to be able to prove that the resulting code works in all, or at least most, test cases. The aim is not necessarily to produce perfect code but to be happy that the code drives the car better than a human. Unlike codes produced by a computer engineer, AI codes produced by AI learning are extremely difficult to *prove*, i.e. they can hardly be reverse-engineered for systematic verification.

6.1.2 *Solving a problem with AI*

As we saw above, the evolution of AI cannot realistically be managed like a zoo of self-modifying executable codes in an unrealistically large user space. Thus the AI can hardly evolve by itself and must be deeply channelized by training strategies. A *learning problem* can be viewed as a set of training data associated with an objective to fulfill. In supervised learning, the objective can be defined by a sequence of a ground truths attached to the training data. For example, for the classic problem of detecting a cat in a picture, the training data is a sequence of pictures and the ground truth would be just a binary label indicating if yes or no, a cat is hiding in the picture. The machine acts as an automaton whose aim is to predict the ground truth from a data (see Figure 6.9). The loss measures the difference between the prediction and the ground truth and can be established under an arbitrary metric. The general objective of machine learning is to minimize the average loss, but since the ground truth might contain some inherent stochastic variations (e.g. when predicting the

Fig. 6.9. Pictures for training cat detection (left) and a very symbolic view of the AI machinery (right).

result of a quantum measurement), it may be impossible to make the loss as small as we would like. Given a learning architecture, there exists a setting (certainly several settings) of this architecture which will provide the optimal average loss. But the optimal prediction might be a non-computable function (e.g. the program halting problem). However, if the optimal prediction is attainable via the execution of an automaton of finite size (e.g. a ground truth which is just the result of a computer algorithm applied to the data has the optimal loss equal to zero), there is still the question of the size of the automaton and the question of the quantity of training data needed to get a suitable convergence to the optimal settings of the learning architecture.

We can note that solving a problem with the AI loop is a kind of summary of the gigantic pyramid formed by the trio data-information-knowledge in a nutshell [114]. The data are the pictures and labels submitted to the machine, the connections between the labels and the picture are the information, and the final weight in the machine after the learning phase are representative of the knowledge obtained after the training. Indeed, the weights enable the machine to make prediction which is the essence of knowledge.

But the representation of the knowledge by vectors of weights may look elusive and would greatly differ from the classic semantic of knowledge as it is taught to pupils and students and might be even impossible to translate in natural language. However, computer-generated knowledge might be as or even more efficient as knowledge expressed in natural language. As a comparison, no one questions anymore the validity of the computer-generated mathematical proof like the four color theorem [24], the proof translated in natural language would be too long to be understood by a human mathematician. However, we see that the representation of knowledge may fail in some cases even when the natural knowledge is very simple. For example, we see that for simple algorithmic problem the machine may completely fail to find a practical solution with a set of weights.

6.1.3   *Taxonomy of learning strategies*

Wikipedia defines "Machine learning as the study of computer algorithms that can improve automatically through experience and by the use of data". The most usual strategy is the supervised learning opposed to unsupervised learning, or semi-supervised learning.

6.1.3.1 *Basic machine learning strategies*

Unsupervised learning: In unsupervised learning, the machine tries to identify the cluster of data or data separability with respect to some metric. One popular method is the spectral analysis on the adjacency matrix of the data. Extracting the k first eigenvalues provides the best fitting reduction of the data in k components. There are efficient methods to extract the eigenvalues, for example, methods based on quantum walk [6]. Other clustering methods such as the spectral analysis of the non-backtracking edge-adjacency matrix [74] can be more process consuming but have the advantage to give the exact number of components as output. It is the number of real eigenvalues which are larger than the modulus of the largest non-real eigenvalues. Similar methods are known under the name Principal Component Analysis.

Reinforcement learning covers unsupervised techniques which are directly inspired by game theory. Adversarial learning is a special case when the machine is taught to mimic a source of data.

Supervised learning: By supervision one means that for each training data we know the ground truth, which we sometime denotes the "label". Therefore, the trained machine will predict the label and greatly facilitate the use of the AI. One example is the "Decision Tree": The machine accumulates the data with their labels in order to build a minimal tree [87] which maps the best the data to the labels. There are many algorithms and heuristic in order to minimize the tree. An arbitrary test vector is submitted as input at the root of the decision tree. It will be routed in the tree toward the leaves, each visited internal node of the tree corresponds to a binary qualification (if the tree is a binary tree) which plays the role of a question on the data contained in the test vector (e.g. some function value greater or smaller than some threshold). The leaves of the tree are the labels. One popular example is the application "Akinator" [126]. The tree structure can be extended to a non-cyclic graph structure such as Dag.

The "Linear Regression" [119] strategy is the simplest version of supervised learning, since it consists of finding the linear function which separates the best the dataset according to their labels. It is equivalent to a decision tree with a single internal node, obtained with a linear function. Linear regression is an historically early learning strategy and has the advantage of being optimized without heuristic either by gradient descent or by global quadratic analysis when the data are submitted by batch. The linear regression [123] should not be confused with the "Logistic Regression" which consists of taking a batch of data and finding the distribution which optimizes the mapping between data and labels. The distribution is to be selected on a pre-existing set of distributions.

Neural networks: Neural networks are made up of artificial neurons called perceptrons. In fact, the resemblance with the functioning of biological neurons is overrated, as the latter is not yet well understood. We can regard a network of artificial neurons as an oversimplification of what was understood about brain function twenty years ago. Mathematically, the artificial neuron receives a linear combination of scalar quantities received from other neurons and retransmits a nonlinear function of the value obtained (this can be the Heaviside function, the ReLU function, etc). The weights of the linear combinations are adapted to each training vector in order to minimize loss and are highly dependent on the neural network architecture and can be complicated. However, in the case of a biological brain, the equivalent process between natural neurons is still largely unknown at present.

Deep learning: Deep learning came as a revolution in regard to the numerous applications it created through its specific neuron training [20]: image recognition, natural language processing, the list is not ending and is augmenting every day. Structurally, it is a neural network; its particularity is that its intimate structure is organized in successive layers which confers some interesting mathematical properties to the architecture. The convergence process is a pure gradient descent because the gradient computation is in this case greatly simplified by the layer structure which allows the back-propagation to be of the same order of complexity than the forward propagation.

Specific improvements of neural networks: Thanks of its simplicity, the deep learning process is robust enough to be associated with other techniques in case they are necessary to improve the performance. For example, in finding a cat in a picture network, the layers must be interleaved with convolution layers in order to foster the training (and so not all positions in the picture need to be trained). The face recognition learning frequently uses a network trained beforehand on cat recognition. Many learning processes need some local and weird techniques in order to be accelerated, or even be made converging, whether or not these techniques have a strong theoretical background. Among these techniques are recurrent Neural Networks, independent RNNs, recursive NN, neural history compressor, second order RNN, Long Short Term Memory NN, Transformer NN, etc.

6.1.3.2 *Machine learning acceleration*

Distributed learning: One of the most obvious ways of accelerating learning in a neural network is to use distributed learning. Many deep learning operations can be performed in parallel on different processing units. By distributing the training data across different machines, learning can take advantage of parallel process processing, and one-off *rendezvous* between these processes is limited to exchanging the weight values of their local neural networks. The optimality of these processes is studied in [55]. But this parallel acceleration does not change the learning strategy.

Adaptive layer multiplicity: As we have seen, there is no universal learning strategy, since even deep learning requires additional adaptation strategies and sometimes fails. The particular case of deep learning is interesting because even its most fundamental parameter, the number of layers, is crucial. If there are too few layers, there isn't enough diversity to capture the essence of the problem. Too many layers and learning doesn't converge (as we'll see with the Fourier transform [66]). One way of solving this problem is to progressively add new layers until convergence is satisfactory. But this can be costly in terms of processing.

Particle systems: The best adaptive strategy known to date is to operate several parallel but different architectures as a particle system. The computing units corresponding to the slowest architectures are moved to the fastest architectures and, at the end of the day, all the particles (i.e. all the architectures) are grouped together on the fastest architecture. This technique delivers excellent performance and can also support non-homogeneous architectures (not necessarily deep learning). However, all architectures have to be tried out, which limits the number of competing architectures.

6.1.4 *Regret MinMax analysis*

There is an obvious analogy between calculability and learnability. There are unlearnable problems, just as there are computationally unlearnable problems. Since a learning automaton is a program, a non-computable problem should also be a non-learnable problem.

One defines the regret of a specific learning strategy related to a specific problem as the distance between the optimal learning and the specific learning strategy both applied on this problem. If a problem induces a regret decaying in $1/\log T$, T being the size of the learning set, then this would be an indication that the problem is hardly learnable, since one should need an exponential number of training data in order to have a significant reduction of the regret. In our analogy, this should be the equivalent of NP hard problems in computation theory.

The example of data compression is an excellent illustration. The optimal compression ratio is the entropy of the text, this could be seen as the optimal loss (we can never compress to zero). The distance between the entropy and the actual achieved compression ratio when applied to a finite text is the *redundancy*. In most universal compression algorithms such as the Ziv–Lempel algorithms, the redundancy is proved to converge to zero on large classes of textual information. The real question is how fast it converges. Indeed, it is proven that over texts generated by Markovian sources the redundancy is in $1/\log n$, where n is the length of the text. It means that the convergence is poor and we need to have exponentially long texts in order to get significantly close to entropy (it is not a problem, in practical application, we just need to get a redundancy of same order as of the entropy).

Let a problem be defined by a source S of features $\mathbf{x}$ and ground truth labels y with a conditional probability $p_S(y|\mathbf{x})$ mapping features $\mathbf{x}$ to the ground truth label y. Given a sequence of T training feature $\mathbf{x}^T = (\mathbf{x}_1, \mathbf{x}_2, \ldots, \mathbf{x}_T)$ and ground truth sequence $y^T = (y_1, y_2, \ldots, y_T)$, we denote $P_S(y^T|\mathbf{x}^T)$ the conditional probability distribution of the ground truth and feature of the considered problem. If the pairs are independent (common hypothesis), then $P_S(y^T|\mathbf{x}^T) = \prod_t p_S(y_t|\mathbf{x}_t)$.

We denote $P_L(y^T, \mathbf{x}^T)$ the conditional probability attained by the learning process in T learning steps t. The ideal learning would be $P_S = P_L$. A learning would be asymptotically optimal if $\lim_{T\to\infty} P_L = P_S$ in a sense we define with the regret.

The simplest definition of regret is the *point-wise* regret which is as follow:

$$\bar{R}_T(y^T, P_S) = \log P_S(y^T|\mathbf{x}^T) - \log P_L(y^T|\mathbf{x}^T). \qquad (6.1)$$

The average point-wise regret is

$$E[\bar{R}_T(y^T, P_S)] = \sum_{y^T} P_S(y^T|\mathbf{x}^T) \log \frac{P_S(y^T|\mathbf{x}^T)}{P_L(y^T|\mathbf{x}^T)}$$

which is nothing other than the Kullback–Leibler divergence of the distributions [33] $P_S(.|\mathbf{x}^T)$ and $P_L(.|\mathbf{x}^T)$, $\Delta(P_S, P_L)$. The Kullback–Leibler divergence is a classical way of measuring the distance between two distributions. It is a semi-distance because it is not symmetrical (to make it symmetrical, simply add the inverted expression). But this is not completely true, because the learning model L may also depend on the training instance y^T.

If we consider a set $\mathcal{S}$ of probability distributions as training sources and a set $\mathcal{L}$ as training strategies' probabilities, we define the average min–max average regret as

$$\bar{R}(\mathcal{S}, \mathcal{L}) = \min_{P_S \in \mathcal{S}} \sup_{P_L \in \mathcal{L}} E[\bar{R}_T(y^T, P_S)].$$

It indicates the performance of the worst learning strategy on the best learnable problem.

But alternatively there is the *maximal* regret:

$$R_T^* = \max_{y^T} \log \frac{P_L(y^T|\mathbf{x}^T)}{P_S(y^T|\mathbf{x}^T)} \qquad (6.2)$$

which we denote as $M(P_S, P_L)$ and the min–max equivalent $R_T^*(\mathcal{S}, \mathcal{L}) = \min_{P_S \in \mathcal{S}} \sup_{P_L \in \mathcal{L}} M(P_S, P_L)$. The question here will be to analyse how far equivalent are the formulations $\bar{R}_T(\mathcal{S}, \mathcal{L})$ and $R_T^*(\mathcal{S}, \mathcal{L})$ for the regret.

One option is to use Shtarkov sums [31,113]. We denote $P^*(y^T|\mathbf{x}^T) = \sup_{P_L \in \mathcal{L}} P_L(y^T|\mathbf{x}^T)$ and $Q(\mathcal{L}) = \sum_{y^T} P^*(y^T|\mathbf{x}^T)$ and $\bar{P}(.)$ the probability distribution defined by $\bar{P}(y^T|\mathbf{x}^T) = \frac{1}{Q(\mathcal{L})} P^*(y^T|\mathbf{x}^T)$. The quantity $Q(\mathcal{L})$ is the Shtarkov sum of the strategy $\mathcal{L}$. We have the inequality

$$R_T^*(\mathcal{S}, \mathcal{L}) \leq \min_{P_S \in \mathcal{S}} M(P_S, \bar{P}) + r_T^*(\mathcal{L}) \tag{6.3}$$

with $r_T^*(\mathcal{L}) = \log Q(\mathcal{L})$. Note that when $\bar{P} \in \mathcal{S}$, the distance is null:

$$\min_{P_S \in \mathcal{S}} M(P_S, \bar{P}) = 0.$$

Note also that the quantity $r_T^*(\mathcal{L})$ does not depends on the set of sources, and we can estimate it in some specific cases of learning strategies, such as the logistic regression [67].

For the logistic regression learning [123], we define $P(y|\mathbf{x})$ by an expression of the form $P(y|\mathbf{x}, \mathbf{w}) = \frac{1}{1+\exp(y\langle\mathbf{w}.\mathbf{x}\rangle)}$, where $\mathbf{w}$ is a scalar with same dimension as $\mathbf{x}$ (in this simplified case, we consider a binary label $y \in \{-1, +1\}$). The objective is to find the weight vectors $\mathbf{w}$ which maximize $P(y^T|\mathbf{x}^T, \mathbf{w}) = \prod_t P(y_t|\mathbf{x}_t, \mathbf{w})$. If $\mathbf{w}^*$ is the optimal value for $\mathbf{w}$, then it satisfies $\sum_t \frac{1+y_t}{2}\mathbf{x}_t = \nabla L_T(\mathbf{w}^*)$ with $L_T(\mathbf{w}) = \sum_t \log\left(1 + \exp(\langle\mathbf{w}\mathbf{x}_t\rangle)\right)$.

An interesting case is when the sequence of feature $\mathbf{x}^T$ takes its values in a finite set $\mathcal{X}$ fixed in advance and we let T tend to infinity. Since $P^*(y^T|\mathbf{x}^T) = \exp(-L_T(\mathbf{w}) + \langle\mathbf{w}^*.\nabla L_T(\mathbf{w}^*)\rangle)$, we get the formula

$$r_T^*(\mathcal{L}) = \frac{d}{2}\log T + \log\left(\int_{\mathbb{R}^d} \sqrt{\frac{\det(\nabla^2 L_T(\mathbf{w}^*))}{T}}d\mathbf{w}^*\right) + O(d^{3/2}/\sqrt{T}),$$

$$\tag{6.4}$$

where d is the dimension of the feature vectors $\mathbf{x}$. We call the function under the integral the *discrepancy* function.

The tensor $\nabla^2 L_T(\mathbf{w}) = \sum_t (e^{-\langle\mathbf{w}.\mathbf{x}_t\rangle/2} + e^{\langle\mathbf{w}.\mathbf{x}_t\rangle/2})^{-2}\mathbf{x}_t \otimes \mathbf{x}_t$ is called the Fisher information matrix; divided by T it is $\sum_{\mathbf{x}\in\mathcal{X}} |\mathbf{x}^T|_{\mathbf{x}}(\mathbf{x})\frac{1}{(e^{-\langle\mathbf{w}.\mathbf{x}_t\rangle/2}+e^{\langle\mathbf{w}.\mathbf{x}_t\rangle/2})^2}$, where $|\mathbf{x}^T|_{\mathbf{x}}$ is the number of elements in $\mathbf{x}^T$ equal to $\mathbf{x}$ and gives a finite value C_d for the integral.

It is important that $\mathcal{X}$ must be finite, otherwise C_d would be of order $d\log T$.

6.2   The Point of View of Turing

Turing (see the portrait in Figure 6.10) has been pioneering computer science before World War 2. During the war, his innovative steps enabled to work out the first mechanical computer specifically designed to break the encoding system of the German military. But he is also known as the one to be the first to promote the feature of artificial intelligence opposed to human intelligence. This is collected in the popular culture as the famous "Turing test": The ultimate goal of AI is to let believe an arbitrary interlocutor that it is a real human. Due to the recent breakthroughs in machine learning, this goal is considered to be almost fulfilled.

In a classical learning process, the AI machine is set up with a $\mathbf{w}$ weight vector. After each test vector (a unit of training data) is submitted to the machine, this weight vector is modified to reduce the loss, i.e. the distance between the actual prediction made by the machine and the ground truth (also known as the label). In the case of searching for cats in an image, the test is a scalar vector $\mathbf{x} = (10, 0, 32, 5, 0, 5, \ldots)$ which is a numerical representation of the image to be submitted (it encodes, for example, the first image displayed in Figure 6.9). The prediction function can be represented by $f(\mathbf{x}, \mathbf{w})$, for example, we can have $f(\mathbf{x}, \mathbf{w}) = \langle \mathbf{x} | \mathbf{w} \rangle$ as the inner

Fig. 6.10.   Alan Turing (1912–1954).

product of the vector $\mathbf{x}$ with the vector $\mathbf{w}$. In this case, the test and weight vectors have the same dimension. But this doesn't rule out the implementation of more complicated functions in the machine, where the dimension of the weight vector is larger than that of the test vector. At this stage of the discussion, we consider the machine as a black box. Since the photo contains a cat, the ground truth y should be equal to 1. If the photo does not contain a cat, in this case a dog as in the fifth photo, we should have $y = 0$. For the four images displayed, the ground truth sequence will be $1, 1, 1, 1, 0$.

The ground truth can be multidimensional. For example, it could indicate also the presence of a dog in the picture, then the sequence would be $(1, 0), (1, 0), (1, 0), (1, 0), (0, 1), (1, 1), (1, 0)$: $(1, 0)$ indicates there is a cat but no dog, $(0, 1)$ indicates there is a dog but no cat, $(1, 1)$ indicates there is a cat and a dog, and $(0, 0)$ would indicate there is none of them.

Back to one-dimensional labels, if for the first image the prediction made by the machine is $f(\mathbf{x}, \mathbf{w}) = 0.65$, the loss would be $\mathrm{loss}(\mathbf{w}, \mathbf{x}, y) = (1 - 0.65)^2 = 0.1225$ (deep learning is mostly using quadratic loss). It is possible to explicitly write the gradient of the loss with respect to the weight vector. The main feature in deep learning is to update the weight vector after each test vector via a classic gradient descent:

$$\mathbf{w} \leftarrow \mathbf{w} - r\nabla_{\mathbf{w}}\mathrm{loss}(\mathbf{w}, \mathbf{x}, y).$$

The symbol $\nabla_{\mathbf{w}}$ denotes the gradient operator with respect to the vector $\mathbf{w}$. The quantity r is a small, non-negative scalar, called the "learning rate". The "$-$" sign emphasizes that the variation of $\mathbf{w}$ is toward smaller values of the loss, since the gradient is naturally oriented toward larger values. The learning rate is low in order to obtain small variations in the weight vector as the gradient descent approaches its stopping point.

6.2.1 *Inside the machinery*

Now it's time to open the black box and examine the neural network. At first glance, the neural network doesn't have much in common with its biological counterpart (see Figure 6.11), but in fact, in some ways it does.

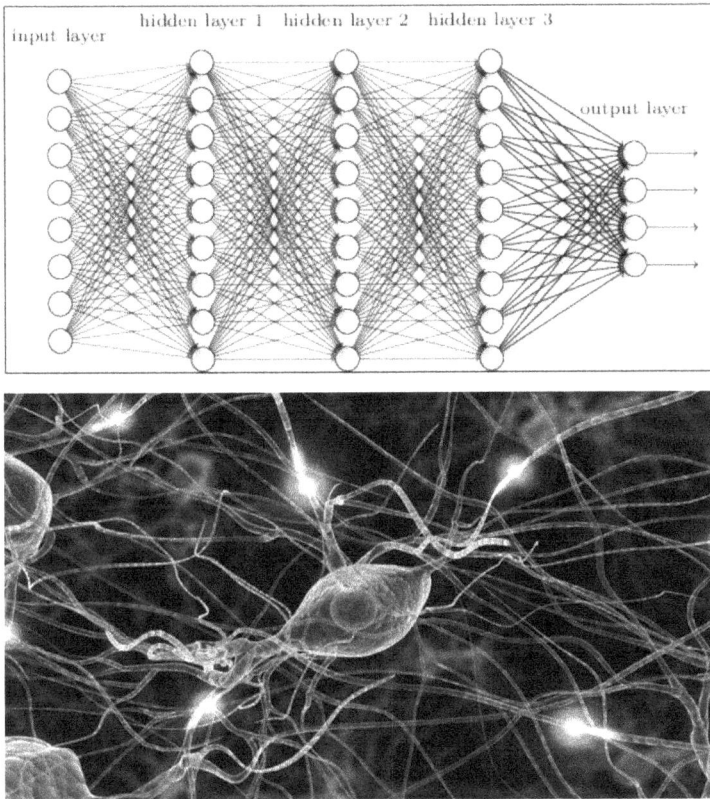

Fig. 6.11. Deep neural network (top). *Courtesy*: Wikipedia; biological neural network (bottom). *Courtesy*: http://www.extremetech.com/wp-content/uploads/2015/07/neural-net-head.jpg.

Unlike biological neurons, which are organized in a kind of disorder, silicon neurons are arranged in layers, making gradient descent calculations much easier. Each circle on the left of Figure 6.11 is a silicon neuron. A silicon neuron has one input and one output. In each layer, the neuron inputs are obtained by a linear combination of the neuron outputs of the previous layer. The zero layer contains the coefficients of the test vector $\mathbf{x}$ for which we want to calculate the machine's prediction. The last layer is the prediction. The linear combination linking the outputs of layer $i-1$ to the inputs of layer i is represented by a matrix $\mathbf{W}^{(i)}$.

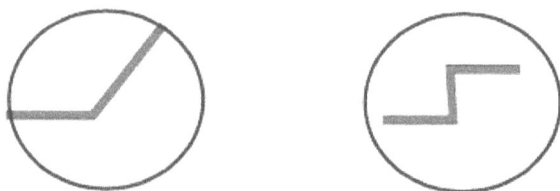

Fig. 6.12.   Activation functions in neuron networks: ReLU (left) and Heaviside (right).

The output of a neuron is a nonlinear function $a(x)$ of the input x, called the *activation* function. If the $a(x)$ function had been simply linear, there would have been no need to divide the network into layers — a single layer would have sufficed. The most common activation functions are $a(x) = \max\{0, x\}$ (equal to x when $x > 0$, zero otherwise), often called ReLU (for Rectified Linear Unit), and the Heaviside function $a(x) = \max\{0, \frac{x}{|x|}\}$ which is 1 when $x > 0$ and 0 otherwise. Of course, these functions can be shifted by a scalar which can be optimized like the other weights (and which would therefore be considered as an additional weight in the prediction function). Figure 6.12 shows the symbol used to represent these functions; we note that it reproduces in a simplified way the respective plot of the functions. This symbolic plot is added in the circle when we want to clearly specify the activation function in the layers in case of ambiguity.

Thus one can represent the computation of the prediction by the following symbolic expression:

$$f(\mathbf{x}, \mathbf{w}) = \mathbf{x}^\tau \mathbf{W}^{(1)} \mathbf{a}_1 \mathbf{W}^{(2)} \cdots \mathbf{a}_{k-1} \mathbf{W}^{(k)},$$

where $\mathbf{w}$ is the vector of all the coefficients of the $\mathbf{W}^{(i)}$s, for $i = 1, \ldots, k$, k being the number of layers. The $\mathbf{a}_i$ terms are the activation function vectors at layer i, and $\mathbf{x}^\tau$ is the transpose of the vector $\mathbf{x}$.

Biological neurons are not organized in layers and are not linked by matrices. The connections between neurons are thin bridges, called synapses. Synapses carry ionized chemical signals that travel from neurons to other neurons, provided they are physically connected. The incoming signal triggers the receiver to fire, i.e. to transmit a signal, or not, depending on whether the incoming potential is above or below a certain intensity. The key factor in silicon neurons is the presence or absence of a physical connection, as many neurons work in parallel. The presence of several incoming connections would equate to a higher weight, but the analogy ends there. "Good" or "bad" connection patterns can be amplified or eliminated according to an unknown reward process during learning, this process has not yet been clearly identified and which certainly has little to do with the gradient descent described above for silicon neurons.

6.2.2 *A school example: Quantum tomography*

As we have seen, one of the key issues in machine learning is characterizing the learning capacity of a problem. Regret is a way of quantifying learning capacity. Quantum tomography is a special case of machine learning where the training set is a set of quantum measurements and the ground truth is the result of these measurements. What makes the case special is the fact that nothing is known about the hidden quantum system since the quantum measurement has no hidden variables (see the chapter on quantum). This differs from classical learning such as cat identification, where the image to be tested is entirely described by a set of pixels. We show that in some cases, quantum tomography is a difficult problem to learn.

We consider a problem related to optical fiber communication where information are encoded in each photon polarization. We show that the learning regret cannot decay faster than $1/T$, where T is the size of the training dataset, and that incremental gradient descents may converge even worse.

Fundamental to this study is the quantum state of the photons used for optical transmissions. Quantum theory is presented in more details in the dedicated chapter of this book.

The learning process after measurements on T training photons will result in an index $L(y^T)$, where y^T is the sequence of measurement results. The index $L(y^T)$ indicates an element of the $\mathcal{L}$ training distribution set. Each index $L \in \mathcal{L}$ defines a complete distribution $P_L(y'^T|\mathbf{x}^T)$ defined over all possible measurement sequences y'^T conditioned by the photon sequence $\mathbf{x}^T$. The difficulty lies in the fact that each learning sequence gives rise to a single measurement sequence y^T, whereas ultimately it should be applicable to any alternative sequence y'^T. So, in general, $L(y^T) \neq L(y'^T)$ when $y^T \neq y'^T$. In the absence of auxiliary information, the learning

process leads to the determination of the maximum likelihood prob-
ability $L(y^T) = \arg\max_{L \in \mathcal{L}}\{P_L(y^T|\mathbf{x}^T)\}$. Our objective is to deter-
mine how close $P_{L(y^T)}(y^T|\mathbf{x}^T)$ is to $P_S(y^T|\mathbf{x}^T)$ as y^T varies.

The distance between the two distributions can be expressed by
the Kullback–Leibler divergence [33]

$$D^*(P_S\|P_L) = \sum_{y^T} P_S(y^T|\mathbf{x}^T) \log \frac{P_S(y^T|\mathbf{x}^T)}{P_{L(y^T)}(y^T|\mathbf{x}^T)}. \qquad (6.5)$$

However, it should be stressed again that the quantity $P_{L(y^T)}(y^T|\mathbf{x}^T)$
does not necessarily define a probability distribution over the
sequences of measurements y^T, even assuming that $\mathbf{x}^T$ is fixed once
and for all. The distribution $L(y^T)$ can vary as y^T varies, making
it unlikely that $\sum_{y^T} P_{L(y^T)}(y^T|\mathbf{x}^T)$ is equal to 1. So $D^*(P_S\|P_L)$ is
not a distance, or even a semi-distance, as it can be non-positive.
One way out is to introduce $P_L^*(y^T|\mathbf{x}^T) = \frac{P_{L(y^T)}(y^T|\mathbf{x}^T)}{S_L(\mathbf{x}^T)}$, where
$S(\mathbf{x}^T) = \sum_{y^T} P_{L(y^T)}(y^T|\mathbf{x}^T)$ makes $P_L^*()$ a probability distribution.
We'll therefore use $D(P_S\|P_L^*)$ which satisfies

$$D(P_S\|P_L^*) = \sum_{y^T} P_S(y^T|\mathbf{x}^T) \log \frac{P_S(y^T|\mathbf{x}^T)}{P_L^*(y^T|\mathbf{x}^T)} = D^*(P_S\|P_L) + \log S(\mathbf{x}^T)$$

$$(6.6)$$

and is now a well-defined semi-distance which we define as the learn-
ing regret $R(x^T) = D(P_S\|P_L^*)$ [67], which we call the Kullback–
Leibler (KB) regret. It should be noted that $D^*(P_S\|P_\mathcal{L}) \le$
$\min_{L \in \mathcal{L}} D(P_S\|P_L)$ and if $P_S \in \mathcal{L}$ then $\min_{L \in \mathcal{L}} D(P_S\|P_L) = 0$ and
$D(P_S\|P_L^*) \le \log S(\mathbf{x}^T)$.

Many applications of artificial intelligence involve physical measurements. The authors of [89] describe a deep learning process on
the physical layer of a wireless network, with the learning applied
directly to the sequence of signals measured on the antennas. The
problem with quantum physical effects is that they are individually
highly unreproducible and, of course, non-deterministic. In our quantum tomography problem, photon polarization is given by a quantum
wave function of dimension 2. In the binary case, bit 0 is given by
the polarization angles θ_Q and bit 1 is given by the angle $\theta_Q + \pi/2$.
The quantity θ_Q is assumed to be unknown to the receiver and its

estimate θ_T is obtained after a learning sequence of length T via machine learning.

For this purpose, the sender sends a sequence of T equally polarized photons, along angle θ_Q, and the receiver measures these photons over a collection of T measurement angles $x_1, x_2, \ldots, x_T$, called the featured angles. They are pure scalar and are not vector ($d = 1$), therefore we do not depict them in bold font as in the previous section which is therefore of dimension 1. The labels, or ground truths, $y_1, \ldots, y_T$ are the sequence of binary measurement obtained, $y_t \in \{0, 1\}$, there are 2^T possible label sequences.

This problem is the most simplified version of tomography on quantum telecommunications, since it relies on a single parameter θ_Q. More realistic and complicated situations will arise when circular polarization, combined with nonlinear noise, gives rise to more complex combinations of polarizations within photon groups. This will considerably increase the dimension of the test vectors and will certainly make our results on the learnability even more critical. However, in the situation analyzed in this section, we show that the system in its simplest parameters is already difficult to learn, i.e. the convergence rate is $O(\frac{1}{T})$, or, equivalently, the regret is $O(\log T)$.

Theorem 6.1. *Under mild conditions, we have the estimate*

$$R(x^T) = D(P_S \| P_L^*) = O(\log T). \tag{6.7}$$

Proof. We assume that the results of the experiment are passed in batches to the learning machine, i.e. the estimate $\theta_t = \theta$ is left untouched for $0 < t < T$. The probability distribution $P_L()$ is a function of θ with the expression $P_L(y^T | x^T, \theta) = \prod_{y_t=0} \cos(\theta - x_t)^2 \prod_{y_t=1} \sin(\theta - x_t)^2$, where x_t is the angle at which photon number t is measured. This x_t sequence is assumed to vary in a finite set. In the experiment, the source distribution is exactly $P_S(y^T | x^T) = P_L(y^T | x^T, \theta_Q)$, so the source distribution belongs to the learning distributions class $\mathcal{L}$, and our aim is to find the value θ^* that maximizes the likelihood probability, in the hope that it doesn't differ too much from the unknown polarization angle θ_Q. We show [69] although with an initial wrong order corrected in [70] that

$$\sum_{y^T} P(y^T | x^T) \log \frac{P_S(y^T | x^T)}{P_{L(y^T)}(y^T | x^T)} = O(1) \tag{6.8}$$

and similarly that

$$\log S(x^T) = \log \left(\sum_{y^T} P_{L(y^T)}(y^T|x^T) \right) = \frac{1}{2}\log T + O(1), \qquad (6.9)$$

where we recognize the Shtarkov sum already analyzed with (6.4). □

Referring to the deep learning method, we study how the gradient descent method reaches the θ_Q value. In this situation, the estimate θ_t varies with time. Due to the absence of hidden variables in quantum measurement, we indirectly estimate the loss by $\text{loss}(y_t, \theta_t|x_t) = (y_t - \sin(\theta_t - x_t)^2)^2$. This is because the average value of y_t is $\sin(\theta_Q)^2$, so the average loss is $(\sin(\theta_Q - x_t)^2 - \sin(\theta_t - x_t)^2)^2 + \frac{\sin(2\theta_Q - 2x_t)^2}{4}$ (minimized to $\theta_t = \theta_Q$). Therefore, the update of the θ_t estimate at each gradient step is

$$\theta_{t+1} = \theta_t - r\frac{\partial}{\partial\theta_t}\text{loss}(y_t, \theta_t|x_t). \qquad (6.10)$$

In Figure 6.13, we display our simulations as a sequence θ_t starting with a random initial θ_1. As assumed in the training process, for all t the transmitted bit is always 0, i.e. the polarization angle is always θ_Q. The learning rate is set at $r = 0.0002$. We simulate nine parallel gradient runs randomly initialized by sharing the same angular test sequence x^T, with $T = 3{,}000{,}000$. We plot the parallel evolutions of the θ_t estimates. Initial points are green diamonds and final points are red diamonds. Although we start with nine different positions, the trajectories converge toward $\theta_Q \pm \pi$. However, convergence is slow, confirming that convergence can't be better than $1/T$. In fact, some initial positions converge even more slowly and, even after 3,000,000 of trials, are still a long way from the target value θ_Q. This is because the target function $\log P(y^T|x^T, \theta)$ has expression $\sum_t \cos(\theta_Q - x_t)^2 \log \cos(\theta - x_t)^2 + \sin(\theta_Q - x_t)^2 \log \sin(\theta - x_t)^2$ which implies several local maxima, as shown in Figure 6.14, where x_t belongs to the set of values $2\pi k/10$ for $k = 1, \dots, 10$. It is highly unlikely that a communication operator would accept such a large number of executions (3,000,000) to achieve correct convergence. However, it would be possible to run the gradient runs in parallel and act as with particle systems to select the fastest convergence.

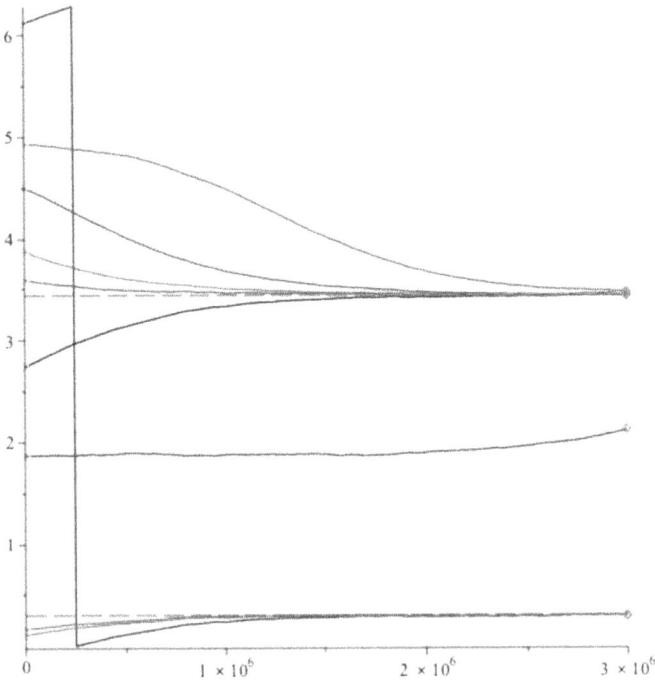

Fig. 6.13.   Angle estimate θ_t versus time of nine slow gradient descents randomly initialized. Green diamonds are starting points, red diamonds are stopping points.

6.2.3   *How universal are neural networks?*

Although the architecture of the deep neural network may seem strange at first glance, it is nonetheless a universal machine. In other words, a neural network is a Turing machine, so any algorithm can be implemented via a neural network. A Turing machine is the first abstraction describing a modern computer, and Turing imagined it in 1936 before any computer had been built. To be precise, the exact equivalence between a Turing machine and a neural network is only complete if neural networks are endowed with a memory, which is the aim of the specific architecture of neural networks called *recurrent neuron networks*, which allows intermediate results to be fed back between layers.

But the fact that any algorithm can be implemented in a neural network does not mean that any algorithm can be easily *learned* by a neural network. Thus, in what follows, we specifically study the

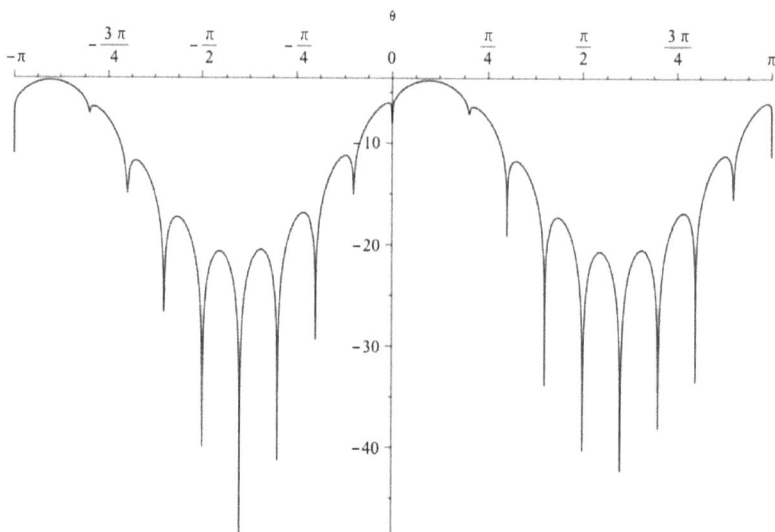

Fig. 6.14. Target function $\sum_t \cos(\theta_Q - x_t)^2 \log \cos(\theta - x_t)^2 + \sin(\theta_Q - x_t)^2 \log \sin(\theta - x_t)^2$ as function of θ.

case where the problem to be learned is a sequence of test vectors $\mathbf{x}^T$ associated with a sequence of labels y^T where for all t: $y_t = f(\mathbf{x}_t)$, where $f(.)$ is a pure algorithmic function.

Of course, it's unrealistic to imagine that learning will converge to the optimal weights. A gradient descent always converges to a minimum value of the loss function but not necessarily to the global minimum (which is zero when the ground truth is a given algorithmic function), but it may stop at the nearest local minimum. Due to the large size of the matrices defining the neural network's weights, learning will always stop on a local minimum, since there will be an overwhelming number of them. In fact, this is not pure gradient descent, as the gradient is also a function of the training vector $\mathbf{x}_t$ submitted to the machine at time t. The $\mathbf{x}_t$ vector is chosen randomly, which introduces a stochastic bias. The correct name is therefore *stochastic gradient descent*. It is expected that the descent will eventually wander around a local minima at a distance of the order of the learning rate r. The key question is whether the local minima are close to the global minimum or too far away. In the former case, the prediction function may have satisfactory accuracy compared to

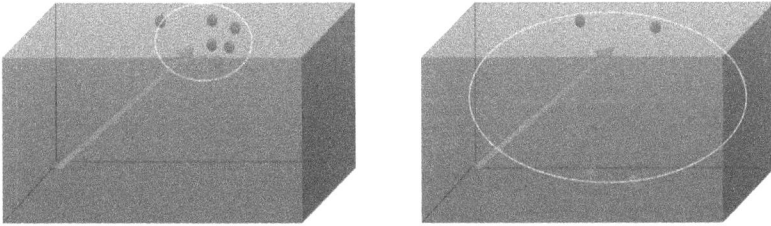

Fig. 6.15.   Gradient descent convergence; good case: the local minima are concentrated (left) and bad case: they are dispersed (right).

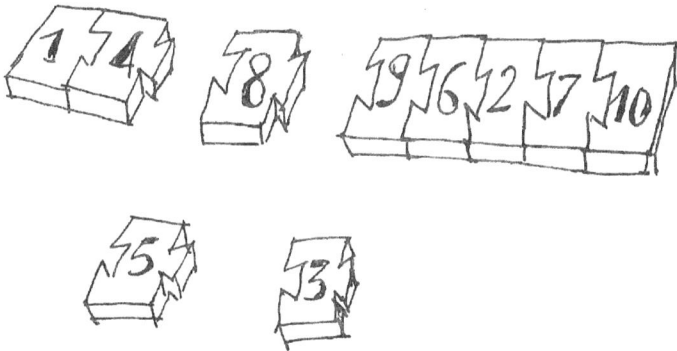

Fig. 6.16.   Toy algorithms.

the optimal function; in the latter case, the prediction function will never have satisfactory accuracy. Figure 6.15 attempts to illustrate this situation, although the display is limited to three dimensions, whereas the weight vectors are of a much higher dimension. The red arrow indicates the position of the global minimum.

We first make few experiments with basic algorithms.

Max finding learning: Any kid can identify a cat in a picture, even when represented in a very symbolic way. Any kid can also easily find the maximum among several numbers as illustrated in Figure 6.16. Therefore, the first tested algorithm is the *max* finder algorithm: finding the largest number among a list of N numbers $\mathbf{x} = (x_1, \ldots, x_N)$. When $N = 2$, the formula $\max\{x_1, x_2\} = \frac{1}{2}[x_1 + x_2]^+ - [-x_1 - x_2]^+ + \frac{1}{2}[x_1 - x_2]^+ + \frac{1}{2}[-x_1 + x_2]^+$ is equivalent to the

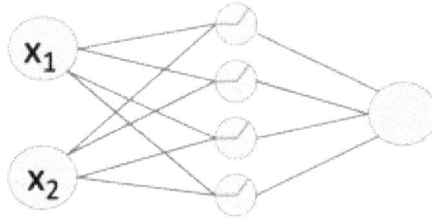

Fig. 6.17.   Max finder network for two numbers: blue for positive weights, red for negative weights.

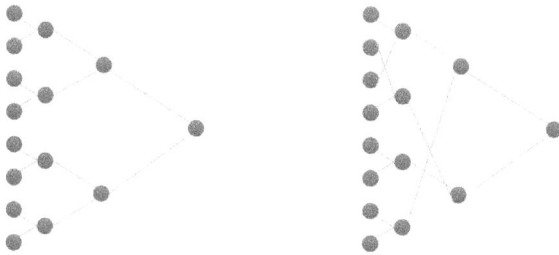

Fig. 6.18.   Two Max finder networks for eight numbers; the elementary element is the network presented in Figure 6.17.

neural network:

$$\max\{x_1, x_2\} = [x_1, x_2] \begin{bmatrix} +1 & -1 & +1 & -1 \\ +1 & -1 & -1 & +1 \end{bmatrix} \mathbf{a}_4 \begin{bmatrix} +1/2 \\ -1/2 \\ +1/2 \\ +1/2 \end{bmatrix},$$

where $\mathbf{a}_4$ is the layer composed of four ReLU activation functions. Another representation of this network is given in Figure 6.17. To extract the maximum from 2^k numbers, simply concatenate k layers of the elementary network of Figure 6.17, as shown in Figure 6.18. Let $\mathbf{w}^*$ be the weight vectors of this optimal network, which is organized into k layers with the ReLU activation function. The left layer is a $2^k \times 2^{k+1}$ matrix and the right layer is a 4×1 matrix, between the k layers a sequence of $2^{i+1} \times 2^i$ matrices of decreasing dimension, for i varying from k to 1. We denote by $f(\mathbf{x}, \mathbf{w})$ the function that relates a $\mathbf{w}$ weight vector and a $\mathbf{x}$ test vector to the predicted label. Figure 6.18 shows another alternative for the max search network with a different branching. There are $2^{k-1}!2^{k-2}! \cdots 2$ possible $\mathbf{w}^*$. In fact, there are many more possibilities, since matrices can also be scaled with arbitrary non-negative numbers. But all these weights $\mathbf{w}^*$ give $f(\mathbf{x}, \mathbf{w}^*) = \max_t\{x_t\}$.

The question is as follows: Will training on signed numbers uniformly distributed on $[-1, 1]$ converge to this max search network (or to an equivalent network)? For this very preliminary study, we have considered nine randomly initialized networks: $\mathbf{w}_0^1, \mathbf{w}_0^2, \ldots, \mathbf{w}_0^9$ trained by classical backpropagation on the same unbounded sequence of training vectors $\mathbf{x}_t$. We denote $\mathbf{w}_t^1, \mathbf{w}_t^2, \ldots, \mathbf{w}_t^9$ as the respective weight vectors at step t. It is difficult to display the evolution of the weights because the dimension of the system is too large, i.e. $2^k \times 2^{k-1} + 2^{k-1} \times 2^{k-2} + \cdots$. What's more, two apparently different weight vectors may in fact give exactly the same prediction. To demonstrate the convergence of the nine learning processes more comfortably, we set two test sequences $(\mathbf{v}_1, \mathbf{v}_2)$, chosen at random but fixed for the duration of the simulation. Calculating the prediction on the two test vectors with the respective weight vectors at time t obtained on each of the nine neural networks provides nine parallel two-dimensional trajectories with points $[f(\mathbf{v}_1, \mathbf{w}_t^i), f(\mathbf{v}_2, \mathbf{w}_t^i)]$ for $i = 1, 2, \ldots, 9$, which we compare with the reference point $[f(\mathbf{v}_1, \mathbf{w}^*), f(\mathbf{v}_2, \mathbf{w}^*)]$, see Figure 6.19. If the formation is indeed

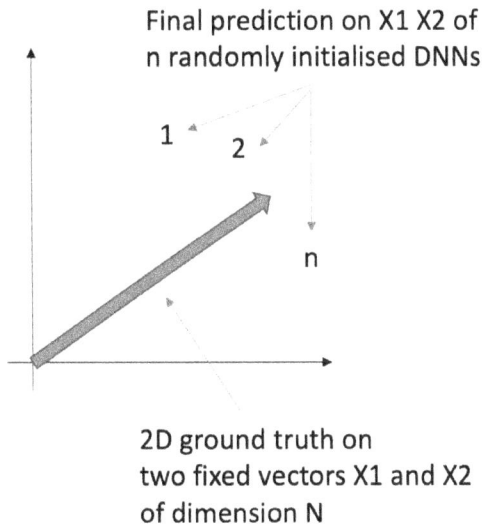

Fig. 6.19.   The principles of the representation of n parallel gradient descents on max finding learning on training vectors of N numbers. Only the final stopping points are displayed.

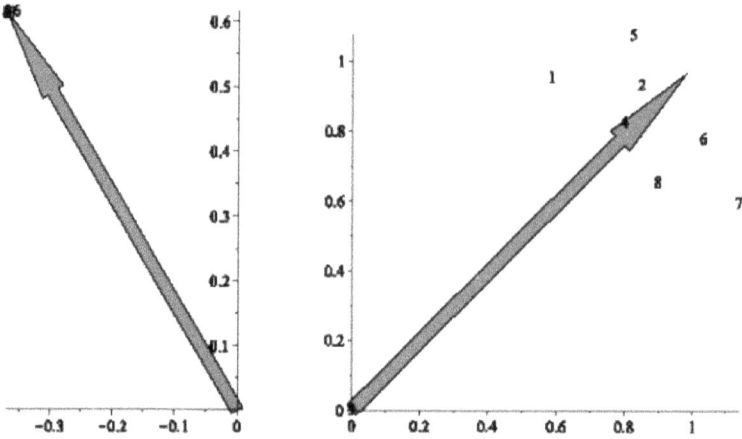

Fig. 6.20. Left: Max finder network trained on sequences of two numbers; right: Max finder network trained on sequences of 32 numbers.

convergent, the plots should converge toward the reference point. Figure 6.20 shows the final position reached by each of the nine networks (listed from 1 to 9) after training. The red arrow shows the ground truth given by the network with weight $\mathbf{w}^*$. We note that for the search for the maximum of two numbers, the final positions of the nine networks are very close to the position of the ground truth, whereas for the search for the maximum of 32 numbers, the final positions are very scattered. In both cases, training is performed on the optimal architecture given by Figure 6.18, where only the weights are indeterminate.

Topological swamp area: If we summarize the results obtained in simulating the learning of the maximum scalar search problem, the conclusion is that there is a learning failure when increasing the dimension of the scalar tuples to be tested. Thus, this problem based on a rather simple algorithm could belong to the class of algorithms that are difficult to learn. In order to find an explanation for this rather surprising phenomenon, we're going to invest in a *toy* model. Since ground truth is a deterministic function of the test vector, we lose any random aspect of ground truth and there are necessarily network weight vectors that are the roots of the loss function. The roots define the networks that actually implement the algorithm. An infinite number of such networks exist, simply by multiplying matrix rows or columns by arbitrary scalars. But if we consider as a single root the roots obtained by continuous distortions, we reduce the set of roots to a finite set [19], each root corresponding to a single row and permutations of rows in the matrices, as illustrated in Figure 6.18. Nevertheless, the number of roots remains enormous; let's note it m. For a network to be trained on a problem involving N numbers, we can expect m to be of the order of $N!$. In the model developed in [19], we abandon the stochastic aspect and use a simplistic loss function. We prove that the gradient descent converges almost surely to a single local minima $\overline{\mathbf{w}^*}$ close to the centroid of the set of roots $\{\mathbf{w}_1^*, \ldots, \mathbf{w}_m^*\}$, even if the centroid is in fact very far from the real roots. The model is far too simple, since it assumes a trivial and very smooth loss function. Experiments on max finder clearly contradict the uniqueness of attractive local minima, since they show a multiplicity of them, albeit topologically distant from global minima. The reason for this division of local minima is probably due to the non-smooth property of the current loss function.

However, for further simplification, we assume that the root set $\{\mathbf{w}_1^*, \ldots, \mathbf{w}_m^*\}$ is obtained by a random point process in the hypercube $[a, b]^n$, where n is the dimension of the weight vector (which must be greater than N) and a and b are free parameters, so that the root centroid is very close to the mean root position $E[\mathbf{w}^*]$, which is $(\frac{a+b}{2}, \ldots, \frac{a+b}{2}))$. The dimension n is in fact of order N^2 corresponding to the order of magnitude of the matrix size, given that the network has $k = \lceil \log_2 N \rceil$ layers.

We now try to import into the real neural network the result of our *toy* model, which indicates that the stopping point of the gradient descent is close to the position of the mean root $E[\mathbf{w}^*]$. We'll confine ourselves to examining the effect on a single layer of neurons. Suppose the coefficients of the b_{ij} layer matrix are chosen at random. Applied to a random test vector $\mathbf{x}$, the function $f(\mathbf{x}, \mathbf{w}^*) = \sum_{ij} [b_{ij} x_i]^+ = \sum_{ij} [E[b_{ij}] x_i]^+ + O(\sqrt{N})$ by virtue of the law of large numbers applied to each row. With k layers, we would have $f(\mathbf{x}, \mathbf{w}) = f(\mathbf{x}, E[\mathbf{w}]) + O(k\sqrt{N})$. In other words, the prediction with any instance of $\mathbf{w}$ is close to the prediction with $E[\mathbf{w}]$ by a relative error $\frac{k\sqrt{N}}{f(\mathbf{x}, E[\mathbf{w}])}$ as long as the position $E[\mathbf{w}]$ is significantly far from the origin (and $f(\mathbf{x}, E[\mathbf{w}])$ value also far from 0 with a high probability, see Figure 6.21 on the left). Conversely, the relative error

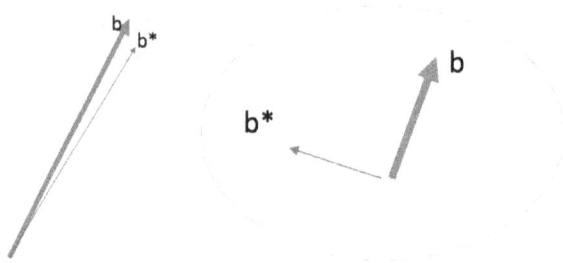

Fig. 6.21.   Left: The relative prediction loss when the average weight vector is non-zero; right: when average weight vector is zero.

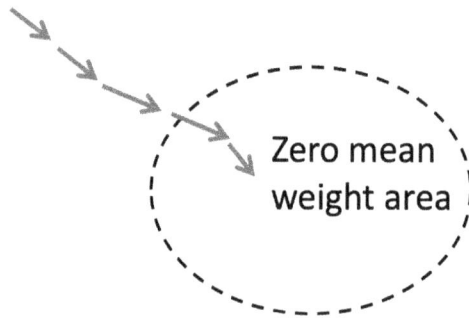

Fig. 6.22.   Swamp area for gradient descent.

becomes too large when $f(\mathbf{x}, E[\mathbf{w}])$ is too small, which is the case when $E[\mathbf{w}] = o(\sqrt{N})$, if learning converged to local minima close to the centroid (without taking into account the possible roughness of the loss function and symmetries between multiple roots), learning would fail under these conditions (see Figure 6.21, right).

We call *zero mean weight algorithms* the following class of algorithms: For each of these algorithms, there exist a set of networks implementing the algorithm, which shows a quasi null centroid. This property can be limited to a subset of layer. We conjecture that this property (possibly with additional intricacies) is fundamental in order to characterize the *swamp area* of learnability (see Figure 6.22). We note that the max finder algorithm shows the zero mean weight property. The Fourier transform algorithm has also the property, but this is a special case which we discuss later. Figure 6.23 shows the convergence of the learning when the original network is generated with random numbers with a non-zero mean (left) and when the

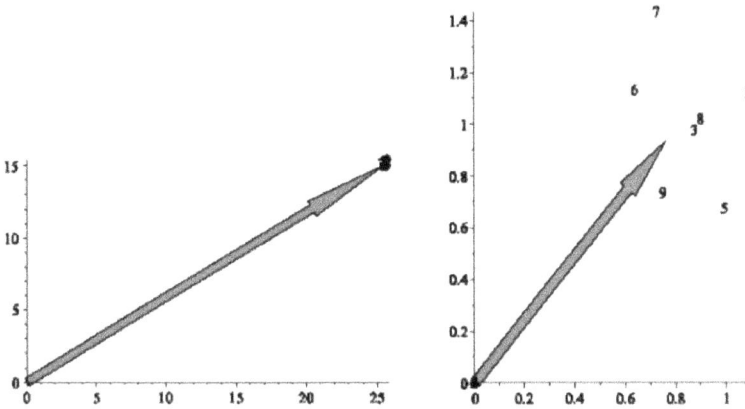

Fig. 6.23. Left: Non-zero mean weight algorithm trained on dimension 16 with four layers; right: zero mean weight algorithm trained on dimension 16 with four layers.

mean is zero (right). We see that with non-zero mean, the learning converges and does not converge with zero mean (the final positions are very dispersed).

Learning the Fourier transform: Of course, the second step is to try the zero mean conjecture on another algorithm than max finding. The Fourier transform is an important algorithm in telecommunication and information theory (see Figure 6.24).

The fast Fourier transform is the most used algorithm in the Web since it is fundamental in image compression and decompression. The Fourier transform is clearly a zero mean weight algorithm but, being also a pure linear transform, it can in theory be easily learned with just a single layer. Thus the zero mean weight property would not define a swamp area when considering single-layered networks. But a learning problem involving the Fourier transform may need in fact more than one layer, for example, when it consists of learning on how to extract transmitted symbols in a sequence of wireless signals. Figure 6.25 compares the performance of training on Fourier transform on one layer and five layers. There are four randomly initialized networks; we have displayed their trajectories before they stabilize. Figure 6.26 explains the plots. The trained networks are defined by their weight vectors when trained on the Fourier transform, but this would be far too large vectors. Instead, the displayed points are the

Fig. 6.24.   Joseph Fourier (1768–1830).

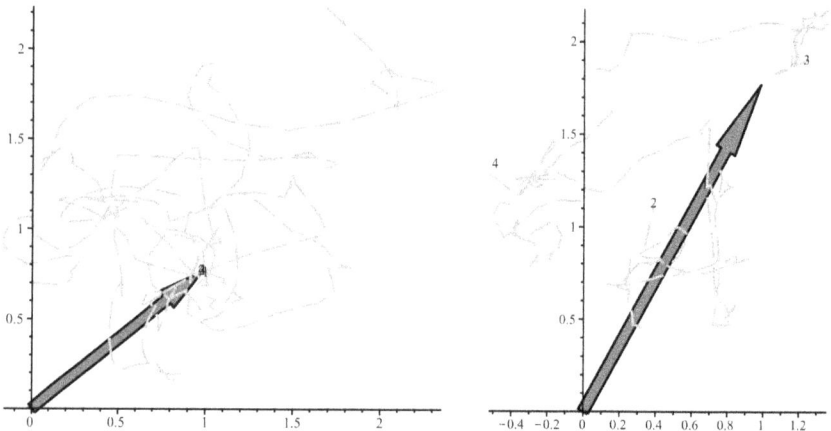

Fig. 6.25.   Left: Fourier transform on 16 numbers learned with one layer; right: Fourier transform on 16 numbers learned with five layers.

application of this trained transform on two fixed random number sequences. With one layer, the networks converge very well to the correct network. With five layers, the networks do not converge any longer and remain far from the ground truth. This may explain why in deep neural training on wireless signals, the Fourier pre-processing of the data considerably boosts the convergence [89], although the

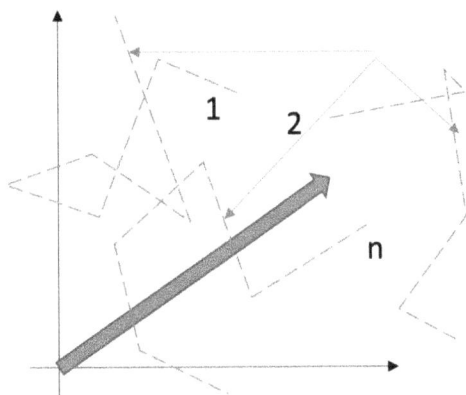

Fig. 6.26. How to display n trajectories of randomly initialized networks, tested over two fixed random number sequences. The red arrow indicates the ground truth.

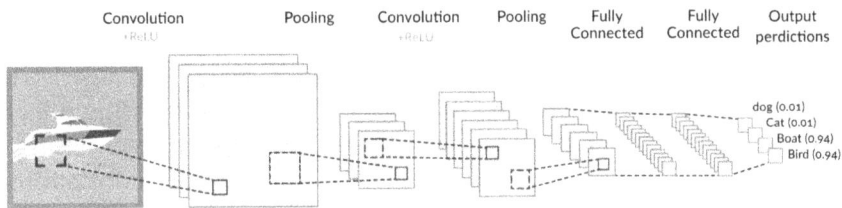

Fig. 6.27. Interleaved convolution operator in the neuron network.

Fourier transform does not reduce the dimension of the training vector when switching from the time series to the phase sequence.

The Fourier transform also plays an unexpected role in learning the cat detection problem. In order to force the network to detect the cat in all possible image positions (top, middle, corner), convolution operations are interposed between layers, since convolution is the classic way of dealing with translations (which result in simple multiplication in the frequency plane). Figure 6.27 shows an example of convolution interspersed with layers of neurons in the dog, cat, boat and bird detection training strategy. Since convolution is based on the Fourier transform and is not trained by learning, the fact that it is interleaved with the layers to be trained indicates that the convolution process is difficult to learn when there are several layers.

6.2.4   *Perspectives for a theory of learning*

As we know, all children are capable of spotting a cat in a photo after a few iterations of learning, and AI can do the same, albeit with some more iterations. Nevertheless, it takes much longer for a child to learn a language or to perform an algorithmic task, such as computing, or to learn to play music. Some are more gifted than others (I'm unfortunately one of the less gifted, as far as music is concerned). On the other hand, we've noticed that deep learning encounters problems when learning basic algorithms, since most algorithms, particularly in arithmetic, have the property of leading to neural networks whose average weight is zero, and therefore their convergence depends strongly on their initialization.

Of course, one could consider as pointless to make an AI learning an algorithm, since it is much more easy to implement the algorithm when we know it than to try to train an AI. AI is here to solve problems for which we don't have algorithmic solutions. Anyhow, the resolution of a problem may go through a pre-processing phase of the training data via a specific and *a priori* unknown, algorithm with zero mean weight property. Skipping the pre-processing might just make the problem unsolvable. One illustration is the need of the Fourier transform in order to train the AI to extract information from a signal sequence.

However, it is probably very bold to compare the training of neurons in the human or just animal brain with the training of neurons in silicon. There is no known equivalent of the zero mean weight property in the synaptic connection. However, the analogy is uncanny.

We end this chapter as we began it, reflecting on a feature from the earliest stages of life's evolution. Olfaction is the oldest sense acquired by living creatures since the appearance of life 3.5 billion years ago. It developed 550 million years ago, shortly before the appearance of eyes and the central nervous system.

The paradox of the *Drosophila melanogaster*: *Drosophila melanogaster* is a common fly (see Figure 6.28) that has the supreme honor of being the most studied animal by neurobiologists. The fly's reproductive rate has made the genetic study of its nervous system a vast playground [46]. The olfactory system of *Drosophila melanogaster* comprises 60 olfactory receptors. Mammals

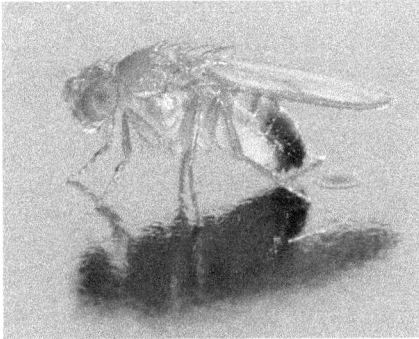

Fig. 6.28.   The *Drosophila melanogaster.*

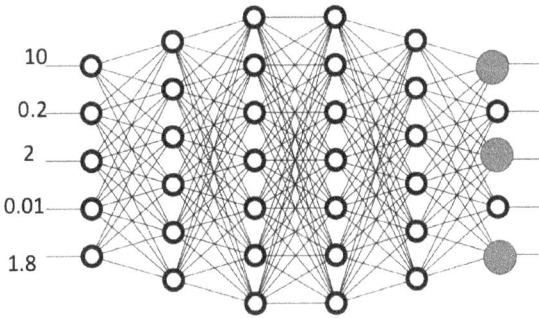

Fig. 6.29.   The neurons' network should able to spot the three maxima of the input signal.

have around 1,000 olfactory receptors (including man, but 30% of receptors are inactive). The olfactory sense is necessary for flies, as it triggers their search for food, breeding sites or escape routes in the event of danger. Olfactory receptors are located in two antennae, enabling them to establish a stereo map of surrounding odors (and detect directions of interest). How odors are encoded in combination remains a mystery, but the fly can recognize up to 1,500 odors that are a combination of stimuli from 60 olfactory receptors. It is thought that odors are encoded by two or three of the most excited olfactory receptors among the 60 signals.

This raises the question of how the *Drosophila melanogaster* neural network detects the position of maxima in the signal sequence, since we have seen that the algorithm for finding maxima is probably one of the hard-to-learn algorithms (see Figure 6.29). Of course, this would implicitly assume that silicon neural networks could have some analogy with their biological counterparts, which is far from being an accepted assumption. In fact, it has been found that the fly uses an open loop to saturate (or de-saturate) olfactory signals. This makes it possible to identify odors even when their signal level is low (or too high). In this case, identifying the maximum simply means detecting when the number of saturated signals is greater than 2 or 3, which is easy thanks to a ReLU activation function. The open loop would play the role of an algorithm unlearnable by conventional methods and would eventually enable the network to converge. This situation is similar to the way dogs identify smells. Dogs' noses don't contain many more types of olfactory receptors than human noses, but their noses have a much larger reception surface. In addition, dogs sniff several times through the mobile nostril, enabling them to make differential comparisons. This mechanical effect could play the role of the missing unlearnable algorithm, but in such an effective way that dogs can detect tiny odors even when they are hidden beneath an overwhelming spicy smell (a tactic tried by drug conveyors).

6.2.5   *Temporary conclusion about autonomous self-optimizing AI*

An AI capable of evolving and optimizing itself to supplant man is far from feasible. Engineers still have a huge amount of work to do in channeling AI to perform well.

Current machine learning techniques are seriously lacking in theoretical background capable of explaining the successes and failures. This is relatively masked by the endless stream of AI success stories, but the failure stories are little known, probably because they would give negative publicity to the huge commercial activity around AI.

Basic algorithms can still help fill in the gaps when AI hits learning hurdles. As far as the development of this chapter is concerned, it seems that the most effective way to break the Turing test is to ask the AI to perform an arithmetic or algorithmic operation.

6.3   Problems and Projects

In this chapter, we divert from the pattern, short exercises and answers, and we try the game of projects where there are much more latitude for imagination.

6.3.1   *The mind reading machine*

Deep neuron networks are very expensive to train. They need zillions of test vectors in order to hope for a convergence. In this set of problems, we address the AI for the poor, namely AI which does not need large quantity of data and relies on simple and provable algorithms, such as pattern matching. We borrow most of our material from [65].

In this problem, we investigate in more detail the implementation of the predictor described in Section 3.4. For optimizing the protocol, we make use of the trie structure [38] which allows optimizing the storage of data in the form of sequence of symbols in a finite given alphabet $\mathcal{A}$. A trie is a tree whose branches contain the access path to leaves which contain the data (one record per leaf). The trie has a branching degree equal to the size of the alphabet $\mathcal{A}$ so that each subtree is indexed by a symbol $a \in \mathcal{A}$. If the tree is denoted $\mathcal{T}$ and $a \in \mathcal{A}$, we denote $\mathcal{T}.a$ the subtree corresponding to the symbol a. When there is no data, the trie is just an empty tree (**nil**) with the following initialization procedure:

procedure TRIE(w)
   $\mathcal{T} \leftarrow$ TRIE()
   RECORD($\mathcal{T}$) $\leftarrow w$
   for each $a \in \mathcal{A}$
     do $\mathcal{T}.a \leftarrow$ **nil**
   return ($\mathcal{T}$)

The records are stored in the leaves of the tree (a **nil** record indicates an internal node). To create a trie containing a single data w, we use the procedure TRIE(w) which creates a leaf with a record equal to w and all subtrees (each of them corresponding to a symbol in $\mathcal{A}$) set at **nil**. To insert a new record w, we use the following procedure:

procedure INSERTTRIE$(w, \mathcal{T})$
  if $\mathcal{T} = \textbf{nil}$
    then return $(\text{TRIE}(w))$
  if RECORD$(\mathcal{T}) = \textbf{nil}$
    then $\begin{cases} x \leftarrow \text{READ}(w) \\ \mathcal{T}.x \leftarrow \text{INSERTTRIE}(w, \mathcal{T}.x) \end{cases}$

    else $\begin{cases} w_2 \leftarrow \text{RECORD}(\mathcal{T}) \\ y \leftarrow \text{READ}(w_2) \\ \text{RECORD}(\mathcal{T}) \leftarrow \textbf{nil} \\ \mathcal{T}.y \leftarrow \text{TRIE}(w_2) \\ x \leftarrow \text{READ}(w) \\ \mathcal{T}.x \leftarrow \text{INSERTTRIE}(w, \mathcal{T}.x) \end{cases}$
  return $(\mathcal{T})$

The instruction READ(w) reads the first symbol of w and erases it afterward. That way the record contains the remaining data. To keep the data in its entirety in the record, it suffices that the read procedure omitted to erase the character after reading.

There is a difficulty when the records are of finite length. In this case, a record may terminate while the tree is still in an internal node. One possibility is to define an extra symbol, a *terminal* symbol which will divert the record to a leaf with record reduced to the terminal symbol.

The *suffix tree* of a word w written in the alphabet $\mathcal{A}$ is the trie built with the suffixes of the word w. Equivalently, the *prefix tree* is the trie built from the prefixes of w. Or it can be seen as the suffix tree of the reversed word w.

An optimization of the suffix tree (also the prefix tree) is to take advantage of the fact that all records are taken in the same word w. Thus the record can be just an integer, called the "rank" which indicates the position in the string where the record should start.

Equipped with this, we can work on the pattern-matching predictor.

Exercise 31

Implement the pattern-matching predictor and simulate it on the time sequences obtained from memoryless sources and on the Markov sources of finite memory. Hint: Use the prefix tree.

Exercise 32

Apply the predictor on natural texts. Extend the prediction on several consecutive characters.

Exercise 33

Assume that the time sequence is made of symbols in $\mathbb{R}$. Using the natural metric and a suitable quantization, find the best prediction in terms of metric.

Exercise 34

Extend the pattern-matching predictor with the possibility of internal mismatches between the copies of the pattern.

6.3.2   *Joint complexity*

The joint complexity between two texts X and Y, $J(X,Y)$ is the number of *distinct* common factors between the two texts. For example, the common factors between "ananas" and "banana" are ϵ (the empty word), "a", "n", "an", "na", "ana", "nan" and "nana". Although it is not the case here, we generally include blank space and punctuation.

Exercise 35

Compute the joint complexity $J(X,Y)$ between "s'il n'en reste qu'un" et "j'y suis, j'y reste".

To enumerate the common factors, a naive algorithm would take a quadratic time. The use of suffix trees would reduce this task to a linear/sublinear time [65], if we combine the two suffix trees. In particular, the internal nodes are common to both suffix trees, but they are not the only common factors to be counted. Indeed, the internal nodes in the suffix tree indicate the factors which appear at least twice in the text. Thus remains the factor which appears only once.

Exercise 36

Implement the computation of $J(X,Y)$ by using the suffix tree. Compute the joint complexity when X and Y are of length n from both memoryless or Markov sources. See the evolution when n increases to infinity.

Exercise 37

Compute the joint complexity when X and Y are natural texts. Compare when the texts are in different language, different topic (but same language). Conclude that the joint complexity is a simple way to make classifications of natural texts without expensive learning and processing. The whole texts can be books.

In the next exercises, we describe a way to make short text classification via joint complexity. Short texts are difficult to classify because they don't carry enough information for an efficient classification via deep learning. However, pattern matching via joint complexity can offer an efficient way. We can note that the classic pattern-matching algorithm might be difficult to learn by deep learning (it would be anyhow useless since the algorithms exist). The methodology we describe in the following is borrowed from [21]. We infer that the higher the joint complexity, the closer the topic of the texts. In [21], the methodology is used to classify short posts in the social medium "Twitter" (now X). It can also be used to classify texts from Facebook, Instagram, etc. It can also be used to classify genoms [68].

Exercise 38

We may want to classify a certain number of posts (tweets). First, we build a primary database of tweets represented by their JSON data. We want, for example, to classify the tweets in function of the following topics: "politic", "sport", "economy", "technology" "entertainment" and "lifestyle". We extract from the bag of tweets the tweets whose JSON texts contain the tagged keywords. Hint: Don't forget to use the decompressed URL addresses which are given in the JSON fields. Of course, this will classify only a minority of tweets, less than 10%. Since the tweets in the topic "politic" won't very seldom mention the word "politic".

Exercise 39

The second step consists of adding to the primary database the remaining tweets. The tweets are taken one by one and the joint complexity with the tweets already present in the primary database. If the tweets in the database with the largest joint complexity are in "politic", then the new tweet is added in the "politic" database. The joint complexity is used here to create a metric between texts. If the metric is not probing, the tweet is put again in the unclassified bag. Note that tweets in various languages and alphabets can be classified that way (e.g. French, Arabic and Chinese) even when they don't contain the English keywords. This comes from the fact that an original tweet in Chinese may contain the keyword in English, e.g. in the URL, and afterward the joint complexity was strong with other tweets in Chinese even without the keywords.

At the end, some tweets may be eligible for more than one bag, since the boundary between economics, technology and politics may be blurred. Some may end as unclassified (not enough joint complexity with the final tagged databases). In [21], we describe a technique using joint complexity to classify tweets from an entire day and using centralities to regroup them. The technique won the third prize based on the some ground truth. The exciting aspect was that the quantity of codes to implement was in the order of few hundreds. The methodology is very interesting in order to build a ground truth database with millions of tweets without requiring human expertize and to submit it to a pure AI which will transform the massive data into training meat.

Chapter 7

Quantum Information Theory

Abstract

In this chapter, we introduce the domain of quantum information. Quantum physics is much more than a perspective for new supports of information taking advantage of nanotechnologies. In fact, the quantum physics enables state superposition which can lead to significant breakthrough in computer programming. For example, the Grover algorithm brings the cost of partial search below complexity $\sqrt{n}$ instead of average complexity $n/2$ with classic computer, where n is the dataset size. Basically, the algorithm performs an inverse function computation when the direct function is given by a black box (i.e. has no particular structural property). The use of Grover algorithm could greatly impair the protection of transactions with cryptocurrencies if the quantum computers were to be generalized. Shore algorithm based on quantum Fourier transform in theory performs the factorization of numbers in $O(n^2)$, where n is the size of the number to be factorized. There is no known polynomial algorithm on classic computers to perform this task. However, the problem is not supposed to be NP hard. Indeed, it is believed that quantum computing cannot make NP-hard problems polynomial. We also look to one of the most spectacular quantum effects which is called the *intrication*. It allows a pair of particularly conditioned particles to be connected regardless of time and space. The intricacy was instrumental to prove via Bell inequality that the quantum measurement is indeed purely not deterministic. Although the connection between intricated particles is not expected to carry information in apparent faster than light transfer, however it leads to science fiction application such as teleportation, already experimented on system of few individual particles.

7.1   Physics and Information

Up to now, we have witnessed two incidental connections between physics and information. Both were related to wireless transmission.

The first connection detected so far was the first Shannon theorem that in presence of a noise $\mathcal{N}$, a signal received over an isolated wireless link with an energy $\mathcal{S}$ (in Joule) over a bandwidth Φ is prone to transmit information at the rate (bit per second) C (see Figure 7.1 (left)):

$$C = \Phi \log_2 \left(1 + \frac{\mathcal{S}}{\mathcal{N}}\right).$$

This is valid under mild condition about the signal spectrum (mostly Gaussian noise). In a multi-user scenario, the cumulated capacity $C(\lambda)$ of several wireless streams issued from a density of user λ toward a single access point is given by the formula (see Figure 7.1 (right)):

$$C(\lambda) = \Phi \frac{\partial}{\partial \log \lambda} E[\log_2(\mathcal{N} + \mathcal{S}(\lambda))],$$

where $\mathcal{S}(\lambda)$ is the cumulated signal from the users and $\mathcal{N}$ the noise received at the access point. In Chapter 5, we have a similar formula, but in this case, for convenience, the noise $\mathcal{N}$ was already included in the term $\mathcal{S}(\lambda)$. In fact, in both cases, the energy is given with an arbitrary unit which plays no role (either killed by a ratio or by the derivative of the logarithm). However, this connection with physics is only an appearance since we have seen in Chapter 2 that

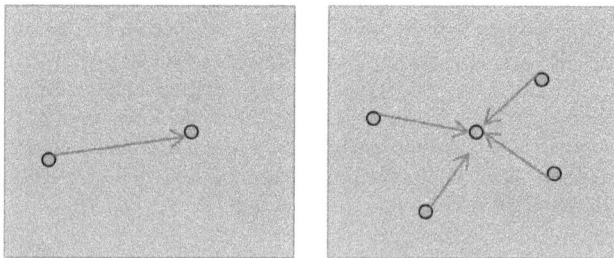

Fig. 7.1.   The point to point single link (left) and the multiple user multiple links (right).

this formula comes from considerations about enumeration and combinatorics. However, the derivation of the second formula when the signal has an attenuation factor of α gives a constant capacity equal to $\Phi\frac{\alpha}{2\log 2}$, involving the radio attenuation factor α, thus physics can be involved in information capacity evaluation, and illustrates the fact that physics introduce limits to information theory.

The striking question is the reciprocal: "Does information theory introduce fundamental limits in physics?" Said in contraposition, can we really manipulate information about physical bodies as if it were an unpounderable, energyless attribute? Interestingly, there is a natural and obvious connection between physics and information which is the concept of entropy. Entropy in thermodynamics is a measure of the disorder or uncertainty over the state of matter, primarily invented by Rudolf Clausius in 1862 [30] for thermodynamics and is strongly linked to the temperature of that body. The second law of thermodynamic states that the entropy of an isolated system of interacting components cannot decrease with time and eventually reaches a maximum entropy state.

The entropy was invented before the composition of gas by free moving individual molecules became an accepted theory. Clausius described his finding in terms of evolution integrals on volume of gas versus its temperature, the gas being considered as a continuous medium. Ten years later, the statistical physics was invented and the definition of entropy was adapted. To simplify, the physical entropy H^P has the following familiar expression:

$$H^P = -k_B \sum_{i \in Z} p(i) \log p(i),$$

where Z is the set of attainable states of the system and $p(i)$ the individual probability of the state i. The physical entropy unit is Joule per Kelvin. The quantity k_B is the Boltzmann constant and is exactly

$$k_B = 1.380649 \times 10^{-23} = \frac{1{,}380{,}649}{100{,}000{,}000{,}000{,}000{,}000} \text{ Joule per Kelvin.}$$

The fact that this constant is a decimal number has just been decreed by convenience. Since the value of the unit depends on the unit of temperature, which is kind of arbitrary, it has been fixed to be a

decimal number to the negligible expense of redefining the Kelvin
unit. Beside the unit, one cannot help noticing that the entropy is
indefinition very, very close to the Shannon entropy which measures
the quantity of information carried by a support:

$$H^S = -\sum_{i \in Z} p(i) \log_2 p(i).$$

This is not quite a coincidence, since John von Neumann himself
advised Claude Shannon to call his measure of information "entropy"
as in physics. An urban legend tells that von Neumann would have
told Shannon that "merging the two concepts under the same name
would not be a real harm", according to von Neumann himself, "since
no one really understands the signification of the original one". If we
take the Stefan–Boltzmann law of energies E at in a gas at temper-
ature T: $p(E) = \exp\left(\frac{E}{k_B T}\right)$, then the entropy is equal to the average
kinetic of the gas particles divided by the temperature.

There are anyhow some conceptual differences, which gave rise
to many controversies in more than a century. The physical entropy
is a way to measure the degree of uncertainty one can have about
a system, e.g. the kinetics of the individual molecules. If one could
measure all the position and speed of every molecules in a gas, the
law of mechanics should in theory make us capable of predicting
the shocks and the future trajectories of the molecules. The system would have lost all uncertainties (at least regarding kinetics)
and fundamentally its entropy would be null and stay as such forever. On the other side, for Shannon, the information is what can be
extracted from the state of a system with the consequence of reducing its entropy. For example, reading a book reduces its entropy since
the book state moves from an unknown arrangement of symbols into
a well-defined arrangement of symbols which makes the text. Sometimes, some physicists refer to information as "negentropy" since
acquiring information reduces the entropy of a system [18,63].

Despite these small twists, the entropy and the information seem
to be very much related. In fact, they gave rise to a famous paradox
called the *Maxwell paradox*. James Clerk Maxwell (Figure 7.2), the
father of the celebrated equations of electromagnetism, imagined in
1967 the following experiment. Imagine a box filled by a gas at a
certain temperature T with enough isolation that no heat will be

Fig. 7.2.   James Clerk Maxwell.

exchanged with the outside. At some moment, a partition is erected to split the box into distinct chambers A and B. The partition does not change the energy repartition of the molecules, and therefore the gases in the two chambers are both at temperature T. Imagine a small mobile sliding door on the bulkhead and a daemon watching it (see Figure 7.3). Every time a molecule with high velocity in chamber A is going to hit the door, the daemon opens the door and the molecule passes to chamber B. When a molecule with low velocity does the same in chamber B, the daemon lets it pass to chamber A. The door remains closed for all the other molecules (low velocity molecules in chamber A and high velocity molecules in chamber A). The consequence of the protocol makes the temperature in chamber A to diminish and the temperature in chamber B to increase, thus reducing the entropy of the whole system. Thus, this would contradict the second law of thermodynamics, since the sliding door can be made massless and energyless. Where does the missing entropy go?

One had to wait until 1949 when Léon Brillouin solved this paradox. The key was to consider that the weak point was not the door itself but the measure of the molecule speed. Determining that speed of a molecule was to be above or below a certain threshold is one bit of information. Collecting and storing this information was not entropyless or energyless. Indeed, storing information modified the entropy of the system. Brillouin [17] assumed that the minimum amount of entropy for storing and deleting one bit of information

Fig. 7.3.   The Daemon of Maxwell.

would be exactly $\frac{k_B}{\log 2}$ and these computations were later confirmed by Rolf Landauer [79] 10 years later. This consequently saved the second principle of thermodynamics, since the entropy stolen to box was in fact moved to the bit storage. Similarly, the minimal energy to do this operation in a temperature T would be $\frac{k_B T}{\log 2}$ Joule. At ambient temperature, this would be an amount of 5.84×10^{-21} Joule, still far below the energy consumption of current memories.

7.2   The Time Arrow

The second law of thermodynamics was in the 19th century the only law which defined the arrow of time. The time brings you from the past to the future (see Figure 7.4), but the law of mechanics in its classic setting (i.e. without quantum twist) are deterministic and time symmetric. If you reverse the time, the laws remain the same. Even gravity is still the same with reversed time. If you reverse the time, the planets will orbit the same ellipses but in retrograded direction. The second law of thermodynamics only accepts the increase

Fig. 7.4.   The arrow of time.

in entropy and consequently makes some physical phenomenon irreversible. In particular, in chemistry, some reactions are exothermic, i.e. creating warmth and energy, and are thus spontaneous, the reverse reaction, endothermic, cannot occur without another source of energy. Without the second law of thermodynamics, the delicate chemistry of life would not be possible. Even the nuclear reactions inside stars, including our sun, won't be possible and the universe would look completely different.

At the surface, life looks like an apparent huge counterexample to the second law since it has evolved from disorganized mixture of organic elements into very structured system of cells and collection of cells as generic constituent of living beings. In fact, when we turn to the elementary chemical reactions which led to this organisation, each of them is exothermic and satisfies the second law of thermodynamics. It is striking to imagine that this is the consequence of some mathematical model about the subjective ignorance on the exact physical state (speed, spin, and position) of each particle in a volume of gas or liquid, i.e. the consequence of information (or lack of information).

This uncomfortable situation got into a climax in 1920 with the emergence of quantum physics. For the first time, nature was showing phenomenons which were both non-deterministic and objectively not time reversible. At the level of fundamental particles, the law of physics accepted pure random effect, i.e. not the consequences of the ignorance about most of the parameters of the state of one particle but because it was indeed impossible to get an exact measure of these hidden variables. For the first time, there was an evidence of a "true" time arrow.

7.3   Information and Quantum Physics

The quantum revolution marks a fundamental step in the understanding of the laws of the universe. It looks like a strange law because it allows pure randomness in the measurement results. Thus, it was lacking the predictive nature what every philosopher was expecting from a law of physics since Newton and Laplace (1810) [80]. Indeed, the Newton mechanics were the consequence of the infinitely large systems, e.g. the solar system, while the quantum physics comes from the observation of the infinitely small systems, e.g. the atom. Although similar in shape, the electrons seem to orbit around the kernel, like the planets around the sun, the two systems are utterly different (see Figure 7.5). Indeed, the orbits of the planets are not eternal and evolve with time (of course not in a human lifetime), sometimes it is due to gravitational interaction between them (Jupiter orbit slowly moved inward after interacting with Saturn orbit), or just because gravity dissipates energy (for example, the distance from Moon to Earth is slowly increasing with time). On the other side, the energy levels of electrons around the kernel are fixed and cannot fall under some minimal values, hence the atoms are not compressible. The fact that the electron does not dissipate energy as planet dissipates energy around the sun was the starting point of the emergence of quantum physics, leading to an energy spectrum with discrete levels (and not continuous). Indeed, a dissipating electron would continuously lower its energy level and get closer and closer to the kernel

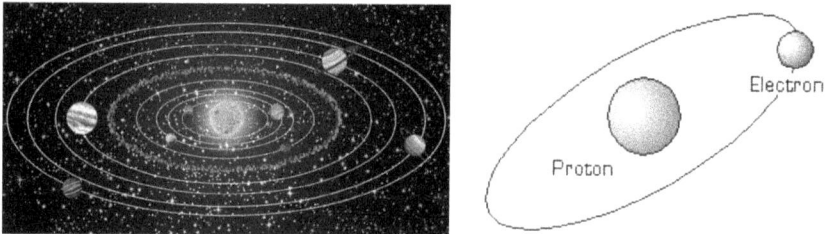

Fig. 7.5.   The infinitely large (left) and the infinitely small (right).

and, sooner or later, would clash with it since it is of opposite charge. The reason why a chair does not collapse under the weight of a sitter is in the fact that the electron cannot fall below a certain quanta of energy and this condition sounds like an apparent repulsive force which keeps the atom stable.

In fact, the quantum physics provides an extremely accurate tool for the determination of the probabilistic distribution about a single particle, which, thanks to the law of large numbers over the huge number of particles (a mole of molecules is approximately $6.0221408.10^{23}$ which is contained in 22.4 liters in regular conditions), offers an unprecedented level of accuracy of prediction for the evolution of large systems of particles. It means that eventually quantum physics and Newton physics agree on large systems. However, these intrinsic randomness at particle level left people very uncomfortable. Albert Einstein was an early opponent of the randomness nature of measurement in quantum theory, although he was the first promoter of the theory of quanta with his interpretation of the photo-electric effect (which awarded him the Nobel Price in 1922). But his deep opinion was that "God does not play dices" and that the current random nature of physical effect was mostly due to hidden physical parameters which were still to be discovered at that time. The dispute was settled later via John Bell theorem [14] and after Alain Aspect experiment [5] in favor of the "Copenhagen" interpretation embracing pure randomness to the exclusion of any hidden variables theory.

The quantum physics is based on some hypotheses, which all revolve around the assumption that the state of a particle is given by a "wave" function $\Psi\colon \mathbb{R}^3 \to \mathbb{C}$ [22]. The wave function can be made more complicated in defining it in $\mathbb{R}^D \times \mathcal{S} \to \mathbb{C}^d$, where $D \geq 3$ and $d \geq 1$ and $\mathcal{S}$ being a symmetry group, and this is just talking of first quantization; the second quantization talks of creation and annihilation of particle operators (thus D would be infinite). In its simplest setting, $\mathbb{R}^3$ stands for the spatial position of the particle and $\mathbb{C}$ stands for the intensity of the wave function. The wave function summarizes the probabilistic properties of the particle. Indeed, for $\mathbf{z} \in \mathbb{R}^3$, the quantity $|\Psi(\mathbf{z})|^2$ is indeed the density of probability of the presence of the particle at position $\mathbf{z}$, under the convention that $\int_{\mathbb{R}^3} |\Psi(\mathbf{z})| d\mathbf{z}^3 = 1$ is called the unitary condition.

Two particles A and B have to share a pair of positions, each in $\mathbb{R}^3$, therefore the resulting wave function will operate in $\mathbb{R}^6$. If the particles are independent with A having the wave function Ψ_A and B having the wave function Ψ_B, the combined wave function will be the tensor $\Psi_A \otimes \Psi_B$ and its value at $(\mathbf{z}, \mathbf{z}') \in \mathbb{R}^6$ is $\Psi_A(\mathbf{z})\Psi_B(\mathbf{z}')$. Of course, we can generalize to an arbitrary number of particles.

It may look weird to consider $\Psi(\mathbf{z})$ as the quantity of interest instead of $|\Psi(\mathbf{z})|^2$. The reason is because the equations dictating the evolution of the wave function are linear on $\Psi(\mathbf{z}, t)$ (now including the time coordinate) and not on $|\Psi(\mathbf{z}, t)|^2$. Indeed, the equations are in the form of

$$i\hbar \frac{\partial}{\partial t}\Psi = \mathcal{H}\Psi,$$

where $\hbar$ is the Planck constant ($1.054571818.10^{-34}$ Joule second) and $\mathcal{H}$ is a self-adjoint linear operator, sometimes called Hamiltonian: This is the celebrated Schrödinger equation [22]. The Hamiltonian can be of the form $\mathcal{H}\Psi = -\frac{\hbar^2}{2m}\Delta\Psi + V\Psi$, where $\Delta\Psi$ is the Laplacian of function Ψ and V is a potential. When $\mathcal{H}$ is independent of time, the general solution is $\Psi(\mathbf{z}, t) = e^{i\mathcal{H}t/\hbar}\Psi(\mathbf{z}, 0)$. The self-adjoint property of $\mathcal{H}$ is based on the Hermitian internal product defined on all pair of integrable functions f and g by $\int_{\mathbb{R}^3} f(\mathbf{z})g^*(\mathbf{z})d\mathbf{z}^3$, where $g^*(\mathbf{z})$ is the complex conjugate of $g(\mathbf{z})$ and indeed, by successive by part integrations: $\int_{\mathbb{R}^3}(\mathcal{H}f)(\mathbf{z})g^*(\mathbf{z})d\mathbf{z}^3 = \int_{\mathbb{R}^3} f(\mathbf{z})(\mathcal{H}g)^*(\mathbf{z})d\mathbf{z}^3$. Since $(e^{i\mathcal{H}t/\hbar}g)^* = e^{-i\mathcal{H}t/\hbar}g^*$ and that $e^{-i\mathcal{H}t/\hbar} = (e^{i\mathcal{H}t/\hbar})^{-1}$, the self-adjoint property implies that $\int_{\mathbb{R}^3} |\Psi(\mathbf{z}, t)|^2 d\mathbf{z}^3 = \int_{\mathbb{R}^3} |\Psi(\mathbf{z}, 0)|^2 d\mathbf{z}^3$; this property is true for any self-adjoint Hamiltonian, even those with time-varying components.

The Schrödinger equation is similar to the equation for the time evolution of a Markov chain:

$$\frac{\partial}{\partial t}\mathbf{p} = \mathcal{P}\mathbf{p},$$

where $\mathbf{p}$ is the vector of probabilities at time t and $\mathcal{P}$ is the transition matrix. But the difference is that the Markov equation operates on probability vectors, not on complex functions. The last point leads to paradoxical situations. If $\mathbf{p}_1$ and $\mathbf{p}_2$ are two probability vector solutions of the Markov equation, then $\frac{1}{2}(\mathbf{p}_1 + \mathbf{p}_2)$ is also a solution, and if on a given state any of the $\mathbf{p}_1$ or $\mathbf{p}_2$ has a non-zero weight, then

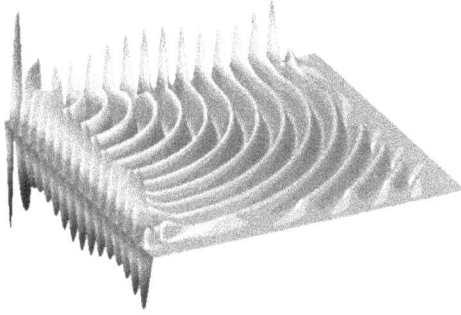

Fig. 7.6. Optical interference.

$\mathbf{p}_1 + \mathbf{p}_2$ has a non-zero weight on this state. With the complex wave function, let two wave functions Ψ_1 and Ψ_2 be the solutions of the Schrödinger equation, thus $\Psi_1 + \Psi_2$ is also solution. Assume both are non-zero on a given $\mathbf{z}$ but $|\Psi_1(\mathbf{z}) + \Psi_2(\mathbf{z})|^2$ can be zero. For example, in a one-dimensional system, we can have locally $\Psi_1(z) = e^{iaz}$ and $\Psi_2(z) = e^{-iaz}$ for some a, then both $\Psi_1(z)$ and $\Psi_2(z)$ are non-zero, but $\Psi_1(z) + \Psi_2(z)$ is 0 at $z = \frac{\pi}{2a}$ and $z + k\frac{\pi}{a}$ for all $k \in \mathbb{Z}$. This is exactly what happened with optic interference (see Figure 7.6) which puzzled the physicists for centuries: "How a superposition of non-zero signals can result in a null signal?" They were even, even more intrigued when they realized that such cancellation also occurs when light beams are replaced by electrons or other particles beams, giving rise to the emergence of quantum mechanics.

7.3.1 The algebra of quantum physics

The eigenvalues of the Hamiltonian operator are interpreted as the energy levels of the particle. Let $E_0, E_1, \ldots$ be the sequence of eigenvalues with respective unitary eigenvectors $\mathbf{u}_0, \mathbf{u}_1, \ldots$: $\mathcal{H}\mathbf{u}_k = E_k\mathbf{u}_k$, the wave functions $\mathbf{u}_k$ are orthonormal bases for all wave functions: indeed, for any Ψ, we have the decomposition

$$\Psi(\mathbf{z}) = \sum_k \mathbf{u}_k \langle \Psi | \mathbf{u}_k \rangle$$

under the convention that $\langle f | g \rangle = \int_{\mathbb{R}^3} f(\mathbf{z}) g^*(\mathbf{z}) d\mathbf{z}^3$. Sometimes, the physicists use the "ket" notation $|\mathbf{u}_k\rangle$ instead of $\mathbf{u}_k$. Taking into account time evolution (assuming a time invariant Hamiltonian), we have

$$\Psi(\mathbf{z}, t) = \sum_k e^{-iE_k t/\hbar} |\mathbf{u}_k\rangle \langle \Psi_0 | \mathbf{u}_k\rangle ,$$

where $|\Psi_0\rangle$ is the initial wave function.

The accurate determination of the energy level of an isolated electron in the hydrogen atom has been the first major success of quantum physics. Indeed, the potential is $V(\mathbf{z}) = -\frac{q^2}{4\pi\epsilon_0 \|\mathbf{z}\|}$ when the proton is centered at the origin, q the elementary charge of the electron and ϵ_0 the permittivity of vacuum. It turns out that $E_k = -\frac{m_e q^4}{2(4\pi\epsilon_0)^2\hbar^2} \frac{1}{(k+1)^2}$ with m_e being the mass of the electron (see Figure 7.7). We note that there is indeed a minimum energy level, below which the electron cannot go. This is a pure mathematical consequence of the equation but it makes the atom incompressible: The chair cannot collapse under the weight of the sitter!

Within this model, the atom is stable: The electron cannot move from one energy level to another level. The probability to move from one level k to another level k' is $|\langle \mathbf{u}_k | \mathbf{u}_{k'}\rangle|^2$ which is zero by consequence of the orthogonality of the basis made by the vectors $\mathbf{u}_k$'s. But in the presence of an external electromagnetic field, the eigenvectors are no longer orthogonal. In fact, they are no longer eigenvectors; the true eigenvectors have to be found with the coupling with the electromagnetic field which, if not too strong, is a perturbation of the atom system in rest. The consequence is that the electron can now jump between the energy levels of the resting atom by exchanging energy with the electromagnetic field.

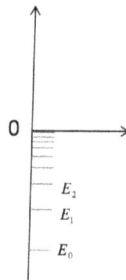

Fig. 7.7.   Energy levels of the hydrogen atom.

7.3.2 Theory of quantum measurement, quantum information

The wave function contains all the elements needed to characterize the probabilities of the parameters of the particle. We can be tempted to say "contains all the information needed to characterize...", but we will refrain to use the word "information" for reasons which are related to the definition of information in the first chapter and which will be explained later. Paradoxically, the wave function cannot be measured; this is not a physical *grandeur*. The wave function can only be indirectly measured via sampling the probabilities of the parameters they represent (assuming that many particles share the same wave function). For example, given an electron wave function Ψ, the probability to measure the electron at level E_k is $|\langle\Psi|\mathbf{u}_k\rangle|^2$. Since

$$\Psi = \sum_k |\mathbf{u}_k\rangle\langle\Psi|\mathbf{u}_k\rangle,$$

the measure can have many possible outcomes, indeed those for which $\langle\Psi|\mathbf{u}_k\rangle \neq 0$. The average value of the measured energy level is $\langle\Psi|\mathcal{H}\Psi\rangle = \sum_k \langle\mathbf{u}_k|\mathcal{H}\mathbf{u}_k\rangle$. The measure has an impact on the wave function, since after the measure, the outcome cannot change, i.e. if we repeat the measure on the resulting system, the result remains the same. Every measured physical grandeur is associated with a self-adjunct operator $\mathcal{M}$ with eigenvalues M_k and eigenvector $\mathbf{v}_k$, for example, the measure of the component on axis x of the speed of the particle is associated with the operator $i\partial_x$. In all cases, the measure projects the wave function Ψ on the eigenvector $\mathbf{v}_k$ corresponding to the measured value:

$$\Psi \to |\mathbf{v}_k\rangle \frac{\langle\Psi|\mathbf{v}_k\rangle}{|\langle\Psi|\mathbf{v}_k\rangle|}.$$

The division by the factor $|\langle\Psi|\mathbf{v}_k\rangle|$ is necessary to keep the system unitary. If the set of wave functions giving the same outcome corresponds to a vector space of dimension greater than 1, then the measure is a projection on this space vectors which is said to be "degenerate". This is the case for the energy levels of the hydrogen atom, the energy above E_0 is degenerate. In all cases, we note that the resulting operation with the normalization is highly nonlinear and irreversible. It is nonlinear because of the normalization to keep the system unitary. It is irreversible because the projection "erases" the components of Ψ on the other eigenvectors. The measurement operation is sometimes called wave function "collapse".

Between two measurements, the wave function evolves linearly: From time t_0 to time t_1 keeping only the time variable, we have

$$\Psi(t_1) = \mathcal{U}(t_1, t_0)\Psi(t_0),$$

where $\mathcal{U}(t_1, t_0)$ is an unitary operator. This is a reversible process. When the measurement occurs at time t_1, we have an irreversible collapse operation Π_1. After k measurement, respectively at times $t_1, t_2, \ldots, t_k$, we get the following time line for $t > t_k$:

$$\Psi(t) = \mathcal{U}(t, t_k)\Pi_k\mathcal{U}(t_k, t_{k-1}) \cdots \mathcal{U}(t_2, t_1)\Pi_1\mathcal{U}(t_1, t_0),$$

as illustrated in Figure 7.8. The presence of the irreversible operations $\Pi_1, \ldots, \Pi_k$ is an indication of the direction of the arrow of time, a direction well anchored in the intimate structure of the physical universe.

Many are tempted to call the wave function the "quantum information". In fact, this would be misleading. At most, we can say that the wave function is a support of information. The information is only retrieved after the measurement and is the result of the measurement. The wave function is not information because it is not a physical grandeur which can be measured. But we could argue that we could perform several measurements on several copies of the wave function Ψ to build statistics about the amplitude and use interference to build statistics about the phase. However, it turns out that it is not possible via the so-called "no cloning theorem" [122].

Theorem 7.1. *It is not possible to make two copies of an unknown wave function* $|\Psi\rangle$ *without violating unitarity.*

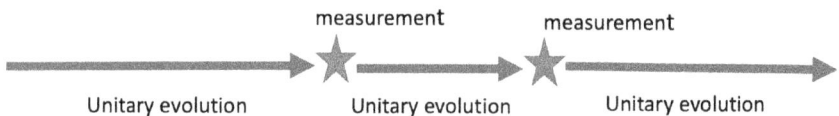

Fig. 7.8. Wave function evolution with measurement interleaved with linear unitary evolutions.

Proof. To simplify, we consider two originally independent particles A and B. The particle A has the wave function $|\Psi\rangle_A$, supposed to be unknown (not fabricated so that duplication would be trivial), and the particle B has the "blank" wave function $|e\rangle_B$ which is supposed to be the vehicle of the copy $|\Psi\rangle$. Thus, the original system wave function is the tensor product $|\Psi\rangle_A \otimes |e\rangle_B$ (to simplify, we remove the sign "$\otimes$" since it is only a multiplication). We want to find a process which leads to the tensor product $|\Psi\rangle_A \otimes |\Psi\rangle_B$.

The process cannot involve a measurement because it would destroy $|\Psi\rangle_A$. Thus, the evolution should be completely unitary. Let $\mathcal{U}$ be the unitary operator which maps $|\Psi\rangle_A \otimes |e\rangle_B$ into $e^{-i\delta} |\Psi\rangle_A \otimes |\Psi\rangle_B$ (we can allow an extra phase shift by an arbitrary real number δ). We know that $\mathcal{U}^*\mathcal{U} = 1$ by property of unitary operators which must keep invariant the dot products.

We suppose two possible candidate wave functions $|\Psi_1\rangle_A$ and $|\Psi_2\rangle_A$. We have $\mathcal{U} |\Psi_1\rangle_A |e\rangle_B = e^{i\delta_1} |\Psi_1\rangle_A |\Psi_1\rangle_B$ and $\mathcal{U} |\Psi_2\rangle_A |e\rangle_B = e^{i\delta_2} |\Psi_2\rangle_A |\Psi_2\rangle_B$ for some real numbers δ_1, δ_2. We have

$$\langle\Psi_1|\Psi_2\rangle \langle e|e\rangle = \langle\Psi_1|_A \langle e|_B \mathcal{U}^*\mathcal{U} |\Psi_2\rangle_A |e\rangle_B$$

$$= e^{-i\delta_1+i\delta_2} \langle\Psi_1|_A \langle\Psi_1|_B |\Psi_2\rangle_A |\Psi_2\rangle_B$$

$$= e^{-i\delta_1+i\delta_2}(\langle\Psi_1|\Psi_2\rangle)^2. \tag{7.1}$$

Since $\langle e|e\rangle = 1$, we get the identity $|\langle\Psi_1|\Psi_2\rangle| = |\langle\Psi_1|\Psi_2\rangle|^2$, which should hold for any arbitrary pair of wave functions. This would imply that either $|\Psi_1\rangle = 0$ or $|\Psi_2\rangle = 0$ thus contradicting $\langle\Psi_1|\Psi_1\rangle = \langle\Psi_2|\Psi_2\rangle = 1$. Therefore, the operator $\mathcal{U}$ cannot be unitary. $\square$

This theorem was originally by James Park in 1970 [122]. We should be careful not to confuse the no-cloning theorem with the fact one can fabricate particles with the same wave function (for example, multiple photons with the same polarization, through a polarizer with a fixed angle). But in this case, the wave function $|\Psi\rangle$ is fully known and the argument above with an arbitrary pair $(|\Psi_1\rangle, |\Psi_2\rangle)$ won't hold. But there would be no information to obtain by measuring $|\Psi\rangle$ since we already know all its parameters. The theorem also applies on partial wave function, i.e. wave functions restricted to one particle parameter, e.g. the spin, as long as the isolated parameter evolves unitarily.

Let $\epsilon > 0$, we call "ϵ-approximate" cloning the mapping of an arbitrary wave function $|\Psi\rangle$ into a wave function $|f_\epsilon(\Psi)\rangle$ such that for all $|\Psi\rangle$: $|\langle\Psi|f_\epsilon(\Psi)\rangle| > \epsilon$. Note that $\epsilon = 1$ is equivalent to perfect cloning (*modulo* a phase factor). We have the easy follow-up theorem:

Theorem 7.2. *For all $\epsilon > 0$, ϵ-approximate cloning would violate unitarity.*

Proof. We assume that the transition from $|\Psi\rangle_A |\mathbf{e}\rangle_B$ to $|\Psi\rangle_A |f_\epsilon(\Psi)\rangle$ is unitary, i.e. there exists an unitarity operator $\mathcal{U}_\epsilon$ such that $\mathcal{U}_\epsilon |\Psi\rangle_A |\mathbf{e}\rangle_B = e^{-i\delta} |\Psi\rangle_A |f_\epsilon(\Psi)\rangle$ for some real number δ. Taking two arbitrary wave functions $|\Psi_1\rangle$ and $|\Psi_2\rangle$ and using the conservation of the dot product of $|\Psi_1\rangle_A |\mathbf{e}\rangle_B$ with $|\Psi_2\rangle_A |\mathbf{e}\rangle_B$ we get

$$\begin{aligned}
\langle\Psi_1|\Psi_2\rangle &= \langle\Psi_1|_A \langle\mathbf{e}|_B |\mathbf{e}\rangle_B |\Psi_2\rangle_A \\
&= \mathcal{U}_\epsilon^* \langle\Psi_1|_A \langle\mathbf{e}|_B \mathcal{U}_\epsilon |\mathbf{e}\rangle_B |\Psi_2\rangle_A \\
&= e^{-i\delta_1 + i\delta_2} \langle\Psi_1|_A \langle f_\epsilon(\Psi_1)|_B |\Psi_2\rangle_A |f_\epsilon(\Psi_2)\rangle_B \\
&= e^{-i\delta_1 + i\delta_2} \langle\Psi_1|\Psi_2\rangle \langle f_\epsilon(\Psi_1)|f_\epsilon(\Psi_2)\rangle .
\end{aligned} \tag{7.2}$$

Thus, we would have $|\langle f_\epsilon(\Psi_1)|f_\epsilon(\Psi_2)\rangle| = 1$ as in general $\langle\Psi_1|\Psi_2\rangle \neq 0$. This would imply that all the $|f_\epsilon(\Psi)\rangle$ are co-aligned with an unitary vector $|\mathbf{w}_\epsilon\rangle$. But in this case, the approximate cloning of a wave function $|\Psi\rangle$ orthogonal to $\mathbf{w}$ will result in $\langle\Psi|f_\epsilon(\Psi)\rangle = 0$ which would contradict the ϵ approximation condition. □

An ϵ-approximate clone of Ψ for $\epsilon > 0$ shares information with the original wave function. Besides, the ϵ-approximate cloning should not be confused with the approximate cloning of quantum system in general. The later can be obtained via unitary operations (see [103]) but with a non-zero failure probability or, in other words, the result of an 0-approximate cloning.

7.3.3   *Spin algebra, photon polarization*

The spin of a particle is a consequence of quantum mechanics [22]. In the first quantization, a particle is defined by its mass, charge and spin. The spin of the electron is a vector which is the quantum model of the rotation axis of the particle, see Figure 7.9. Contrary to its classical model, the spin has the property that its component on any arbitrary axis can be measured only with the values $\pm\frac{1}{2}\hbar$. Indeed,

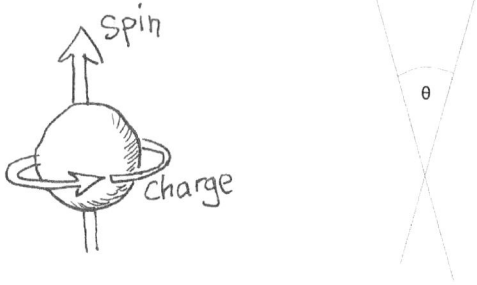

Fig. 7.9. A classic representation of spin (left) and two axes of spin measurement (right).

the measurement aligns the spin on the axis and can have only two values. This effect was also a surprise which led to the birth of the quantum physics.

The quantum wave function of the spin component on a given axis is a space of dimension 2 with the two eigenvectors $|+\rangle$ and $|-\rangle$ respectively corresponding to the value $\frac{1}{2}\hbar$ and the value $-\frac{1}{2}\hbar$: $|\Psi\rangle = \alpha |+\rangle + \beta |-\rangle$ with $|\alpha|^2 + |\beta|^2 = 1$. No surprise here. A particle can be made of k particle of spin $\pm\frac{\hbar}{2}$. In this case, the spin is multiple and its component on any arbitrary axis is the sum of the component of its individual spin. If k is even, the sum ranges over $\frac{\hbar}{2}$ times all even integers between $-k$ and k. In this case, the particles are called "boson". If k is odd, the sum ranges over $\frac{\hbar}{2}$ times all odd integer between $-k$ and k. In this case, the particles are called "fermion". From now we stay on the simplest and common fermion: the electron.

The interesting fact occurs when we move the axis by an angle θ. Let $|\theta+\rangle$ and $|\theta-\rangle$ be the respective counterparts of $|+\rangle$ and $|-\rangle$ on the inclined axis by angle θ. There is the following linear correspondence:

$$\begin{bmatrix} |\theta-\rangle \\ |\theta+\rangle \end{bmatrix} = \begin{bmatrix} \cos\frac{\theta}{2} & \sin\frac{\theta}{2} \\ -\sin\frac{\theta}{2} & \cos\frac{\theta}{2} \end{bmatrix} \begin{bmatrix} |-\rangle \\ |+\rangle \end{bmatrix}.$$

Let $|\Psi\rangle = \frac{1}{\sqrt{2}}|-\rangle + \frac{i}{\sqrt{2}}|+\rangle$. It is a neutral wave function on the first axis because $P(+) = P(-) = \frac{1}{2}$. We note that $|\Psi\rangle = e^{i\theta/2}\left(\frac{1}{\sqrt{2}}|\theta-\rangle + \frac{i}{\sqrt{2}}|\theta+\rangle\right)$; this means that this particular wave function is neutral for any axis.

Measurement
On 1st axis

Measurement
On 1st axis

Measurement
On 2nd axis

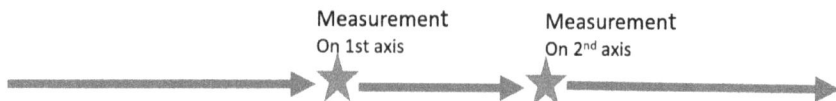

Fig. 7.10.   Two successive measurements on two angles.

If we measure $|\Psi\rangle$ on the first axis, it collapses on the $|+\rangle$ and $|-\rangle$ with same probability $\frac{1}{2}$. If afterward we measure on the second axis with angle θ, the sequence of outcomes is either $(++)$, $(--)$, $(-+)$, or $(+-)$. If we apply the rules of the measurement, see Figure 7.10, that after each measurement the wave function stays on the same eigenvectors, we will have

$$P(++) = P(--) = \frac{1}{2}\left(\cos\frac{\theta}{2}\right)^2,$$

$$P(-+) = P(+-) = \frac{1}{2}\left(\sin\frac{\theta}{2}\right)^2.$$

The polarization of the photon, the particle of light, follows a story similar to the story of spin of the electron. The photon has a polarization which basically corresponds to the orientation of the electric field. The polarization can be detected via polarized glass, which lets pass photon corresponding to the component of the electric field which is parallel to the polarizer axis. If the electric field is orthogonal to the axis, the photon is blocked and absorbed by the polarizer and will be detected by photo-electric effect. However, one can have a non-destructive measure of the polarization by using half absorbing glass. A photon incident to the glass under a certain angle is reflected either when the polarization is parallel to the glass surface or through the glass otherwise. Therefore, the photons take different path depending on their polarization. In this case, the axis of measurement is the direction orthogonal to both the photon beam and the glass surface.

This is similar to spin, but in the case of photon polarization, this is the parallelism to a given axis which is a quantum effect. We denote $|+\rangle$ the wave function corresponding to the parallelism to the axis and $|-\rangle$ when it is orthogonal. The correspondence between two axes of measures separated by an angle θ is

$$\begin{bmatrix} |\theta-\rangle \\ |\theta+\rangle \end{bmatrix} = \begin{bmatrix} \cos\theta & \sin\theta \\ -\sin\theta & \cos\theta \end{bmatrix} \begin{bmatrix} |-\rangle \\ |+\rangle \end{bmatrix}. \tag{7.3}$$

The wave function $|\Psi\rangle = \frac{1}{\sqrt{2}}|-\rangle + \frac{i}{\sqrt{2}}|+\rangle$ is still the neutral state but the conversion in the axis θ is now $|\Psi\rangle = e^{i\theta}\left(\frac{1}{\sqrt{2}}|\theta-\rangle + \frac{i}{\sqrt{2}}|\theta+\rangle\right)$. If from a neutral state we proceed into two consecutive measures on two axes differing by an angle θ, we get

$$P(++) = P(--) = \frac{1}{2}(\cos\theta)^2,$$

$$P(-+) = P(+-) = \frac{1}{2}(\sin\theta)^2.$$

Of course, if we use simple polarized glasses for the measures as depicted in Figure 7.11, the probabilities $P(--)$ and $P(-+)$ will be undetermined because the first measure will be destructive (in this case, $P(-*) = \frac{1}{2}$ (the symbol "$*$" being indifferently "$+$" or "$-$"). Thus, the above expression will be verified on measurement via half absorbing glasses.

7.3.4   *Density operator, von Neumann entropy*

The density operator is an elegant way to describe a system made by a mixture of pure states and quantum states. All the components of the system are described by a positive semi-definite Hermitian operator ρ. If the system is obtained by a mixture of k several quantum

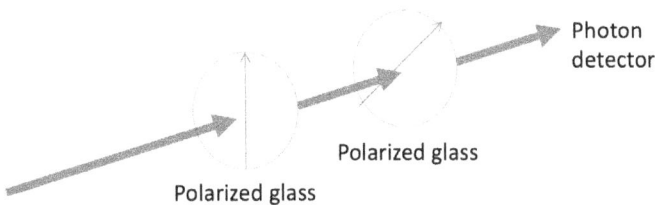

Fig. 7.11.   Two successive measurements of photon polarization on two angles.

states Ψ_i, each of them obtained with a probability $P(\Psi_i)$, the density operator is then defined by

$$\rho = \sum_{i=1}^{k} P(\Psi_i)\Psi_i \otimes \Psi_i.$$

A natural consequence of unitarity is that $\mathrm{tr}(\rho) = 1$. After a measurement, the probability that the system is measured on the observable m is given by $\mathrm{tr}(\Pi_m\rho)$, where Π_m is the projection operators on the eigenspace corresponding to the observable m. Let's take the example of photon polarization. If the system is with probability 1, the quantum space $\Psi = \alpha\,|-\rangle + \beta\,|+\rangle$ with $|\alpha|^2 + |\beta|^2 = 1$, then

$$\rho = \Psi \otimes \Psi = \begin{bmatrix} |\alpha|^2 & \bar{\alpha}\beta \\ \alpha\bar{\beta} & |\beta|^2 \end{bmatrix}$$

expressed in the natural base $(|-\rangle, |+\rangle)$. Under this representation, we come back to the probability that the system is measured on polarization "$-$" is indeed $|\alpha|^2$, since $\Pi_m = \begin{bmatrix} 1 & 0 \\ 0 & 0 \end{bmatrix}$.

If on the opposite we consider that the system is a mixture of pure states $|-\rangle$ and $|+\rangle$, with $P(|-\rangle) = |\alpha|^2$ and $P(|+\rangle) = |\beta|^2$, we get

$$\rho = |\alpha|^2\,|-\rangle\,\langle -| + |\beta|^2\,|+\rangle\,\langle +| = \begin{bmatrix} |\alpha|^2 & 0 \\ 0 & |\beta|^2 \end{bmatrix}$$

and we again find back the probability $|\alpha|^2$ of measuring the system is polarization "$-$". Hence, why a difference?

The two systems differ on their entropy. The von Neumann entropy of the system is the quantity $H(\rho) = -\mathrm{tr}(\rho \log \rho)$ [88]. In the mixture case, we get the classic expression $H(\rho) = -|\alpha|^2 \log |\alpha|^2 - |\beta|^2 \log |\beta|^2$. But if the system is made of the single quantum state $\alpha\,|-\rangle + \beta\,|+\rangle$, then $H(\rho) = 0$!

The reason is that a pure quantum state contains no randomness, the randomness occurs with the measurement. Indeed, after measurement, the system becomes a mixture of pure states and the entropy jumps to $-|\alpha|^2 \log |\alpha|^2 - |\beta|^2 \log |\beta|^2$. Therefore, the von Neumann entropy increases with time because the measurement is an irreversible process. This implies that the measurement theory implies a fundamental time arrow. Anyhow, one must be careful that the von Neumann entropy increase is not the only justification of the third thermodynamic law because there is also an increase even within reversible evolution, due to the "ignorance" effect.

7.4   Quantum Computers

As we have seen, the wave function of a particle can cover several states. Feynman (1918–1988) was the first scientist to imagine that a quantum computer could take benefit of this situation by enabling computations on all these states in parallel and get an unprecedented computation power [36]. This massive parallelism would enable the resolution of NP-hard problems. An NP-hard problem is problem whose resolution consists of finding a data of size n which satisfies whose resolution would basically consists of investigating all possibilities, but each possibility would take almost no time to be checked (in all rigor, each possibility would take a polynomial time to be checked). In fact, instead of all possibilities, it can a fraction of them, which still makes the processing cost exponential in n, or at least not polynomial.

A good example of NP-hard problem is the Boolean satisfiability problems which consists from a general Boolean expression involving n binary variable to find an instance of these n variable which satisfies the Boolean expression. A Feynman computer would take n neutral spins which individually covers the two states $|-\rangle$ and $|+\rangle$, and collectively all the 2^n Boolean values that can be taken by the n binary variables. Such a neutral spin is called "qubit". The quantum computer will necessarily cover the instances of the n variables which satisfy the Boolean expression. The computation of the Boolean can be done via Zhegalkin polynomials which lead to easy representation with quantum circuits.

The issue with the Feynman computer is about the mean to return the solution. A quantum state of qubits is not information and the solution to a satisfiability problem is an information and this information can only be retrieved via a measurement, and a measurement is mostly destructive of the whole qubit chain since it collapses it into a single chain of bits. Thus, there is still far to make a Feynman computer which can give useful information. It is not believed that quantum computer can solve NP-hard problems. Anyhow, there have been interesting proposition such as the following:

- the Grover algorithm which allows us to reduce a search among N into less than $\sqrt{N}$ quantum operations,
- the Shor algorithm which allows us to factorize a large integer of n bits in less than $O(n^2 \log n)$ quantum operations.

7.4.1 *The Grover algorithm*

The foundation of quantum algorithm is the amplification of the probability for the "good" states compared to the "bad" states, and this starting from a neutral wave function, and this via unitary operators.

The Grover probability amplification process works as follow [45]. We assume that there are N elements to be searched indexed from 1 to N. Let σ be the element to be searched (there can be more than one). We assume that there is a function $f(.)$ which characterizes the element to be searched, i.e. $f(\sigma) = 1$ and for all $x \neq \sigma$: $f(x) = -1$. The function $f(.)$ can be the result of a Boolean function as described above.

We assume that we can put in correspondence each integer x in $\{1, \ldots, N\}$ with a quantum state $|x\rangle$. To simplify, we can imagine $N = 2^n$ as a power of 2, in order to use qubits. We denote $\mathcal{U}_\sigma$ the unitary operator defined by

$$\mathcal{U}_\sigma |x\rangle = f(x) |x\rangle$$

or $\mathcal{U}_\sigma = \sum_x f(x) |x\rangle \langle x| = 2 |\sigma\rangle \langle \sigma| - \mathcal{I}$, where $\mathcal{I}$ is the identity operator. The operator $\mathcal{U}_\sigma$ is the symmetry operator around the axis $|\sigma\rangle$. Similarly, let $s = \frac{1}{\sqrt{N}} \sum_x |x\rangle$, we denote $\mathcal{U}_s$ the symmetry operator along the axis s: $\mathcal{U}_s = \mathcal{U} - 2 |s\rangle \langle s|$.

The principle is to start from $s_0 = s$ and then iterate the quantum operations $s_1 = \mathcal{U}_s \mathcal{U}_\sigma s_0$, etc. $s_{\ell+1} = \mathcal{U}_s \mathcal{U}_\sigma s_\ell$ and then measure s_k after a predefined number of iterations.

To watch the evolution of $(\mathcal{U}_s \mathcal{U}_\sigma)^k$, we look at the orthonormal base $(\Psi_\sigma, |\sigma\rangle)$ with $\Psi_\sigma = \frac{1}{N-1} \sum x \neq \sigma |x\rangle$. In this base,

$$\mathcal{U}_s \mathcal{U}_\sigma = \begin{bmatrix} 1 - \frac{2}{N} & -2\frac{\sqrt{N-1}}{N} \\ 2\frac{\sqrt{N-1}}{N} & 1 - \frac{2}{N} \end{bmatrix} \approx \begin{bmatrix} \cos\frac{2}{\sqrt{N}} & -\sin\frac{2}{\sqrt{N}} \\ \sin\frac{2}{\sqrt{N}} & \cos\frac{2}{\sqrt{N}} \end{bmatrix}.$$

We recognize in the last matrix a rotation matrix of angle $\frac{2}{\sqrt{N}}$. Since s makes an angle of approximately $\frac{\pi}{2} - \frac{1}{\sqrt{N}}$ with $|\sigma\rangle$, s_k will make an angle of approximately $\frac{1}{\sqrt{N}}$ with $|\sigma\rangle$ with $k = \frac{\pi}{4}\sqrt{N}$. The probability to get the searched state $|\sigma\rangle$ while measuring s_k will be $(\cos\frac{1}{\sqrt{N}})^2 = 1 - O(1/N)$. If unsuccessful, the process can be repeated (2 would suffice with probability of failure less than $1/N^2$). Thus, we have shown that the Grover algorithm finds the target in less than $\sqrt{N}$ steps.

Of course, if $N = 2^n$, it reduces the search cost in $2^{n/2}$, thus still exponential. But it may have practical impact. Assume that the encryption key needs 3.10^{14} computation steps. With 10 millions operations per second it would take one year of brute force hacking the key. With the Grover algorithm, even with an assumed slow down of the computations at 1,000 operations per second, it would take less than 5 hours.

7.4.2 *Peter Shor's algorithm*

Factoring a large number is a crucial challenge, since some early cryptographic algorithms such as RSA (for Rivest Shamir Adleman Algorithm) are based on the assumed difficulty to factorize numbers with large factors [101]. In fact, number factorization is not expected to be part of the NP-hard problem clique. However, no polynomial algorithm in n on classic computers, n being the number of digits of the number N to be factorized, was known. The best algorithm known is the quadratic sieve, or general number field sieve which factors a large number in $O\left(\exp\left(1.9n^{1/3}(\log n)^{2/3}\right)\right)$.

In 1999, Peter Shor had found a polynomial time algorithm for factoring a number [109]. It is clear that finding one proper factor f of N is enough to easily extract all factors by repeating the algorithm on N/f and so forth. The idea of the algorithm is to find a number a whose exponential order in the group modulo N is even. The order is the smallest r such that $a^r \equiv 1$ modulo N. If r is even, then let $b \equiv a^{r/2} \pmod{N}$. Since $a^r - 1 = b^2 - 1 = (b-1)(b+1) \equiv 0$ modulo N. Since $a^{r/2} - 1$ cannot be equal to zero by definition of the order, if $b+1$ is neither equal to zero modulo N, then $b-1 \pmod{N}$ and $b+1 \pmod{N}$ contains two independent non trivial factors of N. By repeating the algorithm on $b-1$ or $b+1$ will find proper factors.

Ignoring the cost of the order finding algorithm, we can be convinced that the algorithm is polynomial, since the fact that a random number a has an even order r occurs with probability $\frac{1}{2}$. The fact that $a^{r/2} + 1$ is not zero modulo N will happen at least with probability $\frac{1}{2}$. The quantity $a^{r/2} - 1$ modulo N will likely be uniformly spread between 0 and N. In other words, it would take in average 8 tries over the value of a to find a good value with $a^{r/2} + 1 < N/2$ modulo N.

In fact, the magic comes from the order finding algorithm, for which there is no known polynomial classical algorithm. Shor proposes to use n qbits. This quantity is necessary in order to perform the computations modulo N.

The algorithm starts on the neutral state $\Psi = \frac{1}{\sqrt{2^n}} \sum_x |x\rangle$, where x describes all the numbers between 0 and $2^n - 1$. Then use quantum circuit to create the hybrid state $|x, f(x)\rangle$ from the state $|x, 0^n\rangle$ with $f(x) \equiv a^x$ modulo N $\mathcal{U}_f |x, 0^n\rangle = |x, f(x)\rangle$. This can be done with some fixed power of $\log N$ quantum gates:

$$\mathcal{U}_f \frac{1}{\sqrt{2^n}} \sum_x |x\rangle \otimes |0^n\rangle = \frac{1}{\sqrt{2^n}} \sum_x |x, f(x)\rangle.$$

Then we use the quantum Fourier transform where $\omega = e^{2i\pi/2^n}$ is a 2^nth root of unity to map $|x\rangle$ into $\frac{1}{\sqrt{2^n}} \sum_y \omega^{xy} |y\rangle$ via an operator $\mathcal{U}_{FT}$. Thus, we end up with

$$\mathcal{U}_{FT}\mathcal{U}_f \Psi \otimes |0^n\rangle = \frac{1}{2^n} \sum_x \sum_y \omega^{xy} |y, f(x)\rangle.$$

The above sum can be rewritten in

$$\frac{1}{2^n} \sum_z \sum_y \left(\sum_{x:f(x)=z} \omega^{xy} \right) |y, z\rangle.$$

The key is that if x is such that $f(x) = z$, then $f(x + r) = z$ and for all k: $f(x + kr) = z$. In other words, the x are regularly spaced by r (modulo N), thus sum $\sum_{x:f(x)=z} \omega^{xy} = \omega^{x_0 y} \sum_k \omega^{kry}$, where x_0 is the smallest of this x for a given z. In general, the sum $\sum_k \omega^{kry}$ oscillates around zero, except in the close vicinity of the y such that

$ry \equiv 0$ modulo 2^n (which are not necessarily integer), and in this case, the sum shows a large isolated peak. The peaks are regularly spaced like in optical constructive interferences. Any measure on this sum will have a probability amplified on these peaks and give an indication on the value of r. A more thorough analysis over repetitive measurements will show that the ratio $r/2^n$ can be approached with an error smaller than $\frac{1}{2^n}$ and indeed give the true value of the order r.

Collecting all operations, the computational cost of Peter Shor algorithm sums to $O(n^2 \log n)$.

7.4.3 *Quantum computer versus quantum simulator*

The big limitation in the design of quantum computer stands in the hardness to build relatively long chain of qubits which stay coherent for enough long time to make the computations. A quantum state remains magic as long as it is a superposition of pure state, but the magic is gone when it is measured, and the wave function is projected into a single state (we say it collapses), and we can no longer take benefit of state superposition and parallel computation. But for the number factorization, one needs long qubit chains and numerous computation steps. A qubit made of electron spin would collapse as soon as it absorbs enough strong photon which will play the role of a measurement. Any thermal photon would destroy the qubits. This is the reason why many quantum computers require to be cooled at very low temperatures. A number with n digits would require n qubits at least and an order of n^2 of steps (gates); this implies that for the same temperature the quantum computer will statistically fail n^3 times more often than for the factorization of four digit numbers (the maximum attained to date). Knowing that n must be of the order 1,000 in order to have some practical interest, this illustrates the difficulty.

Contrary to quantum computers, quantum simulators (also imagined by Richard Feynman [37]) are not expected to produce results with accuracy beyond the physical size of the parameter. To make it short, the largest non trivial factor of a number N is merely of order $\sqrt{N}$ which is more than negligible compared to N. Therefore, if one has to use an analog representation of N, then the analog representation of one of its factors would be too small to be distinguishable from noise, thus the source of many of the problems.

The quantum simulators use the qubits to build an analog representation of the system to simulate, e.g. system of particles and the quantum gates are used to simulate the interactions of the particle. The resulting physical effect is supposed to be of the same order of magnitude as the size of the simulated system, but the systems are too complicated to be simulated on classic computer. The quantum simulator is perfect to simulate quantum systems, such as the arrangement of quantum states of several particles in various constrained situations which would happen in extreme situations (impossible to experiment in lab).

7.5   Entanglement, the End of the Hidden Variables and the Paradox of the Non-locality

When particles are originated from the same physical event, they are very intimately correlated. This is the example of a pair of photons emitted from a pair of electrons with opposite spins (see Figure 7.12), for example, from a Cesium source). The photons are supposed to be "aligned" in the sense that if one photon polarization is measured according to a given angle, then the second photon will show the same polarization if measured on the same angle, even if the second measure occurs on the other side of the room. This property comes from the fact that the two photons are in Bell state Ψ with $\Psi = \frac{1}{\sqrt{2}} |+, +\rangle + \frac{1}{\sqrt{2}} |-, -\rangle$. The interpretation of the Bell state is that if one photon is measured "+", which happens with probability $\frac{1}{2}$, then the other photon is also measured "+". If it is measured "−", then the other one is also measured "−".

Using the change of base introduced in (7.3), we note that for any arbitrary measurement angle θ, Ψ remains in Bell state: $\Psi = \frac{1}{\sqrt{2}} |\theta+, \theta+\rangle + \frac{1}{\sqrt{2}} |\theta-, \theta-\rangle$. Thus, if the photons are measured on

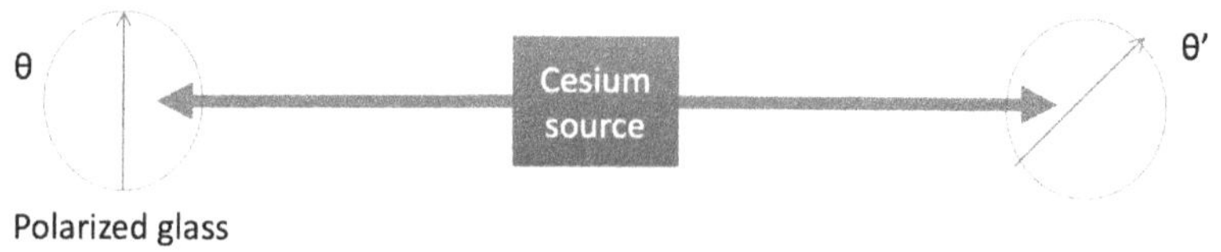

Fig. 7.12.   A pair of correlated photons.

the same angle, they always agree. In the physics in the pre-quantum (classic) mode, this is not surprising since the identical polarization would come as a natural common heritage to both photons. But in the quantum mode, this is a paradox because if the result of a measurement is purely random, then how the second photon is informed of the outcome of the measurement on the first photon, is there any hidden telecommunication between the two particles? In fact, both measurements can occur exactly at the same time, such that the communication should occur faster than light.

This paradox, known as the EPR paradox [32], puzzled the physicist since 1935, the paper published presented the paradox in a different setting, and seemed to be a smoking gun against the hypothesis of the pure randomness in quantum measurement and a strong support to the hidden variables hypothesis. "God does not play dice" was the supposedly Einstein quote on this subject. Figure 7.13 describes the context of the experiment. The letter X denotes the result of the measurement on the photon on the left, and Y the result on the photon on the right. It turns out that the mutual information $I(X, Y) > 0$ even if the measurements of X and Y are simultaneous. We could fastly conclude that a faster than light communication channel exists with non-zero capacity. But this faster than light communication is just an illusion [34] which is permitted by information theory as we see now.

7.5.1 *Causality and information theory*

We can compare the photons as two twin kids who receive from their mother two identical shopping lists X and Y hidden in two envelopes. One kid goes to a mail in the West of the city, and the second kid goes to a mail in the East. Arrived in their respective

Fig. 7.13. The EPR paradox.

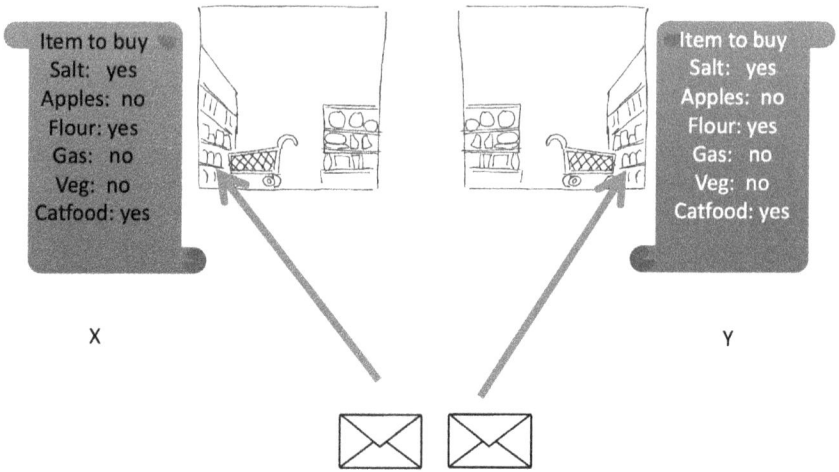

Fig. 7.14. The twin lists.

shopping center, the kids separately open their envelopes and discover their lists, supposedly at the same time. The kids receive the same information at the same time. The fact that $X = Y$ and consequently $I(X, Y) > 0$ hides the fact that the list has been created by the same person, the mother, who plays the role of the hidden variable (see Figure 7.14). In fact, information theory does not care about faster than light information, or even causality. In the expression of a classic channel capacity with X the transmitted codeword and Y the received codeword, the capacity is given by the formula $I(X, Y) = h(X) + h(Y) - h(X, Y)$ which remains the same if we reverse the time arrow so that Y is the transmitted codeword and X the received codeword.

The no telepathy or classical locality postulate: For what does not depend on their common past the twins behave independently. For instance, the common past only contains their identical purchase lists. But the respective orders in which the twins will collect the items can be arbitrary and will be independent.

Indeed, if we denote $\mathcal{P}(X)$ (resp. $\mathcal{P}(Y)$ for Y) the past of the measurement event X as being the set of events created in the backward light cone ending on X (resp. on Y), as shown in Figure 7.15, then the classical locality will imply that $I(X, Y | \mathcal{P}(X) \cap \mathcal{P}(Y)) = 0$. The hidden variables should be included in $\mathcal{P}(X) \cap \mathcal{P}(Y)$. We show that the

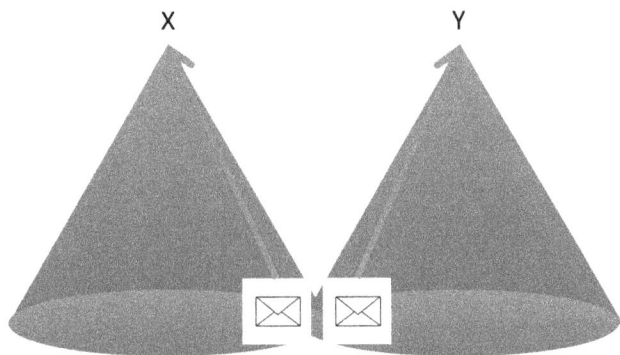

Fig. 7.15. The locality principle.

experiment on the twin photons fails on this law and together kills the hidden variable and the locality. This phenomenon has been coined as the "second revolution of quantum physics" by Alain Aspect, the physicist (Nobel Prize, 2022) who experimentally confirmed the non-locality [34].

7.5.2 *Bell inequalities versus experiment*

The foundation of the hidden variable theory in the twin photon experiment can be described as the "photon conspiracy". Before leaving the cesium atom, the twin photons agree on the answer to give for any possible measurement. In other words, for each possible angle θ of the polarization axis used for measurement, the photons agree on their answer "+" or "−". The answer are of course chosen so that if the two photons are measured within the same angle they will agree (see Figure 7.16). Under this setting, the process in appearance would give up its pure randomness component since the answer is predetermined (however, the predetermined list could be obtained from a random process).

If the measurement was concerning only one photon, then there would be no way to hold or refute the hidden variable hypothesis since only one axis would be involved. But the existence of the pair allows us to get the answer of each photon on two axes. Indeed, it is possible to measure the polarization on an angle θ together with the polarization on angle θ'. For this purpose, it suffices to measure the left photon on angle θ and the right photon on angle θ'.

Fig. 7.16.   The twin photon conspiracy list.

Fig. 7.17.   The whole area of all possible photon conspiracy lists.

The conspiracy hypothesis assumes the existence of a predetermined joint probability distribution of both measurements, and this will be instrumental for the Bell inequalities.

Assume one wants the measure the polarizations over a family of three angles: $\{\theta_0, \theta_1, \theta_2\}$. For instance, we take the set of angles $\{0, \frac{\pi}{6}, \frac{\pi}{3}\}$. Let the plain square illustrated in Figure 7.17 be the universe of all the conspiracy lists built over the answer to the measurement along the three angles.

Figure 7.18 illustrates the various splits of the list universe with the respective answers for the measurement along the angles in $\{\theta_0, \theta_1, \theta_2\}$. Each of the subarea corresponds to the split for the answer "+" and "−". For example, for the angle θ_1 (middle), the answer "−", denoted θ_1-, corresponds to the upper rectangle and the answer "+", denoted θ_1+, to the lower rectangle. The surface of the θ_1- rectangle should be proportional to the probability $P(\theta_1-)$,

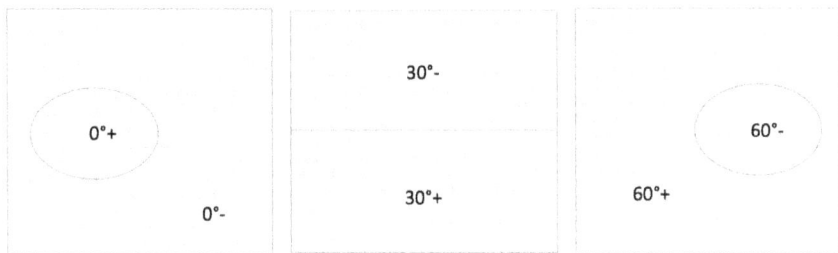

Fig. 7.18. The twin photon conspiracy list universe split with response to angle 0 (left), angle $\frac{\pi}{6}$ (middle) and angle $\frac{\pi}{3}$ (right).

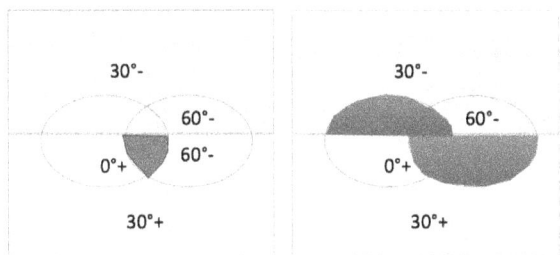

Fig. 7.19. The area covered by $P(\theta_0+, \theta_2-)$ (left) and the area covered by the Bell upper bound $P(\theta_0+, \theta_1-) + P(\theta_1+, \theta_2-)$ (right).

while the surface of the θ_1- rectangle should be proportional to $P(\theta_1+)$ in the same proportion. The surface of the area of the intersection of the area θ_0+ with the area θ_1- should be proportional to $P(\theta_0+, \theta_1-)$, etc.

The idea under Bell inequality is to give an upper bound of the probability occurrence of the list satisfying jointly $\theta+$ and $\theta'-$ with the sum probabilities of the list satisfying other joint requirement.

Indeed, we show that $P(\theta_0+, \theta_2-) \leq P(\theta_0+, \theta_1-) + P(\theta_1+, \theta_2-)$. This is a rather trivial inequality since it consists to cover the area displayed on Figure 7.19 (left) by the areas displayed in Figure 7.19 (right).

The good news is that every term of the Bell inequality can be precisely computed in the quantum framework. Indeed, one has the base identity:

$$\begin{cases} |\theta-\rangle = \cos(\theta' - \theta)\, |\theta'-\rangle - \sin(\theta' - \theta)\, |\theta'+\rangle \\ |\theta+\rangle = \sin(\theta' - \theta)\, |\theta'-\rangle + \cos(\theta' - \theta)\, |\theta'+\rangle \end{cases}$$

which lead to the new expression of Bell state:

$$\Psi = \frac{\cos(\theta' - \theta)}{\sqrt{2}} \, |\theta-, \theta'-\rangle - \frac{\sin(\theta' - \theta)}{\sqrt{2}} \, |\theta-, \theta'+\rangle$$
$$+ \frac{\sin(\theta' - \theta)}{\sqrt{2}} \, |\theta+, \theta'-\rangle + \frac{\cos(\theta' - \theta)}{\sqrt{2}} \, |\theta+, \theta'+\rangle .$$

which implies that

$$\begin{cases} P(\theta-, \theta'-) = P(\theta+, \theta'+) = \dfrac{(\cos(\theta' - \theta))^2}{2} \\[2ex] P(\theta-, \theta'+) = P(\theta+, \theta'-) = \dfrac{(\sin(\theta' - \theta))^2}{2}. \end{cases}$$

Using the above identities makes the Bell inequality to the contradictory

$$P(\theta_0+, \theta_2-) \le P(\theta_0+, \theta_1-) + P(\theta_1+, \theta_2-) \Rightarrow \frac{3}{8} \le \frac{1}{8} + \frac{1}{8} !!$$

Thus, quantum physics just shoots down in flames the locality principle. We indeed have $I(X, Y | \mathcal{P}(X) \cap \mathcal{P}(Y)) \neq 0$, as shown in Figure 7.20. They remained to make the experiment in real and they confirmed to the needed accuracy the violation of Bell inequalities, killing in passing locality and hidden variables. To be sure that the settings θ and θ' of the measurements were not possibly affecting the hidden variables, the two legs of the photon path were made large enough that the angles θ and θ' were randomly determined during the photon flights and could not retroactively affect any possible hidden variable sitting on the Cesium source.

Although the correlations between the photon are faster than light, we stress again that $I(X, Y) \neq 0$ does not actually means that there is an information transfer between the events X and Y confirming what we rehashed with the twin purchase list. To have an information transfer, we should replace X by the setting θ of its measurement and see if there is an information transfer between θ and Y. But we have seen that whatever the value of θ, $P(Y = +) = P(Y = -) = \frac{1}{2}$. In other words, $I(\theta, Y) = 0$. We show in the following chapter that an actually faster than light telecommunication would be equivalent to quantum unitary violation.

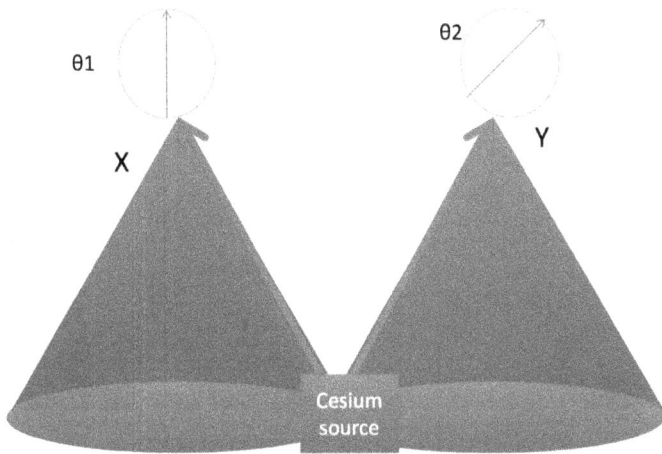

Fig. 7.20. The end of the locality principle.

Other point: After measurement, the Bell state is destroyed and the photons' polarization is projected on their respective measured states and will have independent life hereafter.

7.5.3 *Entanglement against eavesdropping*

One of the applications of entangled photon is perfect end-to-end encryption. During the cold war (again), the "Washington–Moscow Direct Communications Link", nicknamed "hotline" or "red phone", was an encrypted communication line between the US government and the Soviet government. This was supposed to help de-escalation in case of critical situation risking to get out of hands. This was established after the Cuba crisis where the two super-powers were brought on the brink of a nuclear all-out war.

The hotline was no telephone but a teletype equipment (not even painted in red in order to look more dramatic, see Figure 7.21). The secrecy of the communication was protected by a one-time pad. Namely, each text X was XORed with a pre-established sequence of random bits Y known on each end of the channel. The protection is maximal because it is impossible to decipher on random bits. Indeed, $h(Y) = h(X \oplus Y) = h(Y \oplus X | X) = n$, where n is the length of the text to be exchanged. Indeed, $I(X, X \oplus Y) = h(Y \oplus X) - h(Y \oplus X | X) = 0$.

Fig. 7.21.   Early red phone hotline device as a teletype installed between Moscow and Washington in 1963.

Since the sequences were unique and used only once, they were exchanged via diplomatic means prior to each period of communication. In the early times, it could be volumes and volumes of symbols. Now the exchanges are protected via secured computer lines necessitating less volumes of secret keys. But entanglement could have been a perfect solution to this problem.

Indeed, assume Cesium-like atoms are the source of pair of entangled photons; every pair is split so that one twin goes to correspondent A (Alice) and the other twin goes to B (Bob). Alice and Bob measure their received sequence of photon polarization according to a pre-agreed angle θ which makes a binary sequence Y. In the absence of error or synchronization shift, the sequences should be the same on each end of the channel. Let X be the message Alice wants to send to Bob. Alice XORs X with Y and sends $Z = X \oplus Y$ via a regular (unprotected) channel. Bob retrieves the message X by XORing Z with its own copy of the sequence Y (see Figure 7.22). Although the measurement on the entangled photons can be simultaneous, the communication is not faster than light because the channel carrying $X \oplus Y$ is classic.

As the entanglement saves the volume of pre-exchanged keys needed for the early hotline, it does not fully protect from "eavesdropping". Eavesdropping is the action which consists in that a third actor C (Carol) intercepts the flow of secret keys Y in order to capture the

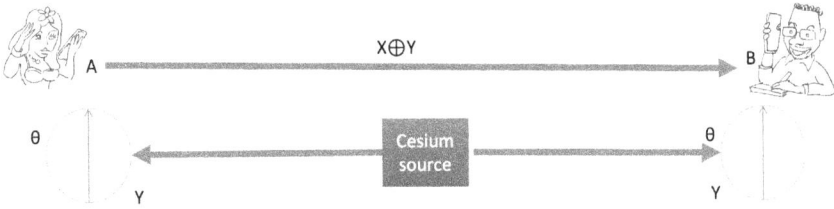

Fig. 7.22. Alice and Bob communicating via a protected channel. Bob reconstructs Alice's message X via the entangled sequence Y.

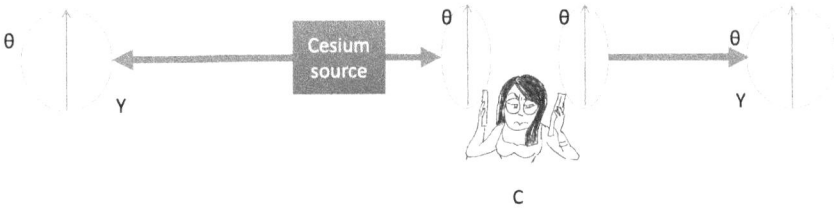

Fig. 7.23. Carol intercepts the entangled flow Y and re-injected non-entangled photons with same polarization. However, the eavesdropping is detected via Bell inequality.

secret message X. In theory, the entanglement is one shot, if one photon is measured in flight to Alice or Bob, then the entanglement magic is gone, the photons are dissociated. But if Carol knows the secret angle θ with which the photons will be measured by Alice and Bob, nothing will prevent Carol from measuring the photon on angle θ and then re-injecting a new photon with the same polarization along with angle θ so that neither Alice nor Bob would detect that an eavesdropping is present.

In fact, entanglement precisely allows us to detect eavesdropping. From time to time, during pre-determined intervals, Alice will interrupt transmitting secret messages and instead send the segment of Y without XORing. On the other end, Bob measures its incoming sequence of photons with a local and secretly varying angle θ'. If the pairs of photon are entangled at both end, then they should violate the Bell inequality. If not, it means that Carol (see Figure 7.23) re-injects non-entangled photons and there is eavesdropping. Bob can notify Alice by an alarm (which could be sent via a return channel over entangled photons).

7.6   Quantum Teleportation

Teleportation is the name of a possible technique to transport objects without moving. It has been popularized in the SciFi TV series "Star Trek". The urban legend tells that it has been introduced as an economical trick in order to avoid the complexity of the special effects required to visually simulate the landing and take-off of a space ship to or from a new visited planet (see Figure 7.24).

In this section, we quickly survey the principles of quantum teleportation. The quantum teleportation consists of implanting in a remote particle the unknown quantum state of another distant particle. Doing so, the quantum state of the distant particle is erased so that the "no-cloning" theorem is not violated. The teleportation takes advantage of quantum entanglement and uses a two-bit classical channel.

The surprise is that two bits suffice in order to transmit a quantum state which *a priori* is defined by continuous scalars, requiring in theory a channel of infinite capacity. In fact, this confirms that the quantum state is *no* information. The quantum state which moves in the teleportation experiment is an electron spin, and the only information it can deliver after measurement is only one bit, therefore teleportation is consistent with information theory.

Anton Zeilinger *et al.* (awarded Nobel Price in 2022 together with John Clauser and Alain Aspect) invented and experimented the principle of quantum teleportation in 1997 [124]. It is based on entangled electrons because there is a need to have particles which can

Fig. 7.24.   Crew teleportation in Star Trek.

interact between each other, which cannot be achieved by photons. The principle of electron entanglement is similar to the principle of photon entanglement with the difference in the trigonometric identities in the base changes ($\frac{\theta}{2}$ instead of θ).

To make the teleportation, we need two entangled electrons: one reaching Alice, the electron A, the other reaching Bob, the electron B. The quantum state of the pair AB is the Bell state $\Psi_{AB} = \frac{1}{\sqrt{2}} |-,-\rangle_{AB} + \frac{1}{\sqrt{2}} |+,+\rangle_{AB}$ since we assume that some subsequent measurements will be done by Bob on a given spin angle θ and Alice knows the angle. A third electron, C, is hosted by Bob with an unknown quantum state $\Psi_C = \alpha |-\rangle_C + \beta |+\rangle_C$. For a change, Bob will be the transmitter. At the beginning of the process, the quantum state of the full system before interaction is just the tensor product $\Psi_{ABC} = \Psi_{AB} \otimes \Psi_C$, see Figure 7.25.

The first operation is to let the electrons B and C interact via their electromagnetic field. The surprise is that the local interaction affects the quantum state of electron A, since the transformation of Ψ_{ABC} results into Ψ'_{ABC} as follows (see Figure 7.26):

$$\Psi'_{ABC} = \begin{cases} \dfrac{1}{2} \left(\alpha |-\rangle_A + \beta |+\rangle_A \right) \otimes |+,+\rangle_{BC} + \\[2mm] \dfrac{1}{2} \left(\beta |-\rangle_A + \alpha |+\rangle_A \right) \otimes |+,-\rangle_{BC} + \\[2mm] \dfrac{1}{2} \left(\beta |-\rangle_A - \alpha |+\rangle_A \right) \otimes |-,+\rangle_{BC} + \\[2mm] \dfrac{1}{2} \left(\alpha |-\rangle_A - \beta |+\rangle_A \right) \otimes |-,-\rangle_{BC} . \end{cases}$$

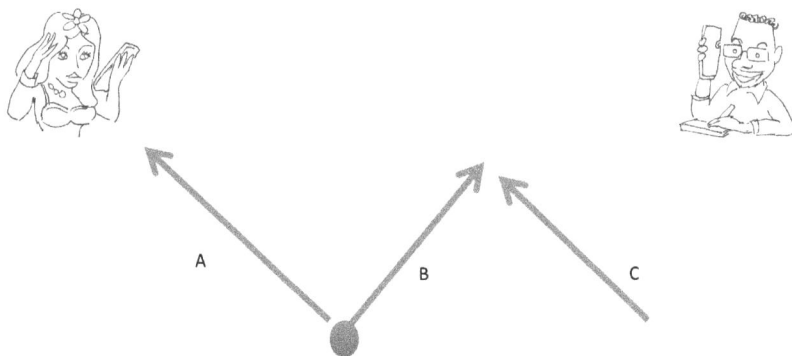

Fig. 7.25.   Teleportation of electron C via entanglement of A and B.

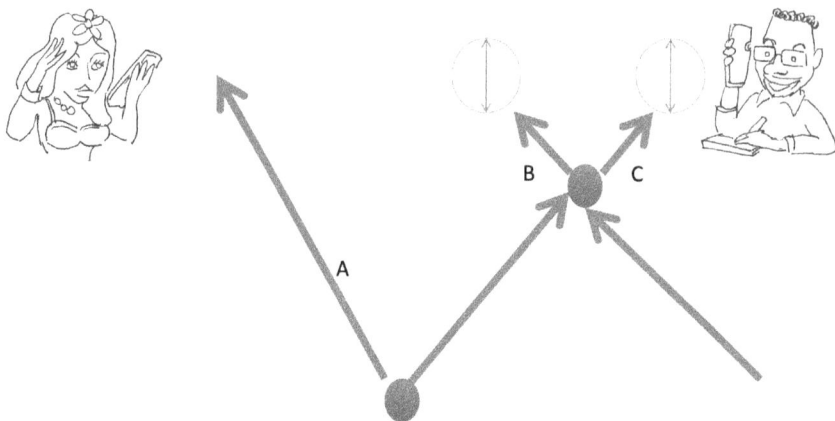

Fig. 7.26.   Local operation via electron B for the teleportation of electron C.

At this level, we note that we can find the state of electron C in the final state of electron A. In fact, it comes as a *fact simile* in the first component in the superposed state Ψ'_{ABC}, the other component being some rotations and symmetries. However, it does not contradict the no-cloning theorem since the initial state of electron C has been erased in the final state of electron C. To remove the ambiguity, Bob will measure both electron B and electron C spins on angle θ and send to Alice the result via a classical 2-bit channel. If the measure spin tuple is $(+, +)$, then electron A state is exactly the state of electron C before the manipulation; this very state has been destroyed by the measurement. If the measured spin tuple is $(+, -)$, one must do a diagonal symmetry on electron A to recover initial C state. For each outcome, there is a rotation or symmetry which recovers C (see Figure 7.27).

Contrary to popular belief, the quantum teleportation does not happen faster than light since the ultimate twist to get the true copy occurs with the help of a classical channel. The no-cloning theorem is not violated because in any case the copy of C is in the forward light cone of electron C after the measurements (the no-cloning theorem allows that a particle is a copy of its past version).

In the exercise, we give some hints about the manipulation on the system ABC via qubit operations as with a quantum computer.

We have described here the teleportation of a very simple electron spin state as it was discovered in 1997. It has been extended to

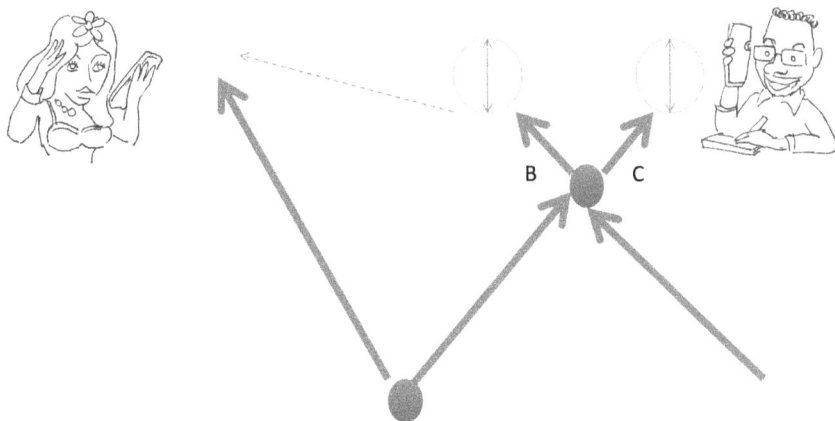

Fig. 7.27. Teleportation of electron C via entanglement finalized by a classical channel.

atoms [12, 100]: in 2004, the full spin state of an atom of Ytterbium (Yb) over one meter. Of course, it is still far from the teleportation of human as in Star Trek. And when it should be available, would people accept it since their body must be first destroyed in the process, in order to satisfy the no-cloning theorem.

7.7 Exercises

7.7.1 *Bell inequality*

An alternative to the violation Bell inequality is to compute the variance of $X + Y + Z$ in the photon conspiracy hypothesis, where X is the measure with angle θ_0, Y with angle θ_1 and Z with a third angle θ_2. This calculation is possible because the Bell state allows us to compute all pairwise angle joint distributions.

Exercise 40

What values can $X + Y + Z$ take? See the Answer of Exercise 40 in Chapter 9.

Now we assume a joint distribution of X, Y and Z.

Exercise 41

What is the value of $E[X + Y + Z]$? See the Answer of Exercise 41 in Chapter 9.

Exercise 42

What is $E[(X+Y+Z)^2]$? See the Answer of Exercise 42 in Chapter 9.

Exercise 43

Find values for angles $(\theta_0, \theta_1, \theta_2)$ which lead to a *negative* variance. See the Answer of Exercise 43 in Chapter 9.

$E[(X + Y + Z)^2] = 0$ which implies that $X + Y + Z \equiv 0$ which is impossible. The hypothesis about the existence of an *a priori* joint distribution is wrong.

7.7.2 *Entanglement*

If we consider that Bob's electrons B and C are qubits but with the distinction that qubit B is entangled with electron A, thus AB is a Bell qubits. We consider the unitary operator $\mathcal{U}_c$ (for cNOT) on Bob qubits:

$$\mathcal{U}_c \left|x, y\right\rangle = \left|x \oplus y, y\right\rangle.$$

Exercise 44

Show that the application of $\mathcal{U}_c$ on the two Bob's qubit gives the following result on the three qubit (meaning that Alice qubit is left untouched). See the Answer of Exercise 44 in Chapter 9.

Exercise 45

Now, we consider the Hadamard operator $\mathcal{U}_H$:

$$\mathcal{U}_H \left|x\right\rangle = \frac{1}{\sqrt{2}} \left(\left|-\right\rangle - x\left|+\right\rangle\right).$$

Show that the application of Hadamard operator on B gives Ψ'_{ABC}, as shown in the teleportation the Answer of Exercise 45 in Chapter 9.

Chapter 8

Non-unitary Quantum Information

Abstract

When Stephen Hawking discovered the evaporation of black holes, he realized at the same time that this evaporation triggered an information loss paradox. To make it short, quantum physics prevents the loss of information because the evolution of quantum state is unitary and consequently reversible. If a support of information is swallowed inside a black hole, and if the black hole completely evaporates, then the information is lost. Thus, the physics around black holes should be non-unitary because they are non-reversible. Non-unitary physics allows some very much magic effects. A non-unitary computer would be capable of cracking any NP-complete problem in polynomial time. More interestingly, non-unitarity allows faster than light telecommunication, even the possibility to transfer information backward in time. We suggest to investigate this very exciting perspective which impacts the common sense. However, such perspective is strictly different than time travel and does not generate the classic time paradox depicted in science fiction literature thanks to the pure randomness of quantum measurement. We show that time backward information transmission is strictly equivalent to non-unitary quantum physics like in the non-cloning theorem.

8.1 The Quantum Accident

One of the most stringent paradox about a quantum system is when it escapes the continuous and linear evolution dictated by the Schrödinger equation, and this every time an observer triggers an accidental measurement. As long as no measurement happens,

measurement

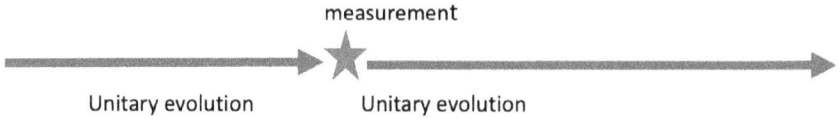

Unitary evolution   Unitary evolution

Fig. 8.1.   The quantum accident.

the system evolves via the continuous application of an unitary linear operator. The operator can be as complicated as we can imagine due to the interaction and quantum entanglements with much larger systems, but it remains, in theory, perfectly reversible. The arrow of time does not impact the nature of the system. But everything collapses when the slightest measurements occur (see Figure 8.1).

The Copenhagen school under the direction of Niels Bohr [54] defined the early theoretical framework of quantum physics. It was admitting that the fact of *observing* (namely, measuring) a quantum system was the cause of the wave function collapse. The following transition shows how the state of spin superposition $\frac{1}{\sqrt{2}}\left|-\right\rangle + \frac{i}{\sqrt{2}}\left|+\right\rangle$ would naturally evolve into a pure state just by the presence of the observer:

$$\frac{1}{\sqrt{2}}\left|-\right\rangle + \frac{i}{\sqrt{2}}\left|+\right\rangle \quad \begin{array}{c} \overset{\frac{1}{2}}{\nearrow} \quad \left|-\right\rangle \\[2ex] \underset{\frac{1}{2}}{\searrow} \quad \left|+\right\rangle \end{array}$$

The idea that an observer with a psyche could be the only cause of the collapse was hard to accept, and this subjective theory of quantum physics didn't fly. When an entangled pair of particles interact with photon or other particles, then the quantum wave function collapses on the measured state and is no longer a superposition on all possible states and *a fortiori* is no longer entangled. There is no need of the intervention of an human observer. This is the reason why it is so difficult to build a quantum computer with arbitrary long chain of qubits. Since the slightest photon would destroy the harmony, this is the reason why ultra low temperatures would be needed.

Since it is hard to imagine that the planet Jupiter would not have a definite position in the sky unless we measure its photon emissions, the problem is what would be the largest size of a system that could be kept in superposition for ever without measurement. Erwin Schrödinger invented the paradox of the dead-alive cat, which everybody knows (in particular this story delights dogs) [104]. As long as the cat remains in its closed box, it remains in a superposed state *dead and alive*; when the box is opened, the state collapses into dead *or* alive. As long as the box does not exchange photons with the outside universe (and as long as there is enough oxygen), the cat remains in superposed state. The paradox was invented in order to make fun of the theory, since no body could imagine that a macro object like a cat could remain in state superposition for a long time. It is amusing to imagine that the cat system stands half way between the planet Jupiter system and the single particle systems in terms of order of magnitude of their sizes. In other words, one needs to pile up the same number of cats to get the mass of Jupiter, as the number of atoms to obtain a cat, that is, an order of 10^{26}.

8.1.1 *The many-worlds interpretation*

The main source of the paradox is how a theory which is essentially linear between measurement suddenly becomes nonlinear when a measurement occurs. The origin of the concept of *many-worlds* is an attempt to unify both situations [35]. Despite a controversial debut, the many-worlds concept remains to date the most solid interpretation of quantum physics which re-conciliates the two contradictory aspects of quantum physics. To make it clear, there are controversies only when one tries to extend it further than a pure theoretical interpretation.

Let's look closer to the sequence of events during a measurement. We consider a "target" electron with the neutral state $\Psi_T = \frac{1}{\sqrt{2}}|-\rangle_T + \frac{i}{\sqrt{2}}|+\rangle_T$. The observer device is made of a tremendous large number of atoms merely one thousand times the Avogadro number, 10^{26}, for a device of few kilograms, and an equally large number of electrons. For example, before the measurement, we can suppose that their wave functions are all equally neutral. Let Ψ_D be the state function of the observer device before the measurement, thus $\Psi_D = \left(\frac{1}{\sqrt{2}}|-\rangle + \frac{i}{\sqrt{2}}|+\rangle\right)^{\otimes 10^{26}}$ and the state function of the pair target-device is $\Psi_T \otimes \Psi_D$ before they interact. When they interact, the spins in the device tend to be aligned with the spin of the target just by electromagnetic interaction. After the interaction, the state function of the pair target-device becomes $\frac{1}{\sqrt{2}}|-\rangle_T \otimes |-\rangle^{\otimes 10^{26}} + \frac{i}{\sqrt{2}}|+\rangle_T \otimes |+\rangle^{\otimes 10^{26}}$. Believe it or not, the transition is fully unitary. The duration of the transition is of the order of the diffusion of the electromagnetic field in the device.

There is a tradition in the quantum theorician world to involve the quantum state of the universe (nothing less). Let Ψ_U be the state function of the universe minus the target electron. The state function of the universe before the interaction between the target electron and the device, i.e. the rest of the universe, is $\Psi_T \otimes \Psi_U$. After the interaction, it is $\frac{1}{\sqrt{2}}|-\rangle_T \otimes \Psi_{U-} + \frac{i}{\sqrt{2}}|+\rangle_T \otimes \Psi_{U+}$, where Ψ_{U-} (resp. Ψ_{U+}) is the state function of the rest of the universe if the measure of the target spin is "$-$" (resp. if the measure of the target spin is "$+$") and the result of the measure has propagated in the universe (or at least in the forward light cone of the interaction event). The result of the measure is first inscribed in the day-log of the lab and then, second, may propagate further to be taught in the university classes. There is no limit of the propagation although a single spin may little change the fate of the universe.

At this level, one interpretation is to consider that the universe "forks" as the measurement result propagates and becomes the superposition of two distinct states: $|-\rangle_T \otimes \Psi_{U-}$ and $|+\rangle_T \otimes \Psi_{U+}$. These states are called "timelines", and there are two interpretations:

- either the timelines coexist by superposition, both have probability weight $\frac{1}{2}$,
- or only one timeline exists and the choice of the actual timeline follows the probability vector $(\frac{1}{2}, \frac{1}{2})$.

The second interpretation leads to the already described *measurement accident*, while the first interpretation is the multiverse interpretation and avoids this accident. The fact that in the first interpretation the two timelines coexist is not an invitation for schizophrenia because as a conscious entity with a single memory, we are only aware of one timeline, and we keep only one souvenir of the result of each measurement. The two timelines don't physically interfere because they are just two different eigenvectors of the same universal Hamiltonian operator. Thus, for an individual point of view, the alternative timelines are always virtual, and finally the

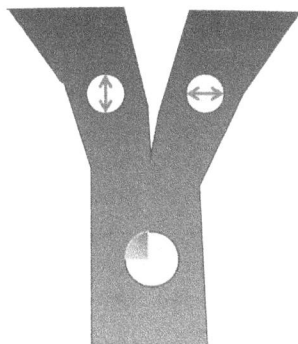

Fig. 8.2. The universe forks on photon polarization: The left timeline contains the "+" measurement, the right timeline contains the "−" measurement. Before the measurement, the polarization is in superposed state.

two interpretations are just a matter of feeling, the computation of measurement probabilities remains the same. Figure 8.2 shows the fork on the photon polarization measurement.

The many-worlds interpretation (MWI) has been proposed first by Everett in 1957 and is at the edge of vertiginous perspectives: If every time a measurement occurs the universe fork, then there are zillions of forks every day. Indeed, if you decide to choose the color of your sockets or the color of your dress as a function of a spin measurement, then every morning when you wear a red dress or red sockets, there is a virtual copy of yourself wearing a blue dress or blue sockets in an alternate timeline. The fact that this alternative timeline is real or not is not important. Indeed, most scientists consider that the many-worlds interpretation is *infalsifiable*, i.e. not testable, thus we can see this interpretation as just a pure curiosity.

The translation of the Schrödinger cat paradox in MWI results in the following sequence of modification of the universal wave function, starting from the original state superposition $\Psi_0 = \left(\frac{1}{\sqrt{2}} |-\rangle_T + \frac{i}{\sqrt{2}} |+\rangle_T \right) \otimes \Psi_{\text{cat}} \otimes \Psi_U$ which corresponds to the spin not yet measured in the box containing the cat:

$$\Psi_0 \to \left(\frac{1}{\sqrt{2}} |-\rangle_T \otimes \Psi_{\text{cat}-} + \frac{i}{\sqrt{2}} |+\rangle_T \otimes \Psi_{\text{cat}+} \right) \otimes \Psi_U$$

$$\to \frac{1}{\sqrt{2}} |-\rangle_T \otimes \Psi_{\text{cat}-} \otimes \Psi_{U-} + \frac{i}{\sqrt{2}} |+\rangle_T \otimes \Psi_{\text{cat}+} \otimes \Psi_{U+}.$$

The first transition is when the spin is measured and its lethal conse-quence if "−" or not lethal "+" has been applied to the cat but the box is not yet opened. In the last transition, the cat box is opened and the result of the measurement propagates to the rest of the universe.

The main problem with MWI is that the sum of the two alterna-tive state functions of the universe implies that least the sum oper-ates on the same common geometric space. It is not hard to have this condition when the superposition is between two states which share the same space-time geometry. But the consequence of the spin mea-surement may affect the distribution of masses in the universe and by virtue of the general relativity, mass distribution affects geom-etry. In passing this problem occurs also for basic quantum theory because the superposition of two state functions of a massive particle in theory may affect the embedded geometry at quantum level in an unknown way.

An argument against the many-worlds interpretation is the zero entropy paradox. But in fact it can be worked out with the funda-mentals of information theory.

8.1.2 *The zero entropy paradox*

A traditional and simplified vision of the big bang theory tells that at its early stage the universe was uniformly and spherical symmet-ric, like the "primeval atom" imagined by Georges Lemaitre in 1931 supposedly constituted of pure energy. The level of energy was so high that the elementary forces should be merged into a single force and no other individual particles could exist. This absence of diver-sity should lead to a zero entropy state. As long as the temperature cooled, the symmetries broke one by one to create diversified forces and particles existed and started to interact.

Every time a spins or polarizations interacted in the early uni-verse, a symmetry broke and the universe became more asymmet-ric and consequently more complexified. But with the many-worlds interpretation, at each interaction, the universe should have forked based on alternative outcomes, in a way that the superposition of all timelines should be fully symmetric and uniform. In a conse-quence, the superposition of all multiverse should have no von Neu-mann entropy and should stay at zero entropy. Like a crushed egg, each fragment shows an intricate shape, but when collecting and reassembling all fragments, we get back to a smooth round egg. Mathematically, since the many-worlds evolve via unitary operations which are time reversible, the entropy remains unchanged. And if the universe starts as a pure quantum states, then its von Neumann entropy is zero and will remain as such because the many-worlds quantum state will remain a pure state.

But this may look not in line with information theory where the quantity of information should permanently increase. This is a new twist between quantum physics and information theory which again confirms that a state superposition can carry information but it is not information *per se*. But as we saw in the previous chapter, the information is revealed only in measurements, i.e. in the many-worlds terminology, the information remains confined in the timeline and no information is shared between parallel timelines. Furthermore, the entropy increase is not only due by irreversible measurements but also by "ignorance" effect even within any reversible evolution, which is indeed the foundation of information theory [75].

8.1.3 *Divorce, quantum style*

In Pietro Germi's movie *Divorzio all'Italiana* ("Divorce Italian Style", 1961, Figure 8.3), Marcello Mastroianni and Daniela Rocca played the role of a Sicilian married couple with a definitively extinct passion. The husband, Ferdinando, is now in love with a much younger woman and decides to plot the murder of his wife Rosalia. The reason is that at this time, divorce was illegal in Italy, hence a deep catholic country. But the husband thinks that the judges will be indulgent if the murder is motivated by a frustrated passion. Thus the husband's plan is to push his wife to have an adulterant relation with an other man in order to justify his future murder in *flagrante delicto* in what would look like an "honor killing". The first part of the plan, the adulteran attraction, is successful, but to her turn the wife might be too attracted by a divorce Italian style. Of course, nothing will happen as expected.

Here we propose the quantum many-worlds style of the divorce. Assume that Ferdinando and Rosalia hate each other so much that none of them can imagine to live in an universe where the other is also alive. The quantum divorce would work as follow. Between the couple (reunited for the last time) you dispose two loaded guns, one

Fig. 8.3.   Divorce Italian style (1961 poster).

aimed to the husband, the other aimed to the wife. A spin is measured and a mechanism makes that if the spin is "+", the husband is shot, if the spin is "−", the wife is shot. It is similar to the mechanism in the Schrödinger cat box, with the difference that whatever the spin measure outcome there is one dead (see Figure 8.4).

There is a probability $\frac{1}{2}$ that Rosalia (resp. Ferdinando) be killed. In the many-worlds interpretation, it would mean that half of the universes contains Ferdinando alive, and the other half contains Rosalia alive, and none contains both alive, which is the expected goal of the operation (see Figure 8.5).

Fig. 8.4.   The quantum divorce mechanism with a double gun triggered by a spin measurement.

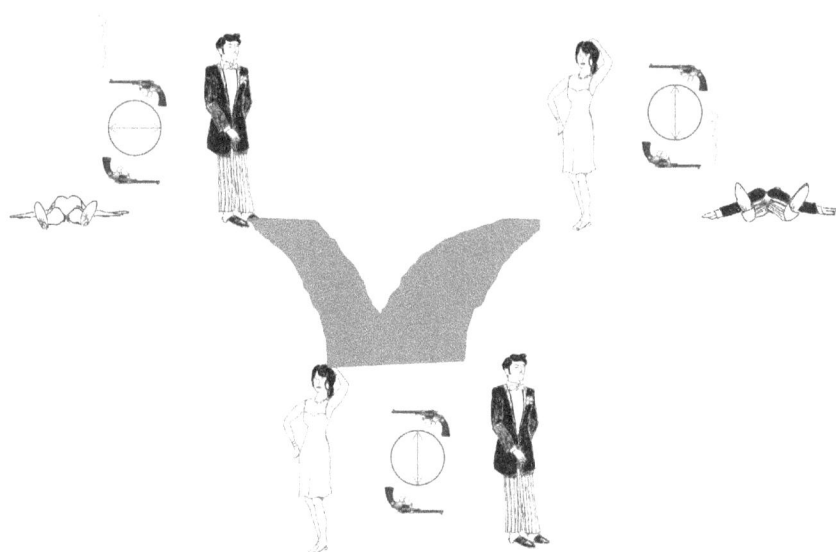

Fig. 8.5.   The quantum divorce with the survivors in different universes.

Of course, if one doesn't believe in the many-worlds interpretation, then it would be good luke, bad luke, if one would survive only in a non-existent timeline. In any case, not something to try at home.

8.2   Black Hole and Information

A black hole is a distortion of the space-time geometry resulting as a consequence of a too large concentration of mass in space. The gravity at the surface of the mass becomes strong enough that escape velocity becomes larger than the speed of light. The consequence of this statement is that a black hole, by definition, cannot emit lights, thus they cannot be directly detected, unless by the effect of their gravitational field at distance. Indeed, a black hole is a greedy stellar object which captures any sort of gravitating bodies in its vicinity, including stars and gas. During their lifetime, a black-hole mass can only augment (if we omit the so-called Hawking radiation, to be discussed later). The german physicist Karl Schwarzschild postulated the existence of black holes and built a theory about their feature based on general relativity [116]. In particular, a black hole should have a mass almost concentrated on a single point, since no known force, including electromagnetic forces, would be able to resist the gravitational pressure. Within a certain distance to the singularity, one can define the event horizon as the theoretical sphere around the mass within which the light cannot escape (see Figure 8.6). A simple analysis shows that for a given spinless and electrically neutral black hole of mass M, the radius R of the event horizon, also called the Schwarzschild radius, has expression

$$R = R(M) = \frac{2MG}{c^2},$$

where c is the speed of light and G the gravitational constant. In passing, we can note that this expression is *twice* the value if would have within pure Newtonian gravitation theory; this is a consequence of the twists brought by relativistic effects. The radius is 0.1 millimeter for an hypothetical black hole of the mass equivalent to the Moon (merely the size of a needle pick), 8 millimeter for the mass of the Earth, and 3 km for the mass of the Sun. The hyper massive

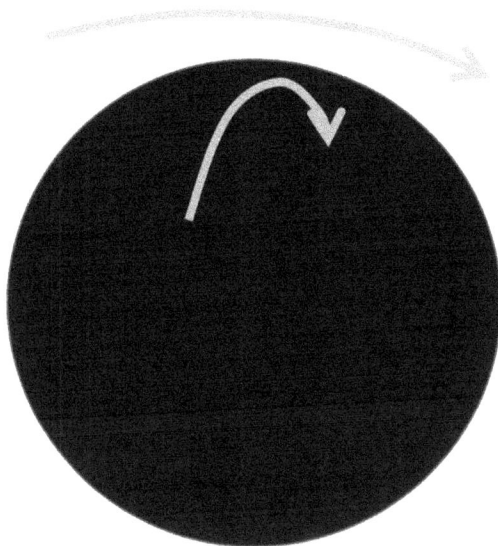

Fig. 8.6. Black hole event horizon where escape velocity is the speed of light.

black hole at the center of our galaxy, Sagittarius A^*, has a mass of 4 millions solar masses and an event horizon radius of 10 millions km.

In 1976, Stephen Hawking discovered that against all intuitions the black holes can emit some radiations due to the permanent creation of pair particles anti-particles on the event horizon boundary [52]. The radiation is so tiny that it is still undetectable with present and near future technology as it can be extrapolated from now. The Hawking radiation should not be confused with the radiation emitted by the matter (gas, stars) when they are circling close to the black hole just as a consequence of their acceleration (an object can loose 10% of its mass in radiated energy just because of the gravitational acceleration when absorbed by a black hole). The much more tiny Hawking radiations are just thermal and implies that the black hole has a temperature $T(M)$ and an entropy $H(M)$:

$$\begin{cases} H(M) = \dfrac{4\pi G M^2}{c\hbar} \\[2ex] T(M) = \dfrac{\hbar c^3}{8 k_B G M} \end{cases},$$

where $\hbar$ is the Planck constant and k_B the Boltzmann constant. We note that the entropy is proportional to the area of the black-hole horizon. After quick numerics, we see that the temperature of a Moon-sized black hole is merely 3 K, not distinguishable from the Cosmic Microwave Background radiation. The power per area unit is in $T(M)^4$. The radiation being localized on the event horizon whose area is in $R(M)^2$, this makes a very tiny radiated power, and it decreases proportionally to the square of the mass M.

8.2.1   *The limit of information storage: The Bekenstein bound*

We denote $h(E, r)$ the maximal quantity of information which can be stored with energy E in a volume of radius r. Jacob Bekenstein has shown in 1981 that for a mass and radius far larger from making a black hole,

$$h(E, r) = \frac{2\pi}{\hbar c} r E.$$

Similarly, if we convert the mass in energy with the formula $E = mc^2$, the limit for a mass m in a volume of radius r is $\frac{2\pi}{\hbar} rmc$. The proof of the bound is surprising since it makes use of black hole of mass M and event horizon radius just larger than r. We just give the proof for the orders of magnitude, the details of the exact constant is left aside. Assuming that the mass m is at distance r to the black hole event horizon, moving the mass within the event horizon, rise the mass of the black hole to $M+m$. Since the entropy of the system mass m and black hole does not decrease, we have the following inequality [15]:

$$h(E, r) + H(M) \le H(M + m).$$

Thus assuming $m \ll M$, we have $h(E, r) \le m\frac{\partial}{\partial M}H(M) = \frac{8\pi}{ch}GMm$. To get the minimal acceptable value M, we just need that $r \le R(M)$, that is, $r = 2\frac{GM}{c^2}$ and we get $h(E, r) = \frac{4\pi}{\hbar} rmc$. In fact, a careful analysis will show 2π instead of 4π; in this case, the bound coincides with the black hole entropy if m is the mass of the black hole and r is the radius of the event horizon.

One side effect of the Bekenstein bound is that indeed the black hole stores the largest concentration of information possible in a volume of radius r. This is not instrumental in Bekenstein proof, since

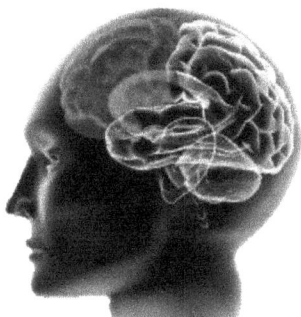

Fig. 8.7. The maximum quantity of information in the volume of a human brain is 10^{42} bits.

it only uses the fact that the entropy of a black hole is a pure function of its mass M, regardless of the complexity of the objects it has swallowed before reaching its mass M.

If we apply the Bekenstein bound to the human brain ($m = 1.5$ kg for 1260 cubic centimeter, thus $r = 6.7$ cm), then we get approximately

$$h(E, r) = 2.6 \times 10^{42} \text{ bits.}$$

Note that this huge quantity of information goes far beyond the memory capability of the human brain; it would also exceed all the complexity of the brain structure down to the molecular and quantum level (see Figure 8.7).

8.3 Time Travel and Information

8.3.1 *Dialogue in a classroom*

Imagine you face several students in a classroom and you ask the question: "Do you believe that time machines exist?"

STUDENTS: No, dear respected professor, this is a pure science fiction invention.

YOU: You're wrong. Look at this image (Figure 8.8): your body is a time machine which travels exactly at a speed of 60 seconds per minute.

Fig. 8.8.   The time traveler.
Courtesy: le penseur by Rodin.

STUDENT: But our body is only moving toward the future, it would become interesting if you could travel at will in both time directions, future and past!

At this moment you show an old remote control of the 1990s for VHS (without its VHS).

YOU: Although it does not look like very sophisticated, it is a "Time Displacement Device" which we will call in short, "TDD".

STUDENT: [*indistinct laughs*]

YOU: If I press on "rewind", then it instantaneously moves all the classroom, in fact the whole universe, 30 seconds backward in time.

STUDENT: [*more audible laughs*]

You press the rewind button, nothing happens. You now show a modern Blue Ray remote control.

YOU: OK this was an early TDD, which had 50% chance of failing. Now this new TDD has 90% success rate. [*You press the rewind button. Nothing happens*]

STUDENTS: [*Mocking*] It is not working! It is not working!

YOU: Are you sure?

STUDENT: Yes we are!

YOU: In fact you may be wrong. I explain: when I press the rewind key, we are all moved backward of 30 seconds and we are back in the situation as we were, half a minute ago, without any memory of what will happen 30 seconds later. Thus we proceed in the same time-line which would end after 30 seconds on me pressing the rewind key and we loop backward again.

STUDENT: But we should be trapped in a time loop? But we are not.

YOU: In fact, we might have looped several times (merely ten times [see Figure 8.9]) until the TDD fails and the timeline proceeds normally. But you would have no memory of the loops.

STUDENTS: But could we have observed that the time of the day have been eroded after so many loops?

YOU: No, we assume the we loop backward, all the universe, including the planets, loops backward.

STUDENTS: Thus your trick is useless, because we may have travelled backward in time, but we have no memory, it is as if it never existed. It is useless!

Here it is time that you acknowledge with humility that your remote controls are not actual TDDs, and you were just pulling the legs of your students. But before they launch rotten tomatoes to your face, you try:

YOU: If the TDDs were *really* working, I will explain how you could observe an effect of the loops.

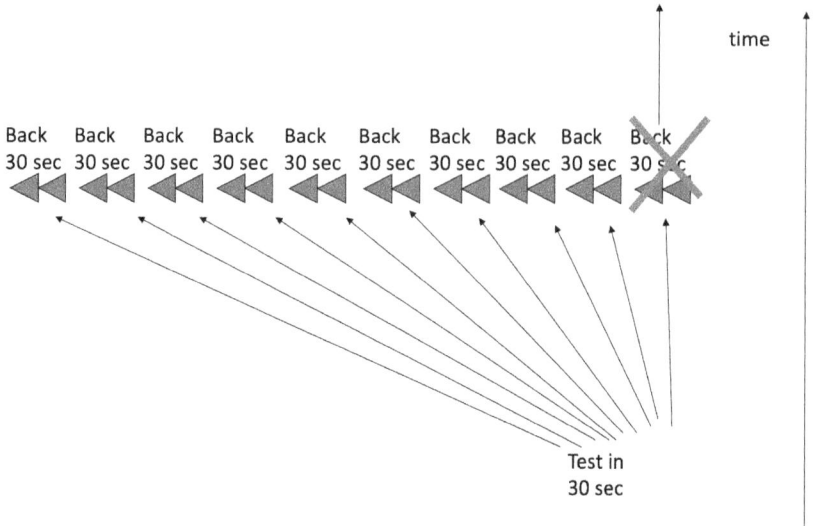

Fig. 8.9.   The repetitive loops before TDD failure.

And now you show a dice together with the TDD.

YOU: Now we imagine that the TDD works and we will play the
       following game: 30 seconds before the time to press the
       rewind, we roll the dice. If it is a 3, we press the rewind
       30 seconds after, otherwise we don't press the rewind. And
       we repeat the operation several times.

Then you draw the following table to simulate the operations while
repetitively rolling the dice.

roll	action
4	
2	
3	◁◁
6	
4	
1	
3	◁◁
5	
...	...

The symbol ◄◄ indicates that the rewind key is pressed 30 seconds later. If we develop the table in a limitless manner, we will see that if the TDD works 90% of the time, then the roll value 3 will disappear 90% of the time from the table since the loop back erases our memory.

All computations done we have $P(\text{roll} = 3) = \frac{1}{51} < \frac{1}{6}$ and $P(\text{roll} \neq 3) = \frac{50}{51} > \frac{5}{6}$. It implies that the effect of the TDD will be very noticeable in the roll statistics. The explanation of this figure is simple: The statistic of the 3 are erased nine times per ten rolls, thus

$$\begin{cases} P(\text{roll} = 3) = \dfrac{\frac{1}{60}}{\frac{1}{6} + \frac{1}{6} + \frac{1}{60} + \frac{1}{6} + \frac{1}{6} + \frac{1}{6}} \\[4mm] P(\text{roll} \neq 3) = \dfrac{\frac{1}{6} + \frac{1}{6} + \frac{1}{6} + \frac{1}{6} + \frac{1}{6}}{\frac{1}{6} + \frac{1}{6} + \frac{1}{60} + \frac{1}{6} + \frac{1}{6} + \frac{1}{6}}. \end{cases}$$

Of course, one must assume that the dice rolls correspond to simple and sufficiently random quantum measurements. A mechanical dice will not work because the feedback loop on the drawing of the number 3 will find the same nervous and muscular schema of the operator and the result of the roll will most probably be a 3. There is no question of using a computer dice either, since the random generators of computers are the product of a chain of purely deterministic arithmetic operations from an initial seed number.

On closer inspection, we see that the choice to go back when the result is 3 can be made just before activating the TDD. Alternatively, and this can be decided at the last second, it can be activated on a 2 or on another number. Therefore, the device can be used to send information backwards in time: from the future activation of the TDD to the last dice roll. As an exercise, the capacity of such retro channel, where X is the number decided to trigger the TDD and Y is the actual drawn number, is

$$I(X, Y) = \log_2 6 - \log_2 \frac{50}{51} - \frac{1}{10} \log_2 10 \approx 2.49125 \text{ bits.}$$

If the fail ratio of the TDD is ρ instead of $\frac{1}{10}$, we have

$$I(X, Y) = \log_2 6 - \log_2(5 + \rho) + \frac{\rho}{5 + \rho} \log_2 \rho.$$

It is interesting to extend to a dice with n sides. In this case, the capacity becomes

$$I(X,Y) = \log_2 n - \log_2(n - 1 + \rho) + \frac{\rho}{n - 1 + \rho} \log_2 \rho.$$

It is interesting to find the maximal value of the capacity $I(X,Y)$ given the loop back failure rate ρ. In fact, the maximum is interesting when at $\rho > 1$, which can no longer be interpreted as a failure probability. In fact, this would mean that in the example with the dice, the TDD is activated every time when the roll outcome is different of 3 instead only when it is equal to 3 as shown in the following table.

roll	action
4	◁◁
2	◁◁
3	
6	◁◁
4	◁◁
1	◁◁
3	
5	◁◁
. . .	. . .

This perfectly mimics the case $\rho > 1$ with TDD failure rate of $\frac{1}{\rho}$. In fact, we see in the following section that the TDD fictitious loop back capability is not necessary for such backward information transfer and in fact everything is based on unitarity violation. Meanwhile, easy computations rise the maximal capacity of the TDD system to be $I(X,Y) \approx \log_2 \rho$ (with an error of order $\log \log \rho$) and is attained with $n \approx \log \rho$ (natural logarithms).

We note that if we press k times the TDD, then the failure rate decreases in ρ^k. At the same time, the capacity of the retro-channel increases by a factor k.

8.3.2 *Unitarity and retro-information*

We see in this section that we can indeed forget about the fictitious TDD and look to the true hero of retro-information, that is, the unitarity violation.

If the dice were a pure quantum object, then its wave function would be

$$\Psi_0 = \frac{1}{\sqrt{6}}|1\rangle + \frac{1}{\sqrt{6}}|2\rangle + \frac{1}{\sqrt{6}}|3\rangle + \frac{1}{\sqrt{6}}|4\rangle + \frac{1}{\sqrt{6}}|5\rangle + \frac{1}{\sqrt{6}}|6\rangle.$$

And its density operator $\rho(t)$ would be at $t = 0$

$$\rho_0 = \begin{bmatrix} \frac{1}{6} & \frac{1}{6} & \frac{1}{6} & \frac{1}{6} & \frac{1}{6} & \frac{1}{6} \\ \frac{1}{6} & \frac{1}{6} & \frac{1}{6} & \frac{1}{6} & \frac{1}{6} & \frac{1}{6} \\ \frac{1}{6} & \frac{1}{6} & \frac{1}{6} & \frac{1}{6} & \frac{1}{6} & \frac{1}{6} \\ \frac{1}{6} & \frac{1}{6} & \frac{1}{6} & \frac{1}{6} & \frac{1}{6} & \frac{1}{6} \\ \frac{1}{6} & \frac{1}{6} & \frac{1}{6} & \frac{1}{6} & \frac{1}{6} & \frac{1}{6} \\ \frac{1}{6} & \frac{1}{6} & \frac{1}{6} & \frac{1}{6} & \frac{1}{6} & \frac{1}{6} \end{bmatrix}.$$

After the measurement just after $t = 0$, it would be (zeroes are omitted)

$$\rho_1 = \rho(t) = \begin{bmatrix} \frac{1}{6} & & & & & \\ & \frac{1}{6} & & & & \\ & & \frac{1}{6} & & & \\ & & & \frac{1}{6} & & \\ & & & & \frac{1}{6} & \\ & & & & & \frac{1}{6} \end{bmatrix}.$$

And after the TDD's activation,

$$\rho_2 = \rho(t) = \begin{bmatrix} \frac{1}{6} & & & & & \\ & \frac{1}{6} & & & & \\ & & \frac{1}{60} & & & \\ & & & \frac{1}{6} & & \\ & & & & \frac{1}{6} & \\ & & & & & \frac{1}{6} \end{bmatrix}$$

reflecting that the timeline of the state $|3\rangle$ is squeezed by a factor 10. The quantities $\text{tr}(\rho_0) = \text{tr}(\rho_1) = 1$ indicating that the evolution is unitary but $\text{tr}(\rho_2) = \frac{51}{60} \neq 1$, indicating a violation of unitarity.

Fig. 8.10. The timeline thinning for unitary violation.

Thus the TDD activation implies an unitarity violation. Figure 8.10 illustrates the fact the timeline of $|3\rangle$ is squeezed by a factor 10.

In fact, this figure suggests that we could read the effect of TDD via the many-worlds interpretation. The wave function of the universe before the measurement (assume the measurement occurs at time $t = 0$) is

$$\Psi_U(0) = |U_0\rangle \otimes \Psi_0,$$

where $|U_0\rangle$ is the universe wave function of everything other than the dice. After the measurement at $t > 0$, the universe splits into six wave functions $|U_1(t)\rangle, \ldots, |U_6(t)\rangle$, corresponding to the outcome of the dice roll:

$$\Psi_U(t) = \sum_{j=1}^{j=6} \frac{1}{\sqrt{6}} |U_j(t)\rangle \otimes |j\rangle .$$

All $|U_j(t)\rangle$ are unitary at $t = 0$: $\langle U_j(0)|U_j(0)\rangle = 1$. Thus $\langle \Psi_U(0)|\Psi_U(0)\rangle = 1$. The action of the TDD makes that at $t = 30$ seconds $\langle U_3(t)|U_3(3)\rangle = \frac{1}{10}$, thus $\langle \Psi_U(t)|\Psi_U(t)\rangle = \frac{51}{60}$. It is not necessary that the evolution of $\langle U_3(t)|U_3(3)\rangle$ from 1 to $\frac{1}{10}$ be abrupt and could be a slow continuous evolution. It could even be only asymptotic, for example, $\langle U_3(t)|U_3(3)\rangle = e^{-t} + (1 - e^{-t})\frac{1}{10}$ such that $\lim_{t\to\infty} \langle U_3(t)|U_3(3)\rangle = \frac{1}{10}$.

By application of the density operator theory, the probability of the outcome 3 is $\frac{\text{tr}(\rho(t)|3\rangle\langle 3|)}{\text{tr}(\rho(t))}$ which varies from $\frac{1}{6}$ at $t = 0$ to $\frac{1}{51}$ when $t \to \infty$ because of the unitarity violation. It is difficult to imagine that the probability of an actual event varies with time since probabilities are eventually obtained after long-term repetitions and statistics. In this context, it would also be difficult to imagine time varying probabilities, which would also depends on the space-time referential.

Therefore, it is natural to consider that the asymptotic value of the probability applies immediately on the time of the dice roll. In other words, the TDDs were not fictitious objects, they should not be limited in time range. Unitarity violation would imply limitless time range for retro-information.

The following theorem is a kind of reciprocal of the previous discussion, indeed that retro-information implies unitarity violation [130].

Theorem 8.1. *The capability to transmit information backward in time implies a violation of quantum unitarity.*

Proof. To prove the theorem, we assume a retro-channel whose reception is at a space time point B. Since receiving data is always based on physical measurements, thus necessarily quantum measurements, we assume a binary outcome 0 or 1 at point B which is affected by some quantum states at space-time point A located in the future of the reception measurement (i.e. in the light cone of B). We assume two transmit states $|A_0\rangle$ and $|A_1\rangle$ which can be located at point A so that

- if the transmit state is $|A_0\rangle$, then the wave function of the reception state is $|\Psi_0\rangle = \alpha_0 |0\rangle + \beta_0 |1\rangle$ with $|\alpha_0|^2 + |\beta_0|^2 = 1$;
- if the transmit state is $|A_1\rangle$, then the wave function of the reception state is $|\Psi_1\rangle = \alpha_1 |0\rangle + \beta_1 |1\rangle$ with $|\alpha_1|^2 + |\beta_1|^2 = 1$;
- $|\alpha_0|$ and $|\alpha_1|$ differ (and so also $|\beta_0|$ and $|\beta_1|$) thus there is an information transfer from the transmission space-time point A to the reception space-time point B.

The selection of the states $|A_0\rangle$ and $|A_1\rangle$ is made on a point E which may not be in the light cone of B (see Figure 8.11) in order to outline the fact that there might be no direct signal from the source of information to the receiver located in the space-time point B.

Till now, the story shows nothing special with respect to the direction of the time arrow. But now we introduce one specific aspect of time backward information transmission, that is, the reception can

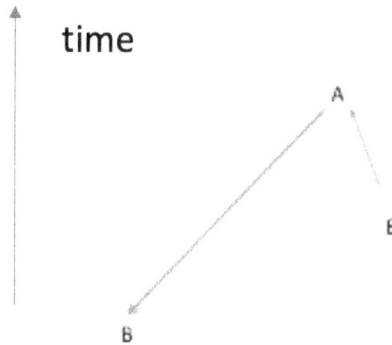

Fig. 8.11.   A retro-channel from A to B.

interfere with the emission, since the later is in the future of the former. We call *forward coupling* the effect of interference of the reception outcome with the emission setting. We imagine two forward coupling scenarios:

- the scenario "+": if the reception outcome is 0, then the emission setting is $|A_0\rangle$, otherwise if the reception outcome is 1, the emission setting is $|A_1\rangle$;
- the scenario "−" is the reverse: if the reception outcome is 0, then the emission setting is $|A_1\rangle$, otherwise if the reception outcome is 1, the emission setting is $|A_0\rangle$.

We denote $|\Psi_+\rangle$ (resp. $|\Psi_-\rangle$) the reception state function when the coupling is "+" (resp. "−"): $|\Psi_+\rangle = \alpha_+ |u\rangle + \beta_+ |v\rangle$ (resp. $|\Psi_-\rangle = \alpha_- |u\rangle + \beta_- |v\rangle$). If we show that $|\alpha_+|^2 + |\beta_+|^2 \neq |\alpha_-|^2 + |\beta_-|^2 \neq 1$, then we would have proven an unitarity violation.

We imagine that at point E the mode of coupling is decided by a quantum measurement between two state $|E_+\rangle$ and $|E_-\rangle$ with uniform probability, i.e. the coupling state is $\frac{1}{\sqrt{2}} |E_+\rangle + \frac{1}{\sqrt{2}} |E_-\rangle$. Let a point C in the light cones of both B and E and such that A is in the light cone of C. We suppose that the state of the forward coupling is decided at point C as a result of the outcome of the measure at B and the measure $\pm$ at E jointly received at C (see Figure 8.12).

The wave function at points B and E before measurement is $|\Psi_{BE}\rangle = \frac{1}{\sqrt{2}} |\Psi_+\rangle \otimes |E_+\rangle + \frac{1}{\sqrt{2}} |\Psi_-\rangle \otimes |E_-\rangle$.

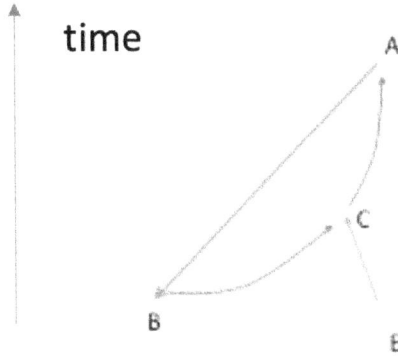

Fig. 8.12. The retro-channel with forward coupling.

The joint wave function at space-time points B and E involves the ancestor of the states $|A_0\rangle$ and $|A_1\rangle$ which we assume to be located at space-time B to simplify. We take advantage of the fact that the uniform weight between $|E_+\rangle$ and $|E_-\rangle$ is like an XOR (addition *modulo* 2) between the binary sequence made of measurements on B and the random uniform binary sequence made of the measurements on E, assuming the state $+$ is interpreted as a "0", and $-$ as a "1". The consequence is that the dependencies inside the non-uniform sequence are lost. Therefore, the wave function at B is exactly as if the states on A, $|A_0\rangle$ or $|A_1\rangle$, were chosen randomly independently of the outcome 0 or 1 on B. Therefore, the joint wave function an the space-time point B is exactly

$$
\begin{aligned}
|\Psi_B\rangle &= \frac{1}{\sqrt{2}} |\Psi_0\rangle \otimes \mathcal{U}^{-1} |A_0\rangle + \frac{1}{\sqrt{2}} |\Psi_1\rangle \otimes \mathcal{U}^{-1} |A_1\rangle \\
&= \frac{\alpha_0}{\sqrt{2}} |0\rangle \, \mathcal{U}^{-1} |A_0\rangle + \frac{\beta_0}{\sqrt{2}} |1\rangle \, \mathcal{U}^{-1} |A_0\rangle \\
&\quad + \frac{\alpha_1}{\sqrt{2}} |0\rangle \, \mathcal{U}^{-1} |A_1\rangle + \frac{\beta_1}{\sqrt{2}} |1\rangle \, \mathcal{U}^{-1} |A_1\rangle ,
\end{aligned} \tag{8.1}
$$

where $\mathcal{U}^{-1}$ is the reverse of the operator $\mathcal{U}$ which maps the states $|A\rangle$ from the ancestor B to the point A. If $\mathcal{U}$ is unitary, the states

$\mathcal{U}^{-1}\ket{A_0}$ and $\mathcal{U}^{-1}\ket{A_1}$ form an orthonormal basis. The couplings $+$ and $-$ give a distinct subset of allowed states:

$$\Psi_{BE} = \frac{\alpha_0}{\sqrt{2}}\ket{0}\mathcal{U}^{-1}\ket{A_0}\ket{E_+} + \frac{\beta_0}{\sqrt{2}}\ket{1}\mathcal{U}^{-1}\ket{A_0}\ket{E_-}$$

$$+ \frac{\alpha_1}{\sqrt{2}}\ket{0}\mathcal{U}^{-1}\ket{A_1}\ket{E_-} + \frac{\beta_1}{\sqrt{2}}\ket{1}\mathcal{U}^{-1}\ket{A_1}\ket{E_+}. \quad (8.2)$$

Collecting the states $\ket{E_+}$ and $\ket{E_-}$ we get the expression for $\ket{\Psi_+}$ and $\ket{\Psi_-}$:

$$\begin{cases} \ket{\Psi_+} = \alpha_0\ket{0} + \beta_1\ket{1} \\ \ket{\Psi_-} = \alpha_1\ket{0} + \beta_0\ket{1}, \end{cases} \quad (8.3)$$

where clearly $|\alpha_0|^2 + |\beta_1|^2 \neq |\alpha_1|^2 + |\beta_0|^2$ yielding an unitarity violation and hinting that the operator $\mathcal{U}$ from B to A is not unitary. $\square$

Corollary 8.1. *Entangled Bell pair of particles cannot be used to send faster than light information.*

Proof.   Let us assume that a pair of entangled Bell particles which make that two space-time point A and B can transmit information (say from A to B). We denote the space-time 3-vector $AB = (\mathbf{z}, t)$, where $\mathbf{z}$ is a 3-vector and t is the time component. We suppose that the vector AB is a space kind vector, i.e. $\frac{\|\mathbf{z}\|}{t} > c$, where c is the speed of light.

We suppose that the property of faster than light information transmission reproducible independently of symmetry, rotation and translation by uniform speed of the vector AB. In other words, a copy of the Bell entangled system over a vector $A'B'$ obtained from AB after an arbitrary symmetry and uniform translation will also be prone to faster than light transmission.

We consider that $A'B'$ in a second Lorentzian referential is equal to $BA = -AB$ in the Lorentzian referential of the first Bell system. We show as an exercise that there exists a speed of translation (slower than light) which makes this possible. In other words, we can make $A' = B$ and $B' = A$ in the absolute space-time: the second Bell system makes possible to transfer information backward from B to A.

Via a small perturbation of the Lorentzian referential, the transmission can be made from a point A' slightly in the future of B to

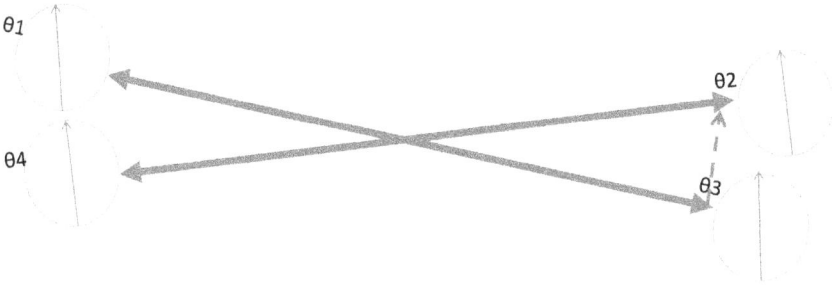

Fig. 8.13. Two entangled pairs in Lorentz opposition.

a point B' slightly in the past of point A (see Figure 8.13), thus it would be possible to transmit information from A to B then to forward it from from B to A' and back to B' in the past of A. This should be in contradiction with Theorem 8.1 since the theory of Bell states is build in a full quantum unitary framework. $\square$

8.3.3 *Retro-information, time paradoxes, and causality*

We coin here the term "retro-information" as the capability of transmitting information backward in time. We have seen that this capability is strongly attached with the possibility of unitarity violation, more precisely with the possibility of differentiated unitarity violation. By differentiated unitarity violation, we mean that there is a difference in the unitary excess or default between two timelines.

Theorem 8.2. *Let A be a space time point where two possible events can take place A^+ and A^-. Let $\boldsymbol{\rho}_+(t)$ (resp. $\boldsymbol{\rho}_-(t)$) be the density operator of the universe at time conditioned by the action A^+ (resp. A^-). Assume that $\lim_{t\to\infty} \mathrm{tr}(\boldsymbol{\rho}_+(t)) = e^{\nu_+}$ and $\lim_{t\to\infty} \mathrm{tr}(\boldsymbol{\rho}_-(t)) = e^{\nu_-}$. If $\nu_+ \neq \nu_-$, then it is possible to send information from point A to any point B in the backward light cone of A, with the capacity $\frac{|\nu_+ - \nu_-|}{\log 2}$ bits.*

Proof. We interpret the fact that the point B is in the backward lightcone from point A by the fact that a classic forward channel can be established from point B to point A. By placing a quantum dice on point B, we are back in the conditions developed in Section 8.3.1

where the TDD is replaced by an unitary violation event. Assuming that $\nu_+ > \nu_-$, we have $I(X,Y) \sim \lim_{t\to\infty} \log_2 \frac{\text{tr}(\boldsymbol{\rho}_+(t))}{\text{tr}(\boldsymbol{\rho}_-(t))} = \frac{\nu_+ - \nu_-}{\log 2}$. This theorem was proven in [129]. □

Contrary to the fictional time travel concept [11], the retro-information concept does not generate violation of causality. Of course, one must have a careful definition of causality. The causality principle assumes that the cause precedes the effect, but one cannot distinguish between cause and effect without requiring a time reference. The simplest definition of causality is the impossibility of time loop: Let X and Y be two space-time events. There is a causality violation if there are two timelike physical trajectories $\mathcal{L}_1$ and $\mathcal{L}_2$ which both pass through the events X and Y (see Figure 8.14 (left)) and such that

(i) on $\mathcal{L}_1$ event X occurs before event Y;
(ii) on $\mathcal{L}_2$ event Y occurs before event X.

Figure 8.14 right shows an example of causality violation with $\mathcal{L}_1$ in blue and $\mathcal{L}_2$ in red.

The example of causality violation created by time travel is in "Back to the future" movie where Doc meets Marty first in 1955 and

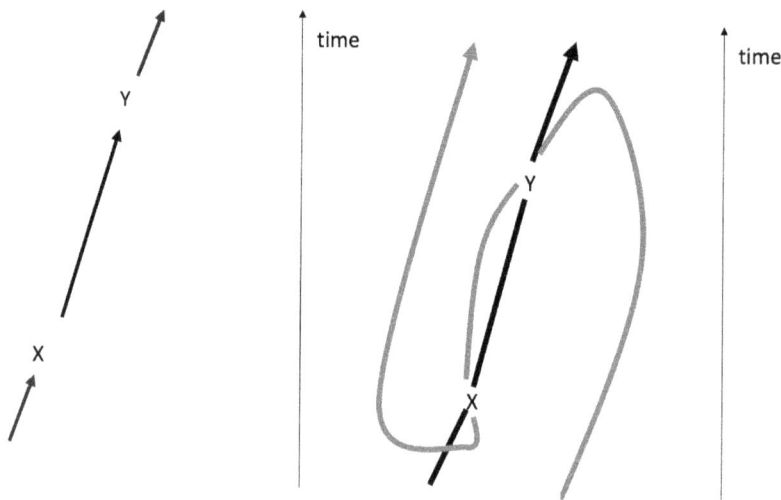

Fig. 8.14. Left: The timelike trajectory $\mathcal{L}_1$ passes through X and Y. Right: The time-like trajectory $\mathcal{L}_2$ passes through X and Y in reverse order causing a causality violation.

Fig. 8.15.   Back to the future causality violation.
Courtesy: Universal Studio.

second in 1985 (trajectory $\mathcal{L}_1$), while Marty meets Doc in 1985 and then in 1955 (trajectory $\mathcal{L}_2$, when the modified de Lorean car reaches the critical speed 88 mph). See Figure 8.15.

The causality violation induces some time paradox; the movie solves them by slicing the universe with alternative timelines, with the classic, and never respected, rule: "Marty, don't meet yourself, or this would be the end of our universe. This is the most pessimistic scenario, I concede, since the annihilation could be limited to not go beyond our galaxy."

The retro-information does not rely on conflicting timelike trajectories, thus it does not lead to causality violation. However, it may be prone to time paradoxes. The grand father paradox [11] tells the

story of a time traveler who by accident kills his grandfather before
he marries. Therefore, the time traveler could not be born and consequently could not travel back in time to kill his/her grandfather, thus
the latter can marry and the time traveler be again in scene. One
possible solution to the paradox is to assume that the time traveler
after having killed the grandfather proceeds in an alternate timeline
and comes back to a present time without father and mother.

The retro-information concept can lead to similar time paradoxes.
The time traveler can just play the role of a time transmitter; she/he
can write to a mobster at the time of the traveler's grandfather and
propose a contract to kill the latter (in exchange, the traveler can
tell the winning numbers in the next lottery). We see that one can
solve the paradox without, this time, requiring alternative timelines.
But first, for the sake of the health of all past and future grandpas,
we will take a less tragic scenario.

Suppose at space-time point B you receive a message from a friend
from space-time point A, one day in the future of B. The message will
just tell you the color of the socket you will wear the next morning.
The message says just that you will have black sockets. Just because
you don't like to receive order (even from your friend), you decide to
take the white socket. But since your friend wants to be honest with
your past yourself, (s)he will transmit the color "white" from point
A to point B. How is it possible?

To simplify, we suppose that the receiver devices at point B is a
spin measurement, whose binary output will be denoted "white" and
"black". If the spin output finally corresponds to the color of your
socket on the next morning, then your friend in the future, knowing
the past color of your socket at point A, will activate the action A^+,
otherwise (s)he will activate A^- because the transmission is clearly in
error. If $\nu_+ \gg \nu_-$, then almost surely the spin output will correspond
to the correct color. This is in the absence of forward coupling, e.g.
you have decided the color of your socket from an external event,
such as your mood inspired by the morning weather. There will be
indeed an information transfer.

But the case we describe is a typical forward coupling, you systematically contradict the prediction. If the spin outcome is black,
then you wear white, and A^- is activated, because the transmission
is in error. Otherwise, if the spin outcome is white, then you wear
black and you will also activate A^-. Since the traces of the density

operators are systematically the same whatever the output of the measurement and the measurement will be white or black with equal probabilities $(\frac{1}{2}, \frac{1}{2})$. Activating A^- in every case would mean that the transmission is systematically erroneous and in fact no information is transmitted.

To be convinced that the forward coupling generates transmission errors, we can imagine that the outcome of the quantum measurement is a sequence of alphabetic symbols which can form the words "black" or "white". If the combination of the letters corresponds to the color of your socket, your friend will activate A^+, and in all the other cases, your friend will activate A^-. By other cases, we understand all the meaningless combinations of five letters. Since the number of meaningless combinations is much more important than the two significant combinations, the contradiction case will mostly produce meaningless combination comparable to transmission errors.

The intimate nature of information theory is that it allows transmission errors. And the occurrence of transmission errors is the very point which allows us to cope with the mess of time paradoxes. Of course, one could imagine that on contrary you are very conciliatory with your friend of the future and that you will systematically wear the color proposed by the prediction. This means your friend will always activate the action A^+, unless the received word is meaningless which will never happen when $\nu_- \ll \nu_+$. Thus, the color of your socket will be white and black with same probabilities. Also, in this case, there will be no real transfer of information, since the quantum device indeed determines the final color. But the difference is that in the conciliatory case, the meaningless combination of received letters will be excluded.

8.3.3.1 *Closed timelike curves*

These would be the consequence of causality violation: A closed timelike curves (CTC) is a timelike trajectory which loops. It can result from the concatenation of the segments $\mathcal{L}_1$ and $\mathcal{L}_2$ between the spacetime points X and Y used for the illustration of causality violation (see Figure 8.14), the result is a closed timelike curve which loops around X and Y.

The General Relativity theoretically allows CTCs. In fact, Gödel, the father of the mathematic incompleteness theory, was able to

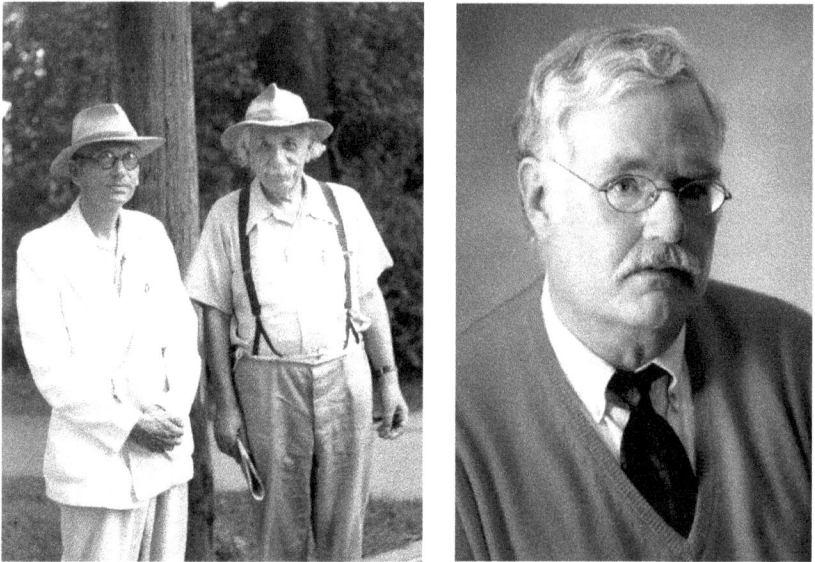

Fig. 8.16.   Left: Kurt Gödel with Albert Einstein. Right: Frank Tipler.

show that the equations of general relativity accept some solutions which describe a cyclic universe [43]. Tipler has also imagined that a infinitely long rotating cylinder will create CTCs [115]. See Figure 8.16 for the portrait of these scientists. These solutions have been criticized by the fact that they would need very unrealistic initial conditions in order to materialize.

Anyhow the CTCs would create causality violation and consequently (since it allows retro-information) would create unitarity violation, and therefore the possibility to send information from any point of the closed curve toward any point in its past, even outside the CTC (with the exception of the Gödel universe which has no external point). In exercise, we discuss how it is possible in an hypothetical CTC.

8.4   The Black Hole Information Loss Paradox

The question is still open about the possibility of unitarity violations in quantum physics beyond the criticized Closed Time Curves and Time travels. Surprisingly, the answer could come again from the general relativity and the black holes. It is known that the black hole thermal radiation, described in 1976 [52], leads to the *Black Hole Information Loss Paradox*. Indeed, an isolated black hole will radiate energy, and even the tiniest photon emission will reduce its mass and eventually lead to its total black hole annihilation after all its mass has been radiated. The power at which the black hole radiates has expression

$$\frac{\hbar c^6}{15360\pi G^2 M^2}.$$

It is worth noticing that the smaller the black hole, the more power it radiates, therefore the evaporation strongly accelerates at the end of the black hole lifetime. Indeed, the time to black hole annihilation is

$$t_e(M) = \frac{5120\pi G^2 M^3}{\hbar c^4}.$$

This time varies in the cubic of the mass and attains 10^{100} years for a hyper-massive black hole of 10^{11} solar masses. For a black hole with the mass of the central black hole in our galaxy, the time is "only" 10^{95} years. For a black hole of the mass of the sun, it is 10^{67} years, for 10^8 tons, the time to evaporation is of order 10^{10} years, the age of the universe. However, no evaporation has been observed so far, although the end-life of a black hole should appear as a very brilliant flash in space.

The table showed in Figure 8.17 displays the figures of typical black holes. Only black holes of the two first kinds (Central galactic and heavy star remnant) have been observed, although indirectly. Their temperatures are by several orders of magnitudes far below of what can be observed, which is the background radiation (2.7 K). Only a black hole of the mass of the moon would radiate on the same temperature of the background radiation. But its section of the size of a needle tip will make this impossible.

The problem which creates a paradox is that a black hole with evaporation becomes mortal. But the Hawking radiations are only thermal and in theory should not carry information. Since by virtue of the principle of unitarity, the information cannot disappear (it would mean that a mixed state could evolve back into a pure state), there is a contradiction (see Figure 8.18). Before the discovery of

Typical mass	radius	Temperature	Lifetime
Central galactic black hole $10^7 - 10^{10} M_\circ$	$10^{-6} - 10^{-3}$ ly	$10^{-16} - 10^{-13}$ K	10^{97} years
Heavy star remnant $10 M_\circ$	30 km	10^{-8} K	10^{70} years
Sun mass 10^{30} kg ($1 M_\circ$)	3 km	10^{-7} K	10^{67} years
Earth mass 10^{25} kg	3 cm	10^{-2} K	10^{52} years
Moon mass 10^{22} kg	needle tip	1 K	10^{43} years
1 km^3 of water	nucleus size	10^{11} K	age of the universe
1 Planck mass	2 Planck length	0.1 Planck temp	15,000 Planck times

Fig. 8.17. Typical figures related to black holes. Planck mass: 2.2×10^{-8} kg; Planck length: 1.6×10^{-35} m; Planck temperature: 1.4×10^{32} K; Planck time: 1.4×10^{-44} s.

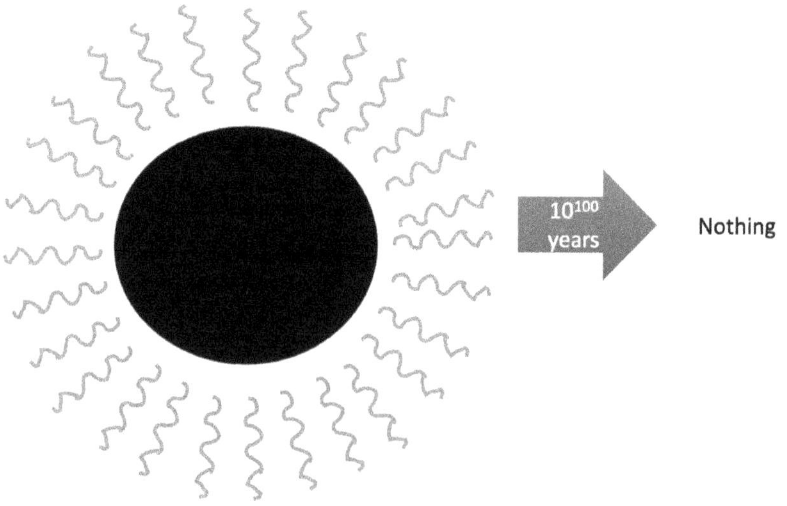

Fig. 8.18. Black hole evaporation leading to complete annihilation.

Hawking radiation, black holes were supposed immortal since no matter and no energy could escape, and the information remaining inside forever.

This paradox suggests a violation of quantum unitarity. Since then, many solutions have been proposed to dissipate the paradox. In fact, collecting all the solutions to the paradox proposed so far is already a challenge. To make it short, the solutions proposed since almost 50 years span over a great variety of ideas:

(a) information could be encoded in thermal radiation, in correlation between the future and the past of the black hole [49,99];
(b) information stays hidden in a Planck size black hole remain [42];
(c) information escapes in a baby universe [96];
(d) an hypothetical firewall prevents the information to enter the black hole or simply prevents the formation of the black hole [82,99];
(e) the black hole has soft hairs which make the event horizon to have fluctuations which allow information to escape [53,81];
(f) a quantum theory of gravity where "graviton" is not massless [41].

Most of the proposed solutions struggle for keeping intact the principle of quantum unitarity. On the other hand, the "anti-solutions", i.e. the solutions which accept the possibility of an actual violation of unitarity, are by far less numerous. The Nobel Prize awardee Roger Penrose [94] has promoted such solutions and has analyzed the impact in long term on cosmology. It seems that the universe may survive some local violations of unitarity, making consequently retro-information possible. Of course, the technological aspects of this perspective are still very elusive.

8.5 Programming a Non-unitary Computer

The retro-information is a fragile concept. We have seen in the previous section with the red-blue socket paradox that the capacity of a retro-information channel may depend on the usage of the information between the reception and emission points. Instead of looking to an endless sequence of paradoxes, one can build within this concept;

we turn to non-unitary computers. We call a non-unitary computer a quantum computer based on non-unitary effect.

A non-unitary computer will work on the following time scheme: Given a problem to solve, the non-unitary computer will work as follow:

(1) *the extraction phase*: a random solution codeword is obtained from a sequence of binary quantum measurements;
(2) *the verification phase*: a sequence of classic computations is applied taking as input the solution codeword and the parameters of the problem to solve;
(3) *the non-unitary phase*: a sequence of non unitary actions is taken as a consequence of the results of the computations.

As an example, we can take the integer factorization. Take as problem to solve the factorization of a large number N written with a large number n of bits. As an extraction phase, one selects a random pair (a, b) of two numbers of n bits each. As a verification phase, one computes the product $a*b$. As a non-unitary phase, we take an action A^+ if $a * b = N$ and both a and b are different of 1. Otherwise, one takes an action A^-. If ν_+ is sufficiently larger than ν_-, one should have almost surely A^+ unless the number N is prime.

In order to better scale with the main parameter of the problem which is the number n of bits which encode N, one should take n action A^+ if $a * b = N$ and both a and b differs of 1 (i.e. (a, b) are non-trivial divisors), and n actions A^- otherwise.

We note that the non-unitary factorization will do much better than the best classic algorithm which is in $O\left(\exp\left(1.9n^{1/3}(\log n)^{2/3}\right)\right)$ and the quantum Shor algorithm which does in $O(n^2 \log n)$ quantum gates computations (see Section 7.4.2). The non-unitary computer does in a single multiplication which takes $O(n \log n)$ classic computation steps with the Schönhage–Strassen algorithm over long integers.

The improvement which brings the complexity in n^2 of the Shor algorithm into a complexity in n looks comparable to the improvement brought by Grover algorithm compared to classic search. Of course, one could note that the Shor algorithm uses quantum gate computations which are expected to be more delicate to operate than classic gate computations, but this is hiding the fact that the non-unitary actions such A^+ and A^- would be far more delicate to design and certainly much more expensive (given one has *a priori* no idea on how to build them).

Let's look into the details of the operations. At the extraction phase, the solution codeword is $2n$ bits long which indicates the cumulated space in bits occupied by the random integers a and b. The density operator as a $2^{2n} \times 2^{2n}$ operator at time 0 is

$$
\boldsymbol{\rho}(0) = \begin{bmatrix} \frac{1}{2^{2n}} & & \\ & \ddots & \\ & & \frac{1}{2^{2n}} \end{bmatrix}.
$$

The density operator is conceptual (thus this huge dimension) since the non-unitary computer works on a single sequence of $2n$ bits. After the application of the non-unitary action, the limit of the density operator $\boldsymbol{\rho}(t)$ when $t \to \infty$ is still a diagonal operator with the diagonal elements corresponding to the pairs (a, b) such that $a * b = N$ without being trivial divisors are replaced by $\frac{e^{n\nu_+}}{2^{2n}}$ and the other diagonal elements are replaced by $\frac{e^{n\nu_-}}{2^{2n}}$. The trace of the limit of $\boldsymbol{\rho}(t)$ is equal to $2(\sigma_0(N) - 2)\frac{e^{n\nu_+} - e^{n\nu_-}}{2^{2n}} + e^{n\nu_-}$, where $\sigma_0(N)$ is the number of divisors of N including the trivial divisors. Thus, the final probability that (a, b) is a pair of non-trivial divisors is $\frac{2\sigma_0(N)(e^{n\nu_+} - e^{n\nu_-})}{2(\sigma_0(N) - 2)(e^{n\nu_+} - e^{n\nu_-}) + 2^{2n}e^{n\nu_-}}$. If $\nu_+ > \nu_-$, it will be exponentially close to $\frac{2(\sigma_0(N) - 2)e^{n\nu_+}}{2(\sigma_0(N) - 2)e^{n\nu_+} + 2^{2n}e^{n\nu_-}}$ which converges to 1 as long as $\nu_+ > \nu_- + 2\log 2$ and $\sigma_0(N) > 2$, i.e. when N is not a prime number. When N is prime, the non-unitary computer will uniformly return any arbitrary pair (a, b). In the case of deciphering an RSA-encrypted message, the number N is the product of two prime numbers, thus $\sigma_0(N)$ is 4, finding the two prime factors help recover the private key from the public key.

If one wants to save on non-unitary actions, one could only extract the factor a, and this over only $n/2$ bits because when $a * b = N$, there is necessarily one factor which is smaller than or equal to $\sqrt{N}$. In this case, the computation phase consists only in verifying whether or not a divides the target number N.

In this case, the operator ρ operates in dimension $2^{n/2}$. The non-unitary phase will be limited to $n/4$ non-unitary actions which is a considerable saving compared to the previous algorithm. After the non-unitary phase, the trace of the density operator is $\lfloor\frac{\sigma_0(N)}{2}\rfloor(e^{n/4\nu_+} - e^{n/4\nu_-}) + 2^{n/2}e^{n/4\nu_-}$. And leaving the exponentially small contributions, the probability of finding a non-trivial factor of N is $\frac{\lfloor\sigma_0(N)/2\rfloor e^{n/4\nu_+}}{\lfloor\sigma_0(N)/2\rfloor e^{n/4\nu_+}+2^{n/2}e^{n/4\nu_-}}$ which converges to 1 as long as $\nu_+ > \nu_- + 2\log 2$ as in the previous algorithm.

The difference with the previous algorithm is that the verification phase is an euclidian division instead of a multiplication, but it is also of complexity $O(n\log n)$.

8.5.1 $P = NP$ with non-unitary computers

An NP problem is a problem whose solution can be verified in a time which is polynomial in the size of the input data [3]. This consideration has to be made regardless of the complexity of the resolution of the problem via a classic computer. It is as if an oracle was capable of guessing a solution to the problem and the user has to just verify that the solution is correct. For example, in the integer factorization of N written with n bits, the problem is to find a non-trivial factor of N. The quantity n is the size of the data. The oracle provides a factor a and the user has only to check that a divides N, which can be done in $n\log n$.

Contrary to a popular belief, an NP problem is not necessarily a problem which cannot be solved with a polynomial algorithm. In fact, NP means "Nondeterministic Problem". Indeed, all problems which can be solved by polynomial algorithms are by definition NP. There are NP problems for which there is no known polynomial algorithms. For example, the integer factorization is among such algorithms since the best known factorization algorithm runs in $O(\exp(1.9n^{1/3}(\log n)^{2/3}))$.

Among the NP problems, there is a special class which contains the NP-*complete* problems. Those problems are the hardest NP problem to solve. In fact, solving one NP-complete problem in polynomial time would solve all other NP-complete problems in polynomial time. Surprisingly, the integer factorization does not belong to the class of the NP problems, meaning that solving the factorization problem in polynomial time will not solve the other NP problems. To have a list of NP-complete problems, one can refer to [39].

The Holy Grail in Complexity Theory is still the very elusive proof that P≠NP, where P is the class of problem which can be solved in polynomial time. It is not believed that P=NP, and it is not believed that quantum computers can solve all NP-complete problems in polynomial time. In fact, the class P on quantum computers is not necessarily equivalent to the class P on classic computers, since quantum gates can perform an exponential number of "parallel" computations (*a priori* a maximum 2^n parallel computations over n qubits), which performed in sequence would not add to a polynomial time. Thus, the NP-complete problems lose their equivalence property on quantum computers, which makes that solving one NP-complete problem in polynomial time solves all the other NP-complete problems in polynomial time.

Among the NP-complete problems, there is the dominating set mentioned in Section 4.3. In fact, the dominating set problem is precisely the following: given a graph with n vertices, and an integer k, to find a dominating set with k vertices or less.

Theorem 8.3. *All NP-complete problems can be solved in polynomial time on a non-unitary computer.*

Proof. It is clear that non-unitary computers if they exist would solve all NP problems in polynomial time. In fact, it is almost by definition: The extraction phase and the non-unitary phase together play the role of the oracle, and the verification phase checks the validity of the solution given by the oracle in a polynomial time. □

Let us formalize furthermore the non-unitary computer. The extraction phase is made via a quantum measurement at time $t = 0$. Let $\rho(t)$ be the density operator for the quantum measurement. Let $\mathcal{X}$ be the set of the output states of this measurement. Ideally, we have

$$\rho(0) = \frac{1}{|\mathcal{X}|} \sum_{x \in X} |x\rangle \langle x| .$$

Formally, $\rho(0) = \frac{1}{|X|}\mathbf{I}$, where $\mathbf{I}$ is the identity matrix. This would imply that the distribution over the states in $\mathcal{X}$ is uniform. In fact, we can depart from this requirement as long as the probability weights ratio remain within some interval (e.g. not greater than twice another weight). We can split the set $\mathcal{X}$ into $\mathcal{G}$, the set of good states, i.e. solution of the problem to be solved by the non-unitary computer, and $\mathcal{B}$ the bad states. We have

$$\rho(0) = \mathbf{G} + \mathbf{B},$$

for instance, $\mathbf{G} = \frac{1}{|\mathcal{X}|} \sum_{x \in \mathcal{G}} |x\rangle \langle x|$ and $\mathbf{B} = \frac{1}{|\mathcal{X}|} \sum_{x \in \mathcal{B}} |x\rangle \langle x|$. After the verification phase, the non-unitary action takes place with respect to the extracted state is solution or not:

$$\rho(t) = e^{n\nu_+}\mathbf{G} + e^{n\nu_-}\mathbf{B},$$

where n is a scaling parameter to take into consideration the predominant solutions after the non-unitary action, typically one should have $n > \log|\mathcal{X}|$. Under this consideration, the component $e^{n\nu_+}\mathbf{G}$ is the main component of $\rho(t)$ being in $\frac{e^{n\nu_+}}{|\mathcal{X}|}$ and the rest being in $e^{n\nu_-}$ exponentially vanishes in comparison as soon as $\nu_+ > \nu_- + \frac{\log|\mathcal{X}|}{n}$. The consequence is that the non-unitary computer selects a state uniformly in the "good" set $\mathcal{G}$ with a probability exponentially close to 1 when n and $|\mathcal{X}|$ increases. Of course, this is provided that the set of good state is not empty, otherwise the output of the non-unitary computer will be uniform on all the (bad) states.

Let us take as example the resolution of the dominating set problem. The problem consists, giving an integer k, in finding a dominating set of size k or less in a given graph with n vertices. We specify the extraction phase as the construction of a random binary codeword as a sequence of n bits with exactly k bits set at 1. There are $\binom{n}{k}$ such sequences and they constitute the state set $\mathcal{X}$. We leave as an exercise the method for extracting such random codeword from an unconstrained binary flow. We map the bits to vertices in the graph and interpret the bit value 1 as a membership and value 0 as an absence, thus the codeword describes a candidate dominating set. If the subset actually dominates, then one activates n actions A^+, otherwise one activates n actions A^-. The density operator $\rho_k(t)$ operates on dimension $\binom{n}{k}$ and has expression at $t = 0$: $\rho_k(0) = \frac{1}{\binom{n}{k}}\mathbf{I} = \mathbf{G}_k + \mathbf{B}_k$, where $\mathbf{G}_k$ and $\mathbf{B}_k$ are respectively the part of the operator operating on the good states, resp. on the bad states.

After the unitary actions, at time t, we have $\boldsymbol{\rho}_k(t) = e^{n\nu+}\mathbf{G}_k + e^{n\nu-}\mathbf{B}_k$, and the trace of the density operator at time t and beyond is

$$\mathrm{Tr}(\boldsymbol{\rho}(t)) = \frac{1}{\binom{n}{k}}\left(c(k)(e^{n\nu_+} - e^{n\nu_-}) + \binom{n}{k}e^{n\nu_-}\right)$$

with $c(k)$ the number of dominating sets of size k (i.e. the number of good states). Consequently, when $c(k) \geq 1$, the probability that the candidate subset is actually a dominating set tends to 1 exponentially as long as $\nu_+ \geq \nu_- + \log 2$ because $\frac{\log\binom{n}{k}}{n} < \log 2$.

8.5.2 *Hard time for non-unitary computers on NP-hard problems*

The minimum dominating set problem is not really a proper NP problem, since there is no way to check in polynomial time that the proposed dominating set is minimal. In fact, the minimum dominating set problem is an NP-hard problem. If we could solve the dominating set problem in polynomial time for any given integer k on a classical computer, then we could solve the minimum dominating set for $k = n, n - 1, \dots$ in descending order in polynomial time until it fails, the last dominating set discovered will be one minimum dominating set. But is it true with a non-unitary computer?

From now, to simplify, we assume that the action A^- is unitary, i.e. $\nu_- = 0$, equivalent of doing nothing and we denote $\nu_+ = \nu > 0$. For the non-unitary computer, the operation consists more or less in solving all the n dominating set problems for $k = n, n - 1, \dots, 1$ and the density operator $\boldsymbol{\rho}(t)$ will be

$$\boldsymbol{\rho}(t) = \boldsymbol{\rho}_n(t) \otimes \boldsymbol{\rho}_{n-1}(t) \otimes \cdots \otimes \boldsymbol{\rho}_1(t).$$

Let k_0 be the size of a minimum dominating set. It turns out that $\boldsymbol{\rho}_k(t) = \mathbf{I}$ when $k < k_0$, thus

$$\boldsymbol{\rho}(t) = (e^{n\nu}\mathbf{G}_n + \mathbf{B}_n) \otimes (e^{n\nu}\mathbf{G}_{n-1} + \mathbf{B}_{n-1}) \otimes \cdots \otimes (e^{n\nu}\mathbf{G}_{k_0} + \mathbf{B}_{k_0}).$$

Clearly, the leading term will be $e^{n(n-k_0)\nu}\mathbf{G}_n\mathbf{G}_{n-1}\cdots\mathbf{G}_{k_0}$ and the remaining terms are exponentially negligible in comparison. For convenience of reading, we have removed the symbols "$\otimes$" between the operator, but these are still tensor products and not matrix multiplications. Consequently, the non-unitary computer is successful to extract a sequence of dominating set of descending order starting from side n, and ending on the minimal size k_0. But instead of starting from n, we could start from the size k_1 of the first dominating set found by the greedy algorithm (see Section 4.3) to save on non-unitary actions ($n(k_1 - k_0)$ instead of $n(n - k_0)$).

The previous analysis assumes that we continue the descent in k even after the first failure. We could write the density operator $\rho(t)$ in the more efficient case when the descent in k stops at the first failure. In this case, we have

$$\rho(t) = \mathbf{B}_n + e^{n\nu}\mathbf{G}_n\mathbf{B}_{n-1} + \cdots + e^{n(n-k_0)\nu}\mathbf{G}_n\mathbf{G}_{n-1}\cdots\mathbf{G}_{k_0}\mathbf{B}_{k_0-1} + \cdots.$$

In fact, the sum stops at the rank k_0 because beyond $\mathbf{G}_k = 0$. The last term is the preponderant term which is $e^{n(n-k_0)\nu}\mathbf{G}_n\cdots\mathbf{G}_{k_0}$ since $\mathbf{B}_{k_0-1} = \mathbf{I}$ and the other terms are exponentially negligible in comparison. Thus, we asymptotically get the same result as with the exhaustive descent.

If we want to save on non-unitary actions, we can opt for the dichotomic search. We first try $k = \lceil n/2 \rceil$ and we divide k by two as long as we get a good state and so forth until we end into a bad state. At that point, we start on the median of the two last value and halve the interval up or down according to the state, etc. There are at most $L = \log_2\lceil n \rceil$ steps in order to reach the minimal state which gives a dominating set. The operator corresponds to a tensor product of L partial operator: $\mathbf{GGBB}\cdots$, where $\mathbf{G}$ is for good state indicating a halving down and $\mathbf{B}$ is a bad state indicating a halving up. Ignoring the fractional parts, in the partial operator, the first operator $\mathbf{G}$ corresponds to $k = n/2$, the second $\mathbf{G}$ corresponds to $k = n/4$, the first $\mathbf{B}$ corresponds to $k = 3n/8$, the second $\mathbf{B}$ to $k = 7n/16$, etc.

To simplify, let us assume that $n = 2^L$, therefore $\rho(t)$ is equal to the sum of all combinations of L factors $\mathbf{G}$ and $\mathbf{B}$. There are 2^L combinations corresponding to the binary encoding to a number $m \leq 2^L$ where the $\mathbf{G}$s corresponds to the "0"s and the $\mathbf{B}$ correspond to the "1"s. More precisely, let $m(i)$ be the value of the ith bit of m. If $m(i) = 1$, let $k(m, i)$ be the integer written with the i first bits of m followed by $L - i$ 0's, then the ith factor in the combination is $\mathbf{B}_{k(m,i)}$. If $m(i) = 0$, then $k(m, i)$ is written with i first bits of

m followed by $L - i$ 1's, then the ith factor in the combination is $e^{n\nu}\mathbf{B}_{k(m,i)}$.

Some combinations may not appear because if $m(i) = 0$ and $k(m,i) < k_0$, in this case, the factor $\mathbf{G}_{k(m,i)}$ is null. In other words, only the combinations corresponding to number greater than or equal to k_0. Clearly, the non-unitary actions make that the combination having ℓ zeroes weight for $e^{n\ell\nu}$. This means that the larger the number of zeroes in the binary expansion, the more preponderant the partial operator. But it is very unlikely that k_0 will have the most zeroes. For example, if $k_0 = 11$ is written as 001011, it has three zeroes, but the combination corresponding to 010000 ($m = 16$) has five zeroes and will have more weight than k_0, although $m > k_0$. The following left-hand side table shows the operations for $m = 16$; we note that the non-unitary computer has faked a bad state at bit 2 while a good state was possible (see bit 6 line). The table on the right shows the optimal dichotomic search for $m = k_0 = 11$; note that we have nevertheless an extra good state on bit 6, due to terminal state border effect.

i	$m(i)$	$k(m,i)$	state
1	0	011111	G
2	1	010000	B
3	0	010111	G
4	0	010011	G
5	0	010001	G
6	0	010000	G

i	$m(i)$	$k(m,i)$	state
1	0	011111	G
2	0	001111	G
3	1	001000	B
4	0	001011	G
5	0	001010	B
6	0	001011	G

The general conclusion here is that the dichotomic search would likely fail on non-unitary computers, since the later may fake bad states just in order to increase the number of visited good states.

The ascending order could be an interesting alternative option since it would stop at the first good state encountered. But we see as an exercise that this won't work since any ascending combination of the form $\mathbf{B}\cdots\mathbf{B}\mathbf{G}$ with an arbitrary number of bad states (more than $k_0 - 1$) would be merely equivalent and would make the discovery of the minimal dominating set (i.e. $k_0 - 1$ bad states) unlikely.

The traveling salesman problem (TSP)(assuming integer distances) can also be solved in descending order but will fail in dichotomic and ascending orders. Note the special case where the distances are rounded scalar in the euclidian plan [90], the TSP is a simple NP-complete problem.

When the distance is arbitrary (i.e. obtained by square roots in the euclidian plan), then the TSP becomes NP hard. In this case, the non-unitary computer will select the minimal path with high probability but the verification phase will last an exponential time, checking all the combinations of smaller costs. This is an interesting perspective where the non-unitary computer gives the result immediately, but the verification phase must run in background afterward and is needed in order to activate the non-unitary actions.

The temporary conclusion of this analysis is that in case of recurrent runs of non-unitary computers, the final output will tend to correspond to a maximization of the number of non-unitary actions.

Another vertiginous perspective is the treatment by a non unitary computer of the program halting problem. The halting problem consists given the code of a program to decide whether or not the program will run forever or eventually stops. The following program is obviously in the second category.

procedure FINITE(n)
  $i \leftarrow 0$
  while $(i < n)$
    do $\begin{cases} i \leftarrow i + 1 \\ \text{PRINT}(i) \end{cases}$

But the following program never stops (unless a physical problem harms the computer, or the printer).

procedure ETHERNAL()
  $i \leftarrow 0$
  while **true**
    do $\begin{cases} i \leftarrow i + 1 \\ \text{PRINT}(i) \end{cases}$

Turing has shown that the halting problem is undecidable, in other words, there is no universal algorithm which determines in finite time whether the program under scrutiny will stop or will never stop [117]. But a non-unitary computer can give an immediate answer. The immediate answer should be contained in the extracted codeword. Then the computation phase is just the program under scrutiny to be run in background. If the later terminates at some future and even far away time, the non-unitary actions will be activated to confirm that the program actually stops. Otherwise, if the program never stops, the non-unitary actions will never be activated and the answer will look like an unintelligible message to be interpreted as negative. This can be viewed as a response from a time beyond infinity. In conclusion, the non-unitary computer can be seen as a surprising way of peering over the abyss of eternity.

8.6 Exercises

8.6.1 *Lorentzian referential*

Let us assume that a pair of entangled Bell particles which make that two space-time points A and B can transmit information (say from A to B) and that the vector AB is of space-time, i.e. the information transfer occurs faster than light. We use the properties of the Lorentz transformation in special relativity. The Lorentz transform is the linear transform which maps any arbitrary pair of space time referential in relative uniform translation (at a speed smaller than the speed of light). Let $\mathbf{v} = (\mathbf{z}, t)$ be a space time 4-vector, where $\mathbf{z}$ is its spatial 3-component and t its time component. The Minkowski norm of $\mathbf{v}$ is $\|\mathbf{v}\|^2 = \|\mathbf{z}\|^2 - c^2 t^2$, where c is the speed of light, is preserved by Lorentz transform.

Exercise 46

Show that there is a Lorentzian transform which maps AB into $-AB = BA$. These mean that one can build a copy of the pair of particle and translate it at a given speed such that it can send information from B to A. See the Answer of Exercise 46 in Chapter 9.

8.6.2 *Closed time curve and retro-information*

An accessible Closed Time Curve (CTC) is when the time curve is finite in size and is accessible from the main larger space-time (like the deLorean time curve in Figure 8.14).

Exercise 47

Show that an accessible CTC can lead to retro-information toward any point in the past of the CTC (in the larger space-time). See the Answer of Exercise 47 in Chapter 9.

8.6.3 *Ascending order for minimal dominating set with non-unitary computer*

Exercise 48

Assuming that $\rho(0) = \mathbf{G}_1 + \mathbf{B}_1 \mathbf{G}_2 + \cdots + \mathbf{B}_1 \mathbf{B}_2 \cdots \mathbf{B}_{n-1} \mathbf{G}_n + \mathbf{B}_1 \cdots \mathbf{B}_n$, show that $\rho(0) = \mathbf{G}_{k_0} + \mathbf{B}_{k_0} \mathbf{G}_{k_0+1} + \cdots + \mathbf{B}_{k_0} \cdots \mathbf{B}_{n-1} \mathbf{G}_n$. And consequently $\rho(t) = e^{n\nu} \rho(0)$. See the Answer of Exercise 48 in Chapter 9.

Exercise 49

Express $\mathrm{Tr}(\rho(t))$ with the constants $c(k)$. See the Answer of Exercise 49 in Chapter 9.

Take a star graph made of a central vertex connected to $n-1$ isolated vertices.

Exercise 50

Show that $c(k) = \binom{n-1}{k-1}$ and that the probability that the ascending order discovers a dominating set of k vertices is proportional to $\frac{k}{n}(1 - \frac{k-1}{n}) \cdots (1 - \frac{1}{n})$ independently of the non-unitary actions. See the Answer of Exercise 50 in Chapter 9.

Exercise 51

Show that the largest probability of the ascending algorithm output is attained at $k \approx \sqrt{2n}$.

We note in this simple example that the non-unitary computer does not always give the value k_0 which is 1 in the star graph.

8.6.4 *Project: Message from the future*

In general, books are written in the obvious purpose to send messages toward the future, for the posterity. On the occasion of the last chapter and of the last exercise of this book, we try the exact opposite, i.e. to use this book as a receiver of messages from the future. Of course, we can't be sure of the result. We've seen that if we were able to act in non-unitary ways, then we could send information to the past. We're going to use a quantum random generator, which is a kind of electron-by-electron spin measurement or photon-by-photon polarity measurement, to extract a sequence of n bits, 0 or 1, of purely random appearance. If, in the not-too-distant future of mankind, this book is still accessible and some of the humans capable of reading it have the power to act in violation of unity, then they could use this book to send messages backward in time. We can suppose that by an action A they can raise unity by a factor $e^{n\nu}$, where ν is big enough. For example, by judiciously sending energy into a black hole. Let's imagine that if the series of bits which we display at the end of this book corresponds to a message they'd be happy with sending in their past they activates A, or doesn't if the message contains no relevant information.

Of course, this may not work for various reasons, ranked from the most important to the less important:

- Non-unitary actions might be physically impossible.
- Non-unitary actions might be physically possible but with too small amplitude.
- The humans of the future may not be highly motivated to waste energy and time in sending information toward humans when the latter live in a too distant past. Indeed, one should not expect the future humans to send us the lottery numbers (although I wouldn't be against it) since very likely they wouldn't care about it.
- Humans of the future may be only motivated only in sending information backward if it turns to be advantageous for them, for example, by sending technological information necessary to make non-unitary actions possible.
- There might not be enough bits in the extracted sequence. The extracted sequence corresponds to 114,688 bits, covering more than 10 pages. I am not sure the editor would allow for more.
- The extracted sequence although large might be based on too few pure quantum measurements, limiting in return the quantity of genuine information it could support.

Exercise 52

We have used the online quantum random number generator of ANU-QRNG (https://qrng.anu.edu.au) which uses quantum fluctuations in vacuum in order to generate high random bits in high rate [111]. Check the bit sequence (displayed in hexadecimal) in the answer of this exercise and try to detect a possible hidden message. Hint: The author has no idea about it.

Chapter 9

Answer to Exercises

Answer of Exercise 1

The classic error is to try to find a fully distributed solution to this challenge. The prisoners elect a leader who will centralize the information. The leader will initialize the lever in up state. When another prisoner visits the room and sees the lever at state up, he does not touch it. If the lever is at position down, he lifts it to position up, if he has never touched the lever before. Therefore, each prisoner has the right to lift the lever only once. The leader's task is to enumerate the number of times he sees the lever up and put it back down, otherwise he keeps it down (don't touch it). When he has visited the room 99 times with the lever at position up, every prisoner has visited the room at least once, and can inform the guardian.

Answer of Exercise 2

This one is tricky. The prisoner decides of an arbitrary order between them. Let x_i be the number stuck on the forehead of prisoner number i (which he can't see). When the prisoner sees the other prisoners' numbers, he makes the sum of all forehead numbers he can see modulo 100, say S_i. $S_i = S - x_i$ (modulo 100), where S is the sum of all the x_is, but he ignores S. The prisoner i will tell the guardian the number p_i equal to $i - S_i$ modulo 100 with the convention that 0 equal 100 (to take into account the fact that the numbers x_i range from 1 to 100). At least one, in fact only one, will have $i - S_i = x_i$ and thus will tell the right number to the guardian. It is the one such that $S = i$ modulo 100. He will tell the number but won't realize that this is the right number (unless the guardian tells him in return).

Answer of Exercise 3

$I(X,Y) = 2 - \log_2(1 + 1/n)$. But codeword set $\{x_1, \ldots, x_n\}$ is OK, thus $k = n$ fits.

Answer of Exercise 4

$I(X,Y) = \log_2(n+1) - 1$, but only $k = 1$ fits.

Answer of Exercise 7

We need the Hamming distance between codewords to be larger than $2n_1$.

Answer of Exercise 8

$I(X,Y) = n - \log_2(n)$.

Answer of Exercise 9

Add $\log_2 n + O(1)$ bit to the binary sequence X_n in order to correct 1 error and recover X_n, the celebrated logarithmic dichotomy.

Let $k = \lceil \log_2 n \rceil$. The $k+1$ last symbols of X_n $r_0, \ldots, r_k$ (or equivalently, those symbols to X_n). We fix r_0 as the checksum of the whole X_n, and for each $1 \le i \le k$, we fix r_i as the checksum of all the symbols of X_n whose position is larger than 2^{i-1} modulo 2^i (i.e. the position written in binary has a "1" as ith bit). The symbol r_0 reveals the presence of an error and the sequence $r_1, \ldots, r_k$ compared to the actual check sum sequence indicates the position in binary of the error. The cases where the error is on the redundant symbol is just a trivial twist.

Answer of Exercise 10

The condition $n_1 < pn$ implies that $I(X,Y)$ is at least linear in n. In fact, the inequalities directly give an upper bound of the decoding failure probability which is $k2^{-I(X,Y)}$ or $2^{-\epsilon I(X,Y)}$. Since $I(X,Y)$ is at least linear in n, the error probability is exponentially decreasing in n.

Answer of Exercise 11

With basic case analysis, we get $q_0 = p_0 p'_0 + p_1 p'_1$ and $q_1 = p_0 p'_1 + p_1 p'_0$. We note that when $p_0 = p_1 = \frac{1}{2}$, we always has $q_0 = q_1 = \frac{1}{2}$ which shows the corollary.

As a supplement, but outside the scope of this exercise, the results in corollary also holds when the bit generation has dependencies and is not stationary. Even in this case, XORing with a uniform i.i.d. sequence results in an uniform i.i.d sequence.

Answer of Exercise 12

We have the identity $I(X,Y) = -\sum_{x,y} P(x,y) \log_2 \left(\frac{P(x)P(y)}{P(x,y)} \right)$ with $P(x) = \sum_y P(x,y)$ and $P(y) = \sum_x P(x,y)$. By convexity of the function $-\log$, we have

$$-\sum_{x,y} P(x,y) \log_2 \left(\frac{P(x)P(y)}{P(x,y)} \right) \geq -\log_2 (\sum_{x,y} P(x)P(y)) = 0$$

since $\sum_{x,y} P(x)P(y) = 1$.

Answer of Exercise 14

We have to be careful to the fact that we ask for the entropy of the whole chain, not the entropy of the final probability vector $\xi_k = \mathbf{P}^k \xi_0$, which will be $h(\xi_k)$. Let $X^k = X_0 X_1 \ldots X_k$ be the chain of length k produce by the Markov transition matrix; we look for $h(X^k)$: We use the identity $h(X,Y) = h(X|Y) + h(Y)$ with $Y = X^{k-1}$, $X = X_k$: $h(X^k) = h(X_k|X^{k-1}) + h(X^{k-1})$. Since $h(X_k|X^{k-1}) = h(X_k|X_{k-1})$ because X_k only depends on X_{k-1}, we get $h(X_k|X_{k-1}) = h(\mathbf{P})\xi_{k-1} = h(\mathbf{P})\mathbf{P}^{k-1}\xi_0$, where $h(\mathbf{P})$ is the matrix made of the coefficients $-P_{ij} \log_2 P_{ij}$ where the $\P_{ij}$ are the coefficients of $\mathbf{P}$. Thus $h(X^k) = h(\mathbf{P})\mathbf{P}^k \xi_0 + \cdots + h(\mathbf{P})\xi_0 + h(\xi_0)$. When ξ_0 is the stationary distribution, we get $h(X^k) = k h(\mathbf{P})\xi_0 + h(\xi_0)$.

Answer of Exercise 15

See Exercise 6.

Answer of Exercise 16

$$P(a_{k-1} + a_k) = P(a_{k-1}) + P(a_k) \geq 2P(a_k).$$

Answer of Exercise 17

In the transition from $\mathcal{A}_{k-1}$ to $\mathcal{A}_k$, the only change in the Huffman tree is the leave $a_{k-1} + a_k$ which is replaced by one root and a leaf for a_{k-1} and one leaf for a_k. The depth of the other symbol a_i for $i < k - 1$ remains unchanged and by the hypothesis of recursion is smaller than $\lceil \log_2 \frac{1}{P(a_i)} \rceil$. The depths of a_{k-1} and a_k are exactly the depth of $a_{k-1} + a_k$ and thus are smaller than $1 + \lceil \log_2 \frac{1}{P(a_{k-1}+a_k)} \rceil$. Since $\log_2 \frac{1}{P(a_{k-1}+a_k)} \leq \log_2 \frac{1}{2P(a_k)}$ which is equal to $\log_2 \frac{1}{P(a_k)} - 1$. Thus the depth of a_k is smaller than $\lceil \log_2 \frac{1}{P(a_k)} \rceil$. Similarly, for a_{k-1} since $P(a_{k-1}) \geq P(a_k)$.

Answer of Exercise 18

At time $t = 0$, the routing table in router A_0 shows a distance to itself of 0, and all other routers show distance ∞. After time 0, the information about the router A_0 propagates one step per time slot. At time t, the routers of index $i \leq t$ show a distance i and the other routers beyond still show a distance ∞. At time $t = k$, all routers got the correct distance.

Answer of Exercise 19

At time $t = T'$, no changes are indicated in the routing table since no TCs have exchanged about the new change of link status. Nevertheless, the router A_1 receives no TC from A_0 and a TC from A_2 giving a distance 2 to A_0. Thus, at $t = T' + 1$, the router A_1 shows a distance 3 to router A_0 via router A_2; the distance in other routers are unchanged.

At time $t = T' + 2$, the router A_1 still shows 3, but the router A_2 shows 4 (via A_1); the distances in the other routers are unchanged.

At time $T' + i$, if $i \leq k$, the distance in router i becomes $i + 2$, the router $i - 1$ shows $i + 1$, and the router $i - j$ shows $i + 2$ if j is even and $i + 1$ if j is odd.

When $i > k$, the router $j \leq k$ shows distance $i + 2$ when $j - i$ is odd, and $i + 1$ when $j - i$ is even.

Answer of Exercise 20

The network shown in Figure 9.1 has a host node A, three neighbor nodes R_1, R_2, and R_3, and six two-hop neighbors: $T_1, \ldots, T_6$. The greedy algorithm will take R_2 as MPR because the latter covers the most on the two-hop neighbor set and then selects R_1 and R_3 because T_1 and T_2 were not covered. Thus, the greedy selection takes all the three neighbors, but in fact the MPR set limited to $\{R_1, R_3\}$ suffices.

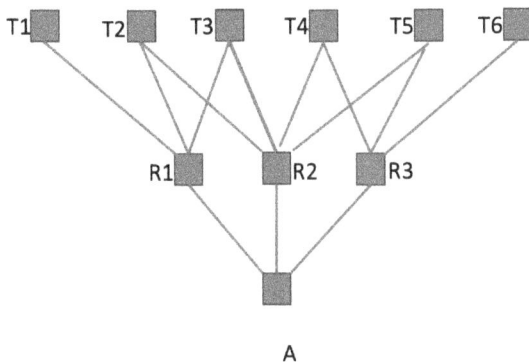

Fig. 9.1. The greedy MPR selection counterexample.

Answer of Exercise 21

$\mathcal{S} = \emptyset$ is a remote spanner. Indeed, for all $v \in V$, $E(v)$ covers all the other vertices of the graph.

Answer of Exercise 22

We know that for $v \in V$: $\mathcal{S} \cup E(r)$ is a connected set which covers all the set V. Since the tree is the smallest connected set and has $|V| - 1$ links, we have the inequality $|\mathcal{S} \cup E(v)| \geq |V| - 1$. Since $|\mathcal{S} \cup E(v)| \leq |\mathcal{S}| + |E(v)|$. Since the average value of $|E(v)|$ is $\frac{|E|}{|V|}$, thus there necessarily exists $v_0 \in V$ such that $|E(v_0)| \geq \frac{|E|}{|V|}$ and we get the demanded inequality.

Answer of Exercise 23

We note that the $\mathbf{S}(\lambda, A_k)$ are independent, thus

$$w(\theta) = \prod_k w(\theta, A_k) = \exp\left(\sum_k \log(w(\theta, A_k))\right).$$

Answer of Exercise 24

At the first order,

- with proba $e^{-\lambda dx dy} = 1 - \lambda dx dy + O(dx^2 dy^2)$, there is no emitter on A,
- with proba $\lambda dx dy e^{-\lambda dx dy} = \lambda dx dy + O(dx^2 dy^2)$, there is one emitter in A located on $\mathbf{z} + O(dx + dy)$,
- with proba $O(dx^2 dy^2)$, there are more than two emitter in A.

Therefore,

$$w(\theta, A) = 1 + \lambda dx dy (e^{-\theta \|z\|^{-\alpha}} - 1) + O(dx^2 dy^2), \quad (5.4)$$

$$\log w(\theta, A) = -(1 - e^{-\theta \|z\|^{-\alpha}}) \lambda dx dy. \quad (5.5)$$

Answer of Exercise 25

We have the Riemann convergence:

$$\log w(\theta) = - \iint (1 - e^{-\theta \|z\|^{-\alpha}}) \lambda dx dy.$$

By change of variable, we have

$$\log w(\theta) = -\lambda \pi \int_0^\infty 1 - e^{-\theta \|z\|^{-\alpha}}) r dr.$$

Setting $\gamma = 2/\alpha$, it turns out that it is $\lambda C \theta^\gamma$ with $C = \pi \Gamma(1 - \gamma)$.

Answer of Exercise 26

We have $E[F^\gamma] = \int_0^\infty u^\gamma e^{-u} du = \Gamma(1 + \gamma)$. We conclude with the identity

$$\Gamma(1 - \gamma)\Gamma(1 + \gamma) = \frac{\pi \gamma}{\sin(\pi \gamma)}.$$

Answer of Exercise 27

We have

$$p(r, \lambda, K) = \int_0^\infty P(\mathbf{S}(\lambda) < \frac{u}{K r^\alpha}) e^{-u} du.$$

By part integration, and change of variable denoting P_S the density function of $\mathbf{S}(\lambda)$, it comes

$$p(r, \lambda, K) = \int_0^\infty P_S(u) e^{-uKr^\alpha} du = w(Kr^\alpha).$$

Thus,

$$\log p(r, \lambda, K) = -\lambda\pi \frac{\pi\gamma}{\sin(\pi\gamma)} K^\gamma r^2.$$

Answer of Exercise 28

Since the contribution of all the plan minus $\{\mathbf{z}\}$ is independent of the signal $s(\mathbf{z})$ from $\mathbf{z}$, we have

$$T(\lambda, \mathbf{z}) = E\left[\log_2\left(1 + \frac{S(\mathbf{z},0)}{\mathbf{S}(\lambda)}\right)\right].$$

Answer of Exercise 29

$$C(\lambda) = \iint T(\lambda, \mathbf{z})\lambda dx dy = \iint E\left[\log_2\left(1 + \frac{S(\mathbf{z},0)}{\mathbf{S}(\lambda)}\right)\right]\lambda dx dy.$$

Note that

$$C(\lambda) = \iint \left(E[\log_2(\mathbf{S}(\lambda) + S(\mathbf{z},0))] - E[\log_2(\mathbf{S}(\lambda))]\right)\lambda dx dy.$$

Answer of Exercise 30

- With probability $1 - dt + O(dt^2)$, there is no additional transmitter on $\mathbf{z}$,
- with probability $dt + O(dt^2)$, there is an additional transmitter on $\mathbf{z}$;
- with proba $O(dt^2)$, there are 2 or more transmitters on $\mathbf{z}$.

Thus

$$L(\mu + \delta_{\mathbf{z}} dt) = (1 - dt)L(\mu) + E[\log(\mathbf{S}(\mu) + s(\mathbf{z}))]dt + O(dt^2)$$

and consequently

$$L(\mu + \delta_{\mathbf{z}} dt) - L(\mu) = T(\mu, \mathbf{z})dt.$$

Answer of Exercise 40

It takes $\{-3,-1,1,3\}$.

Answer of Exercise 41

We have $E[X+Y+Z]=0$ since on all angle the Bell state remains neutral.

Answer of Exercise 42

$$E[(X+Y+Z)^2] = 3 + 2\cos(2\theta_0 - 2\theta_1) + 2\cos(2\theta_2 - 2\theta_1)$$
$$+ 2\cos(2\theta_0 - 2\theta_2).$$

Answer of Exercise 43

$(0, \frac{2\pi}{3}, \frac{4\pi}{3})$ leads to a zero variance!

Answer of Exercise 44

$$\mathcal{U}_c\Psi_{AB} \otimes \Psi_C = \frac{\alpha}{\sqrt{2}}\left(|-,-\rangle_{AB} + |+,+\rangle_{AB}\right) \otimes |-\rangle_C$$
$$+ \frac{\beta}{\sqrt{2}}\left(|-,+\rangle_{AB} + |+,-\rangle_{AB}\right) \otimes |+\rangle_C.$$

Answer of Exercise 45

$$(I \otimes \mathcal{U}_H)\mathcal{U}_c\Psi_{ABC} = \begin{cases} \frac{1}{2}\left(\alpha|-\rangle_A + \beta|+\rangle_A\right) \otimes |+,+\rangle_{BC} + \\ \frac{1}{2}\left(\beta|-\rangle_A + \alpha|+\rangle_A\right) \otimes |+,-\rangle_{BC} + \\ \frac{1}{2}\left(\beta|-\rangle_A - \alpha|+\rangle_A\right) \otimes |-,+\rangle_{BC} + \\ \frac{1}{2}\left(\alpha|-\rangle_A - \beta|+\rangle_A\right) \otimes |-,-\rangle_{BC}. \end{cases}$$

Answer of Exercise 46

A vector $\mathbf{v}$ is of space kind when $\|\mathbf{v}\|^2 > 0$, i.e. the ratio $\frac{\|\mathbf{z}\|}{|t|}$ is larger than the speed of light c. The vector is of forward time kind when $\|\mathbf{v}\|^2 < 0 < 0$, i.e. $\frac{\|\mathbf{z}\|}{|t|} > c$ and $t > 0$, and is of backward time kind when $t < 0$. When $\|\mathbf{v}\|^2 = 0$, then the vector belongs to the light cone.

The property of Lorentz transforms is that they conserve the Minkowski norm, consequently they map space kind vectors into space kind vectors and forward (resp. backward) time kind into forward (resp. backward) time kind vectors. Furthermore, let $\mathbf{v}_1$ and $\mathbf{v}_2$ be two space kind vectors with the same Minkowski norm; there is a Lorentzian transform which maps $\mathbf{v}_1$ into $\mathbf{v}_2$ and another one which maps $\mathbf{v}_1$ into with $-\mathbf{v}_2$.

Answer of Exercise 47

Since retro-information can trivially occur inside the CTC, by Theorem 8.1, it implies that a unitarity violation happens in the CTC. An unitarity violation implies the possibility of retro-information toward any point in the past of the point which creates the unitarity violation. Since the CTC is supposed accessible from the larger space time and the setting of the unitarity violation be affected from there, this could be any point in the past of the CTC, even if the later is a tiny space-time loop.

Answer of Exercise 48

$\mathbf{B}_k = \mathbf{I}$ and $\mathbf{G}_k = 0$ for $k < k_0$ and $\mathbf{B}_n = 0$. Each term $\mathbf{G}_k$ is affected by a factor $e^{n\nu}$ after the non-unitary action.

Answer of Exercise 49

We have $\mathrm{Tr}(\mathbf{G}_k) = \frac{c(k)}{\binom{n}{k}}$ and $\mathrm{Tr}(\mathbf{B}_k) = 1 - \frac{c(k)}{\binom{n}{k}}$. Thus,

$$
\mathrm{Tr}(\boldsymbol{\rho}(t)) = e^{n\nu}\left(\frac{c(k_0)}{\binom{n}{k_0}} + \frac{c(k_0+1)}{\binom{n}{k_0+1}}\left(1 - \frac{c(k_0)}{\binom{n}{k_0}}\right) + \cdots \right.
$$
$$
\left. + \left(1 - \frac{c(k_0)}{\binom{n}{n-}}\right)\cdots\left(1 - \frac{c(n-1)}{\binom{n}{n-1}}\right)\right).
$$

Answer of Exercise 50

In the star graph, only the central node is connected to other nodes. Therefore, a connected dominating set is any subset which contains the central nodes. For a subset of size k, there are $\binom{n-1}{k-1}$ "good" subsets.

Answer of Exercise 51

The quantity $\frac{k}{n}(1 - \frac{1}{n})\cdots(1 - \frac{k-1}{n}) \sim \frac{k}{n}\exp(-(k-1)k/n)$ which attains its maximum for a value close to the root of $1 - (2k-1)k/n$.

Answer of Exercise 52

In hexadecimal, one symbol represents four bits as given in the following table.

Symbol	bits	Symbol	bits
0	0000	8	1000
1	0001	9	1001
2	0010	a	1010
3	0011	b	1011
4	0100	c	1100
5	0101	d	1101
6	0110	e	1110
7	0111	f	1111

The extracted sequence from the quantum random number generator is displayed in 14 placards:

Extracted quantum sequence: placard I

e9783654c08d9c7388ed63e64909830c2ed57c934b3d98d0b7c73363d5fe7f46
bc2c02576d781cd64cd8d87e394a8dcdb369e0ec0839cc55ecac29c47c6614d8
73421984b075d30b338898e4bfebe9aece8ba0e21e6aff0a526ea9783ec939b5f
1afd51717ed0530f66b98bf049bb59c69ad2b8ba26f91e1bb163622b0d7175b
4bead7c08fa0e3f6d970a480b12c13a5e3146a0d0ffb3d1af6d0653d90282550
dfbc54f7f59e59ac7d8061330b5aacdd204a70294c7b33ba783428180d989aa
930c09e8b6a398fea2834e4c4e368717555fd4de0717170b8ad2441ad4aec906
0c6c49d0504e2f643aa560fc8683da9d4cdc96f59e2625a03c7aa116a9eecc30d
b57e5d761efa8be5a7dc8ec191c147593bc0f5541670500c2e8b2f4101486f2ec
9025df740264c798649cf466ed300c873e7218cd619b4efadb800753bd0ad110
637bf8889ff63331cdd87678c6f16a4bf5e0155fce47984546f01af2a88d99eaa8
0bd435f2cca588077182bfddae624c900289aeef084d2b62020f85bfdc69928f5
f266d1a1032448ec0fe4c87f82229ea3225406c3088b0f51da633c269f51ee6070
c79da78a8099b038d6181af8cc03921ea43aad5d0443cbd9826d0dd88793d9c
8660ed04f2e8464e6f4037c96611dd7c771bbbf228e47066a3ec3f8a602acd9b8
e716dd297c42ad31474412b013dc64d2f22c9cb9cec5bb1df115758f5ce44472
61684f3d1b00173d54ede75e78a9b504e6cd0fe7ddbf4fe091945f6326100a12f
7b3e85512cf5acb6cbd98bdf03b2c2ee028390427cad411871cd2cc701b97942
5eba02bd99cffad9696d2bae203620c86fe357591dd9caaee1284caaac3b7a95
15dbaf00239b43f7517d749eb3f0914a66ae680617bd753c3f8334ff3f13eda7ee
fc0961c555eee04b06ca356b263e02be5ed8ad9b789b12d263e25df007aba1af
e15a5c2c64b18f97263dc360f7a75d444bb74bea9db06c23688b7f55f1745cd2
00e502170bb81a1ab30f32b6c02543be9297316bd3c4bc7517a8028f1b9fdf7b
405205c7879b9c341a022e5dd782c630e934a51e0242ec0f475a3d62e724da8a
1c55813bcd649750b353b76eba1b1de9c918b92b0bfabd6d3d16d63b0a8bc0
c442d9ce7c09fe42a846f2d8d4ef05c3b0b401de3b5bc01ddfdb2cecdf4f2a316
460f802704d8449be1c1ed404f1ef35d8ced2d323d1635fca961b81937974095
44437ab38385e3f72cd3b6f48d0ca4a103d8d2739b03b4e7fb6f3706f4cd21ea7
f73c2038e3583d250a93dbca7dfcb5ec1c6e3ccb33763e4fec7cca7bca2fd53cef
cddc200c32e8e06060952f4732bee9ea84da621d74b649022fe77f71e07e2e5fa
47fb0d301a46b08b0960d0ad70a0afdb9c0edc0a646a26b0be57122202c7e04
e5b392a75d66d64628354da9a826fefa66ba9759bb92e3f250d33

Extracted quantum sequence: placard 2

f71290336d4a068076a9948eb9fc3052087257efe6361f16553056e134307395b
670d4334e7a68f0fcf9aa60551ecaa1bb2481906de5a70bdb9c6d352c264b10d
500fafdfa559d6dd18498736254d7ddd74c93400a0d5472f80759d0e0b5a0c58
a0fd8bbbed16b37b4bc755f9ceb27a615a9bda3f7439805e7666713ca670fafc
5281234d8cb12d70b7c7538dc1a194e4b61be06af5a5e13c96e57cfed80b5d05
ff386b65e07dc95d5c5933b711d05b3d358e913ba1c73ff5e0f9ccbcc9c3bb02a
6f8e60b18b9b3ef73b5f4d74703828d681f5ad0e6a43c3f02c1ccef837923c3f18
2f3034932884b52c542a3625431a381b1d76b3fbc4a303663902ecdcd2a8623c
79e8d7d55840d99e510d165a8e59414f285fabcd14b7e1a0e0f4f2ea0ba43fd5d
330b45e40856edf5cc50ffff23d953dcaa6ce228f957878d9b63daa51b1f9adf62
a30144b816bd915e528228c3e9fc690c3f0b456a47cbbf5540a038f21f43f5962
465d5e28b6231ce36134fa5651044d0dafead5761c2108c0cb33bd584c8aac1ff
e030e68aa2c94828d7809ef19a5c7dde51fcbacef50fdf403b507cbf2449e53d42
d624b0684b38283ee3893597156d7825461a97eae7e74585e6d61f5601b85d6
5ac7f504470de330b171fd28a6c78c29697b76ade0d419d79ea831c00bbf1f1e1
9521afff51f448c43cdcf5d5458a9c676a562eae6b3239219e723b291e171a4d5
3db66978a1a05ac43ed4d20e5f8eecba7c4586afb60899e081799f4b64d7fd4ea
38646509042a54b734e2b24cf4bb85b851b1ac917ff6abfdc08cda069a307d35
0e31026e991a999ddf6086e084523555504a6ab929b3182dfeb28154444448d7
ca0b33c4d061fece3fa7328c823ce1696f004e7564bf4265485abce33183fe5f70
986f94ab4f7258ba97741feb54ac37771fb58f9e97e67e842bcd1adde08516103
77fadb9f7de3bdda7b2a4168568a97c7ae673cdbd044b59da34574738ac7f11
7daf13d44c4a46af709871bc13c0c067ab1d335ee8d0d81120277251873f543f
763297bb5c4eb21f646c3c9868a7e9ef8d4f269e15e8003ad20e07db5594ab83
f6353369f511a0efcf45f360842574795393310d36681addcd6bb2981a48b3fd5
963cfa4d278c39eae6dda4d364c9640c879f3b9b1cd4faf584bd5b001716d200
bcc446be61a6c6ef343ed2a81ab78128be5b0912017b5db5e4441a56bc456f68
c964d012bcf88bfbd9ad2357fe0a56571f31e5b8612177ad17dce68873362517
05093b7e2bcca79327cd4588cc54878f90447cbb04a9055dcb0b5008916b662
25f9e71618f944d82150c69e81ba20176d15c105fc1d03b208e868be1508439c
4762f399e88a22c632e3bfe10e277c934152448fbe87be06c2d4ad561a1491f93
c662eb753860cc2456ae95601562a9340027956935f78d70

Extracted sequence: placard 3

dd77ece6ad4afb3696e0dbc25740054b00c6a631e37bb7dc7f298914cab45490
d166fecf650d855d78d779b889a420973ba63d84cc5dae3da4b185d6b3e71a1
5d99b43d37961da3671e93868164267d36d1ef0fa9af24e9fc04caeb0afd5fd37
7b20a89dd12514b920e1542554ed14a753c323ab74dea7d4c66138b9e11c41e
edc14a5f0530134b1e8f04a4061bdabeaa0e043117eb9fd40853d352ab59c927
b3f33738d4d38943082a8a786e46a0aa807152a6d94d6da6d28f3febe7b31462
181ea58ad199b547bee206b8095674ece83a3bc22adf8371f2af67143133acf57
1df8bbfb4d3ea62cb953411205f15e8dea578dab7c8f67ed37b9ccbaaba7168e
4930672644cc32d2de21ed24839878f2f733225892a0e8f1b3ded477adf1e5b46
1f121a3323e75e3bab326473710cfc4984b05fb6dc1511c12b9f99d6da88e7129
fb50c7f9cdf25ba4de9337917fd661c99c3a743a7361b5a54e7255546ce5dbf88
3f3bdc21528a2fc29b65a3b621e0fb075ef53b1c16891e0007fc16d196aebc69a
71b6923cf49cec61ca77275bd6e63995f9d1edae1ed15668b536ea37536c6b76
f5e0931df69bae469bad4de857fd096c7f54ded7a87f4c2b35c8474879bfa1911
78f6c9c21386e65f3a2115bfd6b05cee52bf18199f6aadb143ce4b6447ef3601b
95689046e3114844b5d14c1a454aaa45f45b5ee6f2e2047b785b398ee10a5182
0d796062a78a63a6668b1ba15305cc0429b7b4678925f4ec980aff7a7287936c
1bc043519cdd7422b9179b7e6f308dfc256d1a0dd6bb1e4e0cee0e5e5396b3c8
8206ab69da813f318ecec8d18df767998c6b1c3bf6ddceb02cb4dbf76d33cb4f
32ad1b92c6c52c60b67e7ae8e3f734e9f7617b518d942c5d05a5d0779e57fc89b
8f897622eb12b751c6578093153958564ed694ae7b18bee3e37b0ac74a27f2f2
829bb568d764930ad08692026010fdea5f38736e42ddc8d0b61ea408191d4cb
df3d7da4737b07a3a2c8fa3b52cfd14fe3ca57d6de56b02593bb1fd23f5e7707
f03b8d40726821faf352848c7fb8f98532001e562c379eefb9fdbc0e0db41b019
96970c74545fc12f6eccc410c3ac79b9b95cbafaf7fae79b22b29eec9c4b25f45d
ba70bdb84085259b68a6e3921cb369652263a7466b97176423713ec5e7199dd
dbf758470793d9b704eaff166b3fb803f24299f733db31fbb33218cc61d290057
8cd4c0265b91e63a951791a513be0a35e806bbc6ba35282bb10dbd7ac5db46
f63c2fd0c00b2775a03bf7ed7e92f518afb94bba040058ad2777d4cc0975bb089
23cc3af29eb07f377ed8a6769c2ef167a42580e9bfdc72bd2f96ba3988ddc9937
5c95d71ce2aaaf9ecad0d1582baf383037b1fef5751c274dc62aecdb3355a55e8
de632c716f4ee37812fe52fb11734de77fe1defbab76dfcb74d93a

Extracted quantum sequence: placard 4

8416cfc8b433e0f081af2dbb0a23cfee8becf6f7547000659f5c472e2dd0392fbf5
cfee6500bf02789862c16ff7dbb125d65b2eb2d47f779c5f2e17a44da295093f5c
31f4dfcccf9aefbec5d711a0b2d2b478d756df74d7503765d573d389e1e0e6a87
a58f900a7fee57cadda383ed053ab98903029109537671c9999762274092a492
0c21f00628bc1ba1b546a2355419cf63abcdf55ee3a7c2ac24c0edaa4e30a87af
27e229db203f2c6abd0cb1a0d14b51d3c4a0495e806313fc65f9b9b5a5fba64c
39734fad3722ad3680da1ff6dda87dce66edc6b399122cac5ceb842cd9fb495a1
847e8eea8e5c7721fdf7d583ad365ae174df85ab9ff5e01b7f6ca0aa5e2038512a
782c9fdc9265665a1bb29d70ad67c29790c922bb1d589a6df457d8f4ae9e522c
6956c3f6456a0d8e60d9757520ba8ba5c68cb2751522350026822b99bdfb9c87
e0ce26d515c0ccfbb224fca4315f0ea7e621d09b397b0173fb94792ce3c8e4618
119bfb12e2347b7f1f7d1283728385d0df87ddf65a2368e677e17733becca1c9a
36f629c4e814c73f84f7b6192916d99cef0215a749f6634e4bee196b4bc17a9fb2
0629d3d02feb9aaa82cbc75574a03e984515411f862a8c537f2bfd5205207a7fe
f2fd57a47807071e6ef243fa7d02f5645114d3d6a2b839fd3c791ce50d2bb7874
7f253878bc62a4991d4ad4164cb23b13bdf51e1b96626873805a206afd79a5d3
f764ebf4dccf789723eab91670ce604743db21d188ed98942f32e69b89432a603
bcf7769bf9e05b4b29203259bcd5a71a24093359c26af9421fe7e120375fbca36
a9f285164550b792688c077311d30b7a0dfff7184bf89bc741f2bc6139f030606
da9331ddf81ed071aa71d904aab4c39b196cb55d61f5a4174c7ddc718a64a85
900327a711922ab08e78c340078d1fcdd787d7f05667ff68251848f97d86c0b93
c8a7e421e810df48edba4b21f5cd31c55473d293687481001f47f9cbcc70b2e30
e7b4e07783fe1fa121043ce744a1613889aeeca1b0e92d6d8409c6041912187d
39a3d99683b4869446bc0e15426abef53ba17653fb45d7df44b3eacac3268e06
9fe781884dfbd6eb4747382250fa14ae000aefddcc24597e48608c64cbf212eac
4b0804041202bde23f840eb1a28408e37e53155e7e936b7dca5f4bec95064ecc
b113b963927d635551ce49bb1432f7a9745f47272c3f38dd9806d28b7538eee4
2f42c31ce7da8d2005903982eda47e2b747fd7a9107639c7c9e664b02059e2e1
71c6a5026be099c158e08f4ae3577381b0f1aa63b05a03989e65ca188ff02f44a
4f5e14f83adbc373ac43addf3573aa001cdf950b6727cfcd4dd6efedb81f7c09cf
f0b48860a5be8b48ad76c36030ccde8d4cc9e8db347393c6a42e41fad261d39b
61d777c5f6aec0a47cce9e5b1aaaa30596279df8

Extracted quantum sequence: placard 5

21357a59e37a845335e32b2df66f947a02b9913f43db537d7d355b10affb8806
0618b2b086e028d687f0a1ff728be0a91db3a5a96e72e1903026ac5ff7f64795f9
2370c483dff3167806dad35549285c67f99cdedde5c05bdef8c8f6403230fa483
22a4a0cbeccc3503b144368aeb7fa16f6c8092b38ae6be5687b912fa97248b1a
85d063be9218903029dfc5b96ce3faedc558a472a6b49aea3024c48e4c1ef0554
2b3140f24853dcc55747e316fa595b25da39a438be7df421d35ae4f95f66e714f
d96340675d17465565d63f40352dd795b81d78f6a40e648339c51a16be00c116
61dcb90f0914506b86c4da586d3177c9256d3df7ecbf2abfc49cf0a4d20524e46
ac0640e912672c2156caca26925c804ceb61ec6696afd8d309b1299a39c3cac18
ce2a410f45fee74b2821449ac8a18757666aeafc09ce1f2ee37f12f5ee65bdc4ece
67ce92cb769bf7f1a956104f1dc114c98ae0d542ab5f9a97d6e0ae195dc92db11
a9dca330f3c5d6d2f706cca3d4a478bccba1e8ece642a8ef130a70657fbac5b50
c1b0d9eea02521e416e85a5360f4d685de7913c93f234d9bd0b325099e65bf12
56b109d79f934b0642c0ab0d47a23ebde1c3f6995516dc1c4255f0103519698d
214e5423363b05e2c3f50141170173ae11fc157e13c2c77baeb4f3742e470ebc7
18f8407ea33f8189a22aa891573d5d7742de83ffa67aa274e05ce998423fc921c5
ebc1b5ec1c5523a9e3e5bc6c95e01f9fa4c6f80d5103d46e5a6592f3d12358546
54db47e8455a0f43fbb9f90d639b2750a89cef6f973771acc52b59b9b3a20a3a2
d63b7a6b004216e7f499112cb48d55a3cecbe971705bea832d6b5164b9c6f96
d41025ae971807db49e5d88ca0b74e099b867169ea0fa0488958127a1d42787
2e4912fb3bb4a0b975866b93c723af594daf90c0c351fdc2fb376c50768e3c610
8a3d8e96816b1ec29da4a73eae7c8d06b4e4a356726554c8a73a78034982fa75
4a73319be1ef761a30622a54ede4879ff2bc52dc502f9a9653633332b18434385
ef24ef206962f5a05e3fff1b02c70e37e0a76363b2fc0d30d79ee2b6d17435dcbf
dee4c92dafaf8832432bf646972bad050c79a546d9267651fd95fee587054022c
53276304dea6ca1479671167c24ba1f09cbf53888245ba404ca9d88ed909ceaf
2929e276a162915c2957e31672fe57a58e354e610ff594c92200b09a43ff7d72f0
43d1c6db4940c3063d8f565f7a61bf3a24cdc6cd01603ca4090db958289f0dc9
7258a2746b320a68e3f8296fd5a96f8ae5a1d464e366ae5ed19a7423ee8b221b
8e3aeecd2bcb0772a9460846207f7747a955cde6608e4b93f316ae3c14e13bed
4df8c72b8fc5b77785024ea55b8b7816c3a0e29d01a03f8e742de11eecefa071f5
c2fb1628ea69f1ac88cd104033ce85e709ba6308

Extracted quantum sequence: placard 6

0591d466bbf024ee3120318d7b38a014b32ac0ff75c4fc4b61451f9f2084ba864
ea3dc2af2d686809481a7add22a56a13a47adbc0a71631517e5b1f50efeb3fde
092b9572fb480dddff7d5e27c97feef348711a014ce6937d8fcf4fee87306858c0
49328c4b8c039bd6d804b90afd50dcc5adc672ee1aed4c56e191a1080c9d7e1
4e793967c37949b65f00074d58aa110271490a38c1279a8031c5f048657feff1b
275ae249691fb743a431fa600291a018bbeb4147e3858f2a0d716ea549799cba
0cd3f2838f6b6830392a9d68eab98d30129db049f36c3c197fe67e470798012b
9304a4f424ec17bb1f3e3ab17c5910dc564f28265796d87f9cc330c4db473d6fa
c87d9c284c3e1cd0fa34435c23a9aa9462e7c84575871e4bfb78f506d6d74a07a
d7ceb5c7017fdcfbbe16fe835bb95dd82636a264d00be5d7ca71995d57ca04c7
899b8451e578e8e1877d6c133e7b27d421371eb06ba84682b1646b22677a12c
def1369e8257626321173abbeec36c2e1ba6757dc04edb568cf8cbe0fd822b00
a36b016fbe21e5f818234af89a49dca56e4c01b6bc6aceb52b8a71d42679f1a3f
75f4822660414b4f845ef4bd2842092567aa62e95ddb6d526a736ba8997f9603
4283b946301f5e73f2858e3d6d06bde673ea7fda2e216c7c3671d63265130b74
9b8e5857b50467ace21b9510fabbbe67db2130b661428fca1b14f63e5128e1c4
ddc5b772057d79594e0f04f3abcc79bafb1656fb8503f8e66cf45aa43a12aedd7
ef87a0246023d44574e5c33e991b41b0343180f8ee8d01c13c0ef89d8377a6f10
18dfc7f5e457231178044e5c7a97f3dd3fa57bcbedd0029acdccaf49ad0b87ac8
206221fef62b400f48258005769a9b0b594a8ac9594d98089fa68d92dc56bb75
ca7976708f24bb6ea88f8337b9a2ccd325cb94e2da53d848fc906b8284664976
e926b244f037a9e7557e92aff27877367751af69d3a2a4cc569f1b2469b2996c20
03107f0e7d20e42af231d16a6a33a08cd4ecd67d4f67ba3cf1b24170c4cab2f3d
af32d20084ccb6f41dc4fb162252b532e104b938844edd6ed19ed048426a0239
8bf627797b783d2e43f0ea9fd4e993b4ca3eddbf096df9caf6bc66458a5c814d6
6beda2736e890d10f07d2bccfe9ca4f8be2a1d3c5c064561c92aadc0f6df0a92e
30bef83b11fa77673fdd27fd45ee47ab6c90122a3147423c5eeb8464432c83551
6b723693733e1233f9e4a8167d8142e21f06867e2a50e0be353304620dbb9f82
c6c530ec4edcaf303bfb8f7b76aafad9802a633ee26be69aa71a90ed413905994
7426787c88e43a9e1b2dc71e85b65e86640fdeb7fc0ebb60b2605fb965128501
8c7e2d82179af94745d1444f2923cf18c71b64903122f6b6a65c092f39b363f91
2e93fd252b6bd26594709997c0fa43b91b8c801d92793d54

Extracted quantum sequence: placard 7

4a3131d68a5ed5c74870abfedd84513895eff34fd16bdaa73912a48560a1c8af2
8724faf140af83a87c0deb24435b6fbae149a8f63a51eeaf86cbe90bb17bf8d9b3
84b8d22054d6b55c912251da2fe325bcf719b717d852e1d3cd3da7060e3cb2ce
6622c4f6d34f789d9235ae67e5c61fd98586f1398cf70cb37cdb7e7503c7859a3
1adedcea504980ebf2b27b1c7ed113dedcc353bc4f1a360742aa7bf4f46592491
d68ca27bf3c2e32acd9afdf796acb9cadcc87a4fefd387b77517e6fd39ca83df38
3a3ef08f8303bb062594e02e03fecb526cdca4319619f9f3005e322050e5f35787
192a040fb0aa6630927390bb0f809707e1e62ca2ada7fbcffa671c8a6c24f15be
03030dcd78f3475a48f8833cd6622d5302db6ddc3ebfae88af5d7068bcbb6899
81da2a5701efea0ac29f9b4761760231d852fd98959d23c51b1fce63469f07cc4e
70423028d2a6c5d66ef644eb2985f57698821e139b8822ff44541c9ce2a6af384
d795ef4d76fcb34d8d7badafa768b6e552c8236136e44ee5291b73c4ab0cfec09
65e120d8c2682c1137fe7fadab5b46a5f3396f767a31261221dd1f13b14299eb5
50c464a91265909406fa048f15e0eba9eeee7e31b4a76794a9b73094675c4d54f
09e82a6eb5e86fe1ed575ba77d9973a39b78e10d969d49f69d12556a6a6d7b09
866cf6df81fdfa8132805676edb1c9eb6949e6a48be8bdae8f03e6647373fd296
373de320e790e44d8007faad63dcb51fd6ad00cd3a9a6babfd54ffff0cab0d7f1e
cbbc0f55a0c8d7503e2160de23c9ef70566449a6bdfac7119af4fed1af9aef07cc
9db7fb77be2a02b23b087542c0e3d118d8ac3e6e4554fbdbd64adc2038c94a1
3b37962acafa3e2652a534e8c4053a94f00e1876953241c856817ee932802b0cd
dac2e913fbf6bdc5ae277bce6c6ae1b1519bdc32938fd5581fd2d42cb22343c7
026092a6a19f7c5857fc4b9726b401fed3b09dcce204277fe89df1ec7f1fa6142c
b5cd566148814e4f254fe50316ff16fe6c377d468060f2b437a270917f6f3e6b3a
4469619f249c46749eeb6ee4dd9ff2a304c2a0193cbd1b13c58097d751047fcfb
ddaa78c7d2c542c69822d3eacc73ee35a08decaea565281fa9d1fd89f3aab806
d7ecbaba60c2fc840320378f9fef80c4e762aae60b5bd6f1203565a9aa3ac40fee
395abc7836ad26189498987bd0117ee088c34e1a3dbb86f5ea66b5455729b60
610c19871137b307abaf083faca053b8be4b54b608dbe814e0ebab2edf63c16e
93419a68c5610cdcb5b2e6931777c58258a6c6d494d7a572d5b7785dd553d46
d1a437ef66d807268e9f59a465fd384a945b78190068f8b7e3762725d3804fcfc
6da499af43202dafa72c7a2a1f00855562a880df89c6dca5bdf1cd63ef7f431c4d
b8a503e4f52b94aff41fa890f1ffa4f7d1e

Extracted quantum sequence: placard 8

732d863a3602bd474728bdfd3cbbe9914ae29960a3e8fdeb613c99db885f0438
7719289d1e09d1a55cb8a9a533e67dc2f580067204ae4ac3e6fc43d7cb7b91db
7d2f4c0956876f4fdbf1647550a812a48f1b298072b1a24e893be00db487e33e1
24a061a3970bc721043aca900a98610d9aee9f650d86b6475d654fa4927f138d
1b3540b9b93ec357c50019242a0613b004cc6397fe10fa106b88fe7089d34a420
c6ae970cbcad56ecd903738e71dd911659dbac6b4fd4b8afef970da163b49c39
3e01294e4b02bde27287bce43c5b5b865ef1462f070941475cf56ff1318bea5de
d43ba62478dd5a4fa961768248322ff0848186666c24d22f21bb7d5363c1869a
0b1e98ec62148f91e1f42a5ad069820827b28875334251f87760b91186a36c0d
ebf0e9139fa255e0f4b6605115ae7dc03fe353e7ffd74d66c6c7a0e44b7c0d5ae1
4229a172a06450067b0c7b9fc96ca2616e65890e8b57f7daf2c659ffcf77706210
5891745fe6676e856ac0526e443b663500a91ea615f4ddaa6b778a36c7de92d2
e1397db5e01a0d57f026c8dc77655b53f524ae120e94b11c06db1e76340411a2
5da1770e69b3bcdf6bf5d854918b6e6b27c8978c89fc03b546f82ebe4ce7773e2
ed7327a02438d0c872c80d4ffe27c45983c354c459c9226a7a2f0339fe697aea1
2890a914e99366d924335e6b0a1b39a30a37e2b8fab2f940a472a36e7a0ea635
150f5458db5cbd1a19a7bea1f0d6794c7cc5bfa920e6f54078517170eb34d9f88
bb65aef6ebf1fda210882cc4c2871af5284892df852b263ef9ea362c608bf487c5
08680d2da98149b79e1617ed3975fb2bd87dc654fdd73c4ce1771a1ea96f4823
4e0cd445480be7932d6fb76aa07791f1803d837731164a88ce7e61714b88b973
301e97cd3c4ccd4604c167c1566d0668ab8dc0065e02abc748c06c6c9df759d2
2327fbf02200c5a845f659ca463ebe8ae44ccfe294c67b7549543559da3f9e5e87
a4ea6e24e7839a80d9969d6b98a1bc0e02685570ff9bd45ff1c8e227c9e9ecfc60
c15b40753a6380d2ebbf9f7dd30ced3afe41476b5641d1d0f990fd048f0d84e5f
f8d17d6da05a48a020fcdd7ec80b0fd9e1a6bcc91c5e80cb7964080b133bedb0
266838905dc2d6aa5545f50817bacce3467a71c2da58e55fb6d6b5dacd938832
ddda2ccccda7c6ba89b80f2447711792657f98e948046b5814582d95236e2458
3cc67b79066e3e265c64a9644ec8a8bf9b3d6e2b71fe1c2d53d3b8926793d554
737fd35271077f6ce02f0a5084e5dd05afe66cac0b32a6e43ecfd9c50dfa0a3b79
4d433eaf5ce17218dd5f79a43bff4e5b3a142fe20ccd7aed7e880ce153cfd5f7a0
44427087dbb2590f615562aad7669f1bc34b4883e4556bfc6cb9559c55bf16b7
aa3c93e2cf9c1c2ae8944970f5724263b69b537b18bc

Extracted quantum sequence: placard 9

113649c8b43756466d49b96b70fb8b2266bd5d916beb9f5f12caa5c09ecebf3d
9ef91d2747c1104af2be24194a5b88cf997bf6892cd48e329ae0684c9a34e3016
7caa573fec0cb54504cd12c02efe5a064eed8b973cee500e6e30a1682d8850c5fd
7df738ccd5f02763d5af51de55521259558a7b04e3a6b3f2d77825e35f107adb6
128dd5558ef6f07ffaa4f1b5d2b037e39cac6c2cf2d498684190a78edbebc4ffcc
152387ccbdb6f5954f3de01d547e6eb578886b15ef4395268cad144f1824a623
94ba5ecefb6d90b5bdbd4078a7b859729298f14195645226281e2fa79922464f
a7867a03c38eb0f777e71142216664438966bbc8757ddc070aa6613ff72cf894b
9fdd244f0c99c5ba922ddbf30406f89f800a870b2022c0b5dcc23dae20d0abd0
48ca3e1ab37efd2bcab6f1175e46d63d7c7e5320190c6ec2695fc2e611ed63d2b
0f743cd1c3c3cacd535211741c5d32de4e585d2bbd064cfc088edc1c40d30d8d
a6c5cfa1c5f9fcefda528d74450b94a10479d850cf4b81c394ec992c17ff033041
bd9493241edfe5c0df0dfeaa7e60dea9cba83e73af8fd9b22722f3a001697459e
e059110e23d3c2325db5810a3be81653239f9e966bdeaca65296e855733c79c5
afbbd383442bf6e54f9d7fc1d001ffd703a1f6336b53a3d53f2187d304cc5c7283
dd0659c349ae7858703a2a7a3ba2ce4d5f8cd2111c5f2013ede18decc111d24e
4cb117385c886ef87b5346a7838975a9ee8f2cb7a60720349cdd28b0c042732e
b3028578b5c09a27784f7103bb00e0df0438326ab1ce6e26b49bdf58fee82e20
aa312dde9285dd3cf6952a7eaee0c3edd68f8b223507bf0e70a37bf0815f65c07
d767e459fdca1371db485248d19f461a09bb0839189bdd054f66db9b2a19235
2e123604595e9190a6991c181d7f5b10450bc29d0764cf87024f78a9095812c6
eb03e7fb88fc7ae8d23854ef2fed1dcb5eda5e3d614a0773033d041ee9e22be0d
aa9888a9af402ca4b928c2f74306937bfbbf13fbbcc019f00b44bf7899c9975e8
2dde62a2d78ee7f6150c12c1e575a6259688a19b131fc13655a2c6ea6ba38752
5cee8839f7ca9f0e276495542969156e90da5db8b5d53280b1ab8688305a203c
f6be39e8c24c6801c7700ed757f4cfaa0a5bb58ba27993ca8f1169e9f7d2bbbc5
e98b2bdc601e451658f3d6f8b4d0d3c9ec48502e03663005ef8a779ed3b6a943
f6748884fda0d314ca2b80cceba57e5260e86b76b80b87aa9b909b06266aebcc
0bce2bfeaab9c7a43ca252fcc94c2e451bf746f347ab18d7101e3dee83263339c
546b71a854d4fcb824c4e9b1c281f03eae92b171c9914895bb2fdc09f7e26bf93
4134520b1a48cd18cab760d1b06db309cc3670d8bf08f362f808cdf995154614
05d7d5760eda223f3814f976f8c8073cd8a6bdc000ef33

Extracted quantum sequence: placard 10

f3bd6cefa900a507312d1bfc54dfedf2ca8e0f1ea53f937c03786711a2c41b7ff06
804de41c018109ef5f0a4030527c171334e90a06c6e556cde5f4c60e892835aa6
ef2610a1f3f9ca88a7edad54a0b96107e983de0e3717726381f366a92377a46c6
722b191d45153a1c81d7a2fc344fb36b93eb585986d3e2eca0284d190e00dea2
75d58b1f040f48b92e405b2d66d7b86af939ba0014f58c7208400ae0c4c877df0
04e0435eeb980d79adfc2522388795147732ad54eeeae01ba12d15113be6d926
ff5d338082f8970e8a90c2696fa02b6e69c4ab0baba92f8952de9876c23d0d311
e5ff1e62fe85429381bc97ff5e3da52bc1f3e64970f31211e24916a5f1f5d161b15
7e25abf71d25bc06135eda67ef23824ddcf53e865a987339ef044dd5775f31689
9d08fe3bf44f911e28d99dc03113dfcca4aee0bdf1adf024517f09b6bd0a1c4f66
a57ef11c2ebddd880a68724b802b7f82bffdd4c6b0c62cc9ada613be360fbcd9
b351afb08372d657afa76823e4dd286b112d7b5259140ef96583d35b06fd68cc
15ba089eefc38d610a3e792781cd8414b72345540bfa490b75b9c95c9ac7b616
dc56a27bd6a98f1fc2de890d73013ae799e120284dfa72eb97f0cc00fc0f54b052
4e5c0e1d0ef4f079440c86f2213f9586875e524d011f8c5761ae3b88a084d7b9a3
40c5fd9488ed77e1e90a44f22a60b80451f6b89b403840eaf5bed0b79a29a2a2
99761af8e10ebf101259f64901500b5479ee96a3c80a634d862fa53a02f6b58d8
db5d445bdee36dd0bf1553cfc3c10dc84564e9af0c55eb02d7e6deb1da6d10e6
288a97fbb97d55f3384b0ea72089e03e2d5c3bcbca0759412a9d6d1217607aa
deb789a6479cc8c8dbb421d571a904a806c92299f06070eaddb3401e4326362
b4a9ada2361be2f98b9e7a80112c8b88b2f55e9d498c5209d1f2924c05e49880
03927589ab45afcd775ff06d045f748fed504ba21b1dc346fb8ff4527e449ab27e
4b30547a8a000439a03d4287f56de6981ccec99108b0102485dcc31de57d08ff6
f292661bc131fd358c3d083dd091d43d73f2440d4f3b9b49155d853927f655f3
7723f2ab121a03be432d3cdfa976b5c2914c18f5c129846c61610476c566a5667
2aa7621c6cece6a59af06dcddee7b12d52affa2397637798575d4fc07f7b47e47f
42f25ba96a2dfb264db849e275b8a20256b03835ec16d14134111bd0f73527f8
4380eeb293e29b451cf329021eccdc155da968d825ec5d18df15ceff3483980d8
9362dc0581fe32e1e22b30be4527b0c7ee3c137a24fac275a3fd50bd9353f69bb
36c1443017c30c6feb29b8c4249ec61d8ec25165a1e17287fa2ce71195484c340
3876da424e4fde98c80224b772bb58519a54e8c93fd3a321ba8a0e0777e0473f
c914fff43c401ae60d2fc2e9ff369e226d4bd90

Extracted quantum sequence: placard 11

9b359c54bcd58d6234d60e0da54c35be0eef8ca84a5412a17f13a21014655729
381d70b746001475bd274b165e82ac2f46dde6414137d06196ad887332bcb69
c3eb6046607e820bc08a3054a09ffaef9ec1e172c3fcaab1dc8f368ca89900dbb0
e181924222cb0805ee1022c7c4f175864519f6ed77cd7f5ebc43100c5297ab6bb
a811d475ea76f3ba02aa0c92a7ce3dee5e44510042699f9240d0cdb9a9cbef616
be02cc86f187b7f305d762de63ddcc35d921b7bf19f38a721e9bc42760aa2989e
365ce2a6a8c79b68282636b886eae69615f758acdb935feda3b3504c67ecd0c2
160e193be7bd0b135eabf1e8c16b22b43097343dc0a905f6ab5bdc3fd1757911
49db4b2570fb7864862c4ca6e6071084575dbbe81dee6cb857f1f7dc00f7cd267
19d8b2cf6c3b169c964cacb5e3f616ed099da3520db4a6a0f82b8e0e1031e0e4a
c7780cf5f07c20c773f87e00dee4ef2fd4059122fb30270fe44c586cc21b6238da8
4ddc645bcf2d48d456cc27f854593b99d6f78f781de4eab049d6aca9b0319a22
739fe4606f083d61536c978988139abb772ef80f863e40df5bdb4ee5bb09bde0
4e4ba55832232387b865033dddf93d74052a90c37a0f3ed7c3ec966fb0c10699
f9fec9fb1a2fec2beab0858ff2d7c824b9ad9a9435d7ad0fb108bcbfb755de5201
6edfd08433ac4828ae80b2ce33d8c1f25789d55a6dd9b8e73314c6377b877dea
298f5bc5e14642fa140ebf82c96d470c99ced922a875af382548ae47fef7563b63
a24c372aabb6c8cf9a96a7a5ee6cf7e7c9f11d917f291549802985378870467aa
febdeb6477d369271416cb27dc3c520261613f8137347956d942934ef5e1ff96d
0227721add50c6184ada6dadec583e8899272a3d211747fd58ced69e1f9112cf
3741f12910d03302ae25c7795508312fdf76e2e1908b8efb7d11f1787ba2318bd
7722cdbbb6124197a79eeda66972b44210894806ef928b8d5ef132c91e41fbb3
4c4cb1f29d0fe02e9714ceec3983242a8a20a4e72a2e737574a7fcb7aefeba1cd2
00f5049f222bdcc69f6a548577d44dbf5abf0fe6ba004629d8c993b97e3cffb841
6505e3d189ec82865e927d261e4959c0fa1ac5e5d2c5fbcf7aa3b4c294cce028c7
d52d30f187674b2c80d830f87209b7cc9cbd93c3e23262c12a97b8a73d666b9c
d075885df87fc381f5958b305a6e592c146286d0551a3594d4592f545753b434
b90be1ce389a1e289fc3e8e6f5a8842ac7dfb9ec4cbf7277a9b3bb6f0faff71258b
07c119a8b6ceea5805ca20afe88d7f485529861df93933307dc7fbf7c1a920248
ccbc545e30c90f031f97dc7726ca18798340516ab14687a3eb9311eb0455336b
583d9b747de4edb8683f7f31197bf13991746ef9cde42ff59700e27e3448930e17
5d99a863ccc24c5fe056014ee5ccc5caad70

Extracted quantum sequence: placard 12

c34c37cd6919d0be2a2ae0227f3099dd6bc2adfa2d4abc3c5242bdd8b352faab
e9a26c6352d771e025d9c4b9cfe97ead97ae3d22f25cc191930cfc43facbc75a
5ed8d705cb341571a06200e08ed55209ab9e9a03d2745942d2b6ec36fb9c2d74
67094937af975f3f116475a1d66cd8fe04a8ca8e393cdd688567b0271eb2cec3
26f5ebbca7ac1601fd36d5ba744abc78b1623afc590a45bf8c141f844fe8c451
d5b6d634050bec7031c485bd3cb09285300fef96b2340c965dcb2f1978f69fd3
a3b889abd552f4268e636deefb57b9c870fdc3275cce2dfd70637057e6ccf7307
3b6e650988d96aa1013e0ccb6ef9d81e36814041f3b4257c18454360e89de2cd2
af9328fceaef50db7e273c6cbf8ccf4354aeba24143445dfdea98a7ec01930b576
c6ec05cfd6ee29149a2719580f0cd0f67e43766b9d38104a0686bc06d4584cc352
fb1e8d4d58c7c742c45af63fea98a4ddde1932981e766302f2468e388f3d7bfc48
f9e8c062480a54971267f1ba1f96fa0db5a7dc76c8eb1b6fc7be7b378c0d944903
e93129cb7c1e6e3d3bc6a2ddd5006aadd26f9f49631d07ee16d515046da8209ef
479fbb7054a62732dec76253a5f1f4544ed5396036acaf04221e9a79d8d046e7b
c12cabef2f08f5c18d3fd11f1c2e018475e9e3e69909cb4c4303238971c4ea92fefa
29085d07d2116c979fa5e8ee16c0c9bf3928b13b2ad100bf27c5f33077bfa4af00
fe25127cfe39292c9d25fd75c2912d4820e55af4151e4ccd8e014f6e4717e56fde
9761fc55600999ac18d9559dba9b5af60a8ef239f6564a7e26f610388b64e7a426
c6e416968d722779744eab6d8a935cd24eea271ab8461156831ce82e0bfb0d46
4c76dc6fbfe3d669ba9ac2963fa466be4c1238dd5f0376091f22263749ff72e8c35
8480d1a9b1ec3e20848beb7595f81284d263bda6f6aeeadaa2cfce4797c4a16d4
e1157dd8638b5eaf0a2ec2f376d1838f9aeb856ca147c5444e35e8021e94e9e83
ae6e11cfd8dbd21d2f8fe15fd36729e5f8d7327a3a68c882aec1d009807f042d99
2cebd3e5d2213670999fc583ea3e4b8d6ed12de1d5485f110dbcc0f1a8dd10f972
8836e292ecac9ec13054d8f8ee39f52c8b1b3580fc6fafc0aa784ca5c15791a6bc
bf8a760f074ce7af3a2180f829524ecae3c82befa7237c7237a7bcbcdc01ac9000
a9d9a6c30668a64bca1d7415ce63c57c5f9478caf5d0b29c3de510ff4c6e375f54
75e5e118d36b2ed013381a13db984723ed3c5508d822238a409a596a9e17a6b
5f19f109401f7f4866580fb74ffb9f13aa8b47a5ef01c76ed059c308179dacaaf27
76abd0c463eb85f3fc63b2c4ba0686b4a9002283864410673c16ef63276cf793e
8555084046fee7165096206af3943b75e955311bf03009745585ec51b541dd496
39cbf8ab8218fa22a04e7a2

Extracted quantum sequence: placard 13

6a2ea0a6c131e8ceb001fbaae115e1bb9a0be047727ff8b053b78d16e443accdb
dfc26bedd946986aced538a7ee595c13be7483ec72bfbcad27a80b45b55652d7
1622b84b34aa3c9324775ae30f71b724d7fcd87e9bc5011a85b8aff14fb0ac32d6
929bfe9962a6be5c4ba62601f1591a8ea462e153eebb4d1ec92cc8c48e750bf02
36ce84a7ba3e961685dad81aaa61c4ad14745f127aade66bf0cac1f3e8677ea30
0380d09dc811312f741a5ef790d7e4e07ba28fe8426d5aae29b2909f5e6c6740d
892da41226da73b8a246b1a17c63532b57a306a9f590452ee68467c35a70378f
070b81e239a141c480e02b05584e29ee51f0676b5df8818a635135525ebca177
ca359f7caac1c0fab3e25f1f6f86171fcdfd32d24a1b6506d301a5804af58661401
b6edf4537d83504c4dbf172ee19b2c95c36ac9e2f0c39bf29314dcf2634d4cb62
bb5bf41168f24bffae8487253b1e463a2287f7fb30b11c557caef6f3bc526ed65
f1aef9a0fdc3acab67594f50f6a71f00595ba1839b96e77298c44943fa8885ea6
8609835959c6a027468e3eefa369d848b18190e777f1f4575ee5a6dd33d841549
ca3c3b87f3cf1cf07130ca781b1f0b5f8e04d463ac0400b58d66cd01b4aaccf0a
676318df1972e23bd35e072e2257bb399618f5972040c60edffb4cad1e9b11011
d5af7acb237bf1bbf9adba5f56a655ab0c7441d8fa98b2a7b9ec58efb85754261
a5756cafddf6477cfffa278ee76e1a8d6ae967dacfe593881136678404361112a9
662f6cd6d31451c95e822fa22d1c143dcada299000860771c62f8ec9dca9d1f558
0eafcc65f11f6845cba45968795bc936acce959a3bd55f1fc5c09b77e80327c28c
bacd486e9b321ac49d2e23165dc4597260f9d882f0647c0904d8a7c71c64d4647
fa8d08cbf80df6e108d60c3738dcf43d7b19ddfbc57f8fbcdc7ca44c064648f80a6
55f2de56e56f711a75a15a2cd31845f92855b140310c7b571c3b42e04acf572ea7
3ade3520ccffc77797e5d6904240bef3e7f6f02dbde9f1c769d305b1d642dce4a8
dd435d376e7c0d66d2ad5d15cb80f6e918bee9aff715b680a3c1a9ff33545aa20a
d76a30a632f197bf2d19f66ee3efd84027113fe34c44ba8d39fb36ec59f026607f
61d1a879d2acff0e5f05bfabc43bf89ebe5d59c7abd07951ef43380e3c5510f905
1fd9d0e74b8b13a5aa6ef4201eda46830e441d271b280f00951794f6f3a3bfceb5
88cc55e953d6dbc0e8e7dc0013a3af784848a0869844296f84349fecfe6076c45b
deaae083b6a48555763df0c20b042ef130c8ce60e235d60491a3d33b99809298
5898a6804d25d526544a2a21bb936de973e485a82cff287ad4c19abfa012e2c3
6183734ba89535fc8cfb8ba6fb3cdc9d2df5c3790511da070a6467a4a1e39cd9
5dadbb8238b90b9eef22d9d0

Extracted quantum sequence: placard 14

dfcfbfcaded65a43efe13a4c1853c8ea9168be09978986b5ae008b8f324fce2e7d
2cd13edcab0a0450946a445332bf7003d96cb75f799270e23cb6f2277fb512c61f
f9ce10e1d2847c19c95c02f646e17c9e2606914404d03eafca222bff9c5aa64181
e3c5017575cf7df53e6fe4e306a08ca6e6564366080da334a00dd2c182c69cdaa4
9773d07b7d530b2a8d22814e4d45606461d6f73592813b1cc9bf6cd5405a90c9
5e3e3d465ff7acf32fa06a39fdff57a7a361d52d472b5b0afdefc1574aaa4508480a
2e97276b8e52335a4050b97d315aa7bcf98dc972eac6d3d5f9da59079e7490d9
ba98c4b075fac5c24057de2e086baed1dc5c99ff1c890077c56b7c92ff65209f4d
6460eafb5e34213018545ede07c6e9983a626126050e8e0b4824bee7bbd8390f6
1d0edd83b0c88caf67958d685f763e92537ccf27c320d3d77bc2c3e5598eacb6fc
e68639ac35b8138c18e4a4b14ca8a645ecac9fecbbd029ae3397ce0ce8b45a277e
ab7921b73f8d0f4e6c767e7762f7b51a2b10c809e50063ba610d7776320f67f4c5
5e68a35c737843ca7e2e4d8565be01e0430c9383eb034065439e6e7f51e2ea3b2
87900bcc7bc3cb8eae94b4feaf66179ca868cce209e0e92be3c6f6e67419d16b20
ff20b4d83953505cacfb9ff6f45b3e5ac7f3321c03955e42c0cd53c212aeccb7af8
9f6902dc151e70aa59f8c33a4b2292bfb850c069dc20371313357dbbabba1c79c
c420873861d2659f965fe4ce5429cf89fdc7a68c8ec3c55db9ee47b8934796263
be34f115e84b030230f477f23a218360db2147a52498792774890d53c85525ba9
ddd0a396ff40289d00a48288954313147c93b23e268de4da990249aa60b7e99c2
70a77dbc6df922b8ac6ef0e9046653cfb8e37e0fd37e46d5dca035afd557c98ea
9597590250fefed9694e16d6d48466dffe7a6c21b42b265d5fcdc4436ef0e6866
a445e44b7e73e8828f04b2eca598e5cd4450fea2b0ff1ce3e31baa4bcfe240ea6
3d050b41a403277c4ffab6701b4d2470cf6263ddd756d43d4bf79650ee24df736
07be06180cbdca5a1cc38cdd8742e61fb6cb6228cab6c95bedd5ca25a8ee4629
2487c3254b8bdb674beec63cf915c031295e3051b8cea36ed2a0810866948892
43ee7443d0273ef4a1ca30a6c6c3f4c327c369d8fe74e088d0493ef41e5af0f187b
629511d6fd67fc41b91cf3dcc18d7ba4eb36651a4f5eae1acc1e50f80edf41380
9e41182361e83a68cf6c6fe084d547331c915c8d9ab7448a46d5443c5ca991eb
1de766aab76d13f875f2d2db1d2b6568b2245219b24068b4d88db8926e7692b
3d749c546350074a31123ced8323b7f626a0bb4cdda4e8d608b39722805cf903
9c9d2b461990f95bd85e6f8c0fdb0a4cad994d7982b908bdb69145cb6066513a
ae9cbba77237086afd2f474b71

Bibliography

[1] Adjih, C., Jacquet, P., and Viennot, L. Computing connected dominated sets with multipoint relays. Dissertation INRIA, 2002.

[2] Rolf, B. Graphs to Prof. Sommerfeld's attenuation formula for radio waves. *Proceedings of the Institute of Radio Engineers* 18(3), 391–402 (1930).

[3] Aho, A. V. and Hopcroft, J. E. *The Design and Analysis of Computer Algorithms*. Pearson Education India, 1974.

[4] Arnold, H. W., Murray, R. R., and Cox, D. C. 815 MHz radio attenuation measured within two commercial buildings. *IEEE Transactions on Antennas and Propagation* 37(10), 1335–1339 (1989).

[5] Aspect, A., Dalibard, J., and Roger, G. Experimental test of Bell's inequalities using time-varying analyzers. *Physical Review Letters* 49(25), 1804 (1982).

[6] Avrachenkov, K., Jacquet, P., and Sreedharan, J. K. Distributed spectral decomposition in networks by complex diffusion and quantum random walk. In *IEEE INFOCOM 2016 — The 35th Annual IEEE International Conference on Computer Communications*. IEEE, 2016.

[7] Whitman, G. M., Kim, K.-S., and Niver, E. A theoretical model for radio signal attenuation inside buildings. *IEEE Transactions on Vehicular Technology* 44(3), 621–629 (1995). DOI: 10.1109/25.406630.

[8] De Prycker, M. *Asynchronous Transfer Mode solution for Broadband ISDN*. Prentice Hall International (UK) Ltd., 1995.

[9] Baccelli, F., Klein, M., Lebourges, M., and Zuyev, S. Stochastic geometry and architecture of communication networks. *Telecommunication Systems*, Springer 7, 209–227 (1997).

[10] Baccelli, E., Jacquet, P., Mans, B., and Rodolakis, G. Highway vehicular delay tolerant networks: Information propagation speed properties. *IEEE Transactions on Information Theory* 58(3), 1743–1756 (2011).

[11] Barjavel, R. *Le Voyageur imprudent: roman extraordinaire (Future Time Three)* (1 Vol., 255 p.). Paris: Denoël, 1944.

[12] Barrett, M. D., Chiaverini, J., Schaetz, T., Britton, J., Itano, W. M., Jost, J. D., Knill, E., *et al.* Deterministic quantum teleportation of atomic qubits. *Nature* 429(6993), 737–739 (2004).

[13] Leiner, B. M., Cerf, V. G., Clark, D. D., Kahn, R. E., Kleinrock, L., Lynch, D. C., Postel, J., Roberts, L. G., and Wolff, S. S. The past and future history of the internet. *Communications of the ACM* 40(2), 102–108 (1997).

[14] Bell, J. S. On the Einstein Podolsky rosen paradox. *Physics Physique Fizika* 1(3), 195 (1964).

[15] Bekenstein, J. D. Universal upper bound on the entropy-to-energy ratio for bounded systems. *Physical Review D* 23(2), 287–298 (1981).

[16] Benettin, G., Galgani, L., and Strelcyn, J. M. Kolmogorov entropy and numerical experiments. *Physical Review A* 14(6), 2338 (1976).

[17] Brillouin, L. Maxwell's demon cannot operate: Information and entropy. I. *Journal of Applied Physics* 22(3), 334–337 (1951).

[18] Brillouin, L. *Science and Information Theory*. Mineola, N.Y.: Dover Publications (1956).

[19] Bouillard, A. and Jacquet, P. Quasi black hole effect of gradient descent in large dimension: Consequence on neural network learning. In *ICASSP 2019-2019 IEEE International*.

[20] Bronstein, M. M., *et al.* Geometric deep learning: Going beyond euclidean data. *IEEE Signal Processing Magazine* 34(4), 18–42 (2017).

[21] Burnside, G., Milioris, D., and Jacquet, P. One day in Twitter: Topic detection via joint complexity. In *SNOW-DC@WWW*, 2014.

[22] Cohen-Tannoudji, C., Diu, B., and Laloë, F. *Quantum Mechanics, Volume 3: Fermions, Bosons, Photons, Correlations, and Entanglement*. John Wiley & Sons, 2019.

[23] Darwin, C. *On the Origin of Species by Means of Natural Selection, or the Preservation of Favoured Races in the Struggle for Life* (1st edn.). London: John Murray, 1859.

[24] Detlefsen, M. and Luker, M. The four-color theorem and mathematical proof. *The Journal of Philosophy* 77(12), 803–820 (1980).

[25] Diacu, F. The solution of the n-body problem. *Mathematical Intelligencer* 18(3), 66–70 (1996).

[26] Forouzan, B. A. *TCP/IP Protocol Suite.* McGraw-Hill Higher Education, 2002.

[27] Black, U. D. *IP Routing Protocols: RIP, OSPF, BGP, PNNI, and Cisco Routing Protocols.* Prentice Hall Professional, Hoboken, New Jersey, USA, 2000.

[28] Chukin, V., Mikhailova, D., and Nikulin, V. Two methods of determination of ice crystal fractal dimension. *Science Prospects* (2012).

[29] Clausen, T. and Jacquet, P. Optimized link state routing protocol (OLSR), IETF RFC 3626, Fremont, California, United States, 2003.

[30] Clausius, R. *The Mechanical Theory of Heat.* Macmillan, London, UK, 1879.

[31] Drmota, M. and Szpankowski, W. Precise minimax redundancy and regrets. *IEEE Transactions on Information Theory* IT-50, 2686–2707 (2004).

[32] Einstein, A., Podolsky, B., and Rosen, N. Can quantum-mechanical description of physical reality be considered complete? *Physical Review* 47, 777–780 (1935).

[33] Van Erven, T. and Harremos, P. Rényi divergence and Kullback-Leibler divergence. *IEEE Transactions on Information Theory* 60(7), 3797–3820 (2014).

[34] d'Espagnat, B. A la recherche du réel. *Le Journal de Physique Colloques* 42(C2), C2-99 (1981).

[35] Everett, H., Wheeler, J. A., DeWitt, B. S., Cooper, L. N., Van Vechten, D., and Graham, N. In DeWitt, B. and Graham, R. N. (eds.), *The Many-Worlds Interpretation of Quantum Mechanics.* Princeton Series in Physics. Princeton, New Jersey: Princeton University Press, 1973.

[36] Feynman, R. P. Quantum mechanical computers. *Foundations of Physics* 16(6), 507–532 (1986).

[37] Feynman, R. P. Simulating physics with computers. *International Journal of Theoretical Physics* 21(6/7), 133–153, (2018).

[38] Fredkin, E. Trie memory. *Communications of the ACM* 3(9), 490–499 (1960).

[39] Garey, M. R., Johnson, D. S., and Stockmeyer, L. Some simplified NP-complete problems. In *Proceedings of the 6th Annual ACM Symposium on Theory of Computing*, 1974.

[40] Le Gall, D. MPEG: A video compression standard for multimedia applications. *Communications of the ACM* 34(4), 46–58 (1991).

[41] Geng, H. and Karch, A. Massive islands. *Journal of High Energy Physics* 2020(9), 121 (September 2020).

[42] Giddings, S. Black holes and massive remnants. *Physical Review D* 46(4), 1347–1352 (1992).

[43]   Gödel, K. An example of a new type of cosmological solutions of Einstein's field equations of gravitation. *Reviews of Modern Physics* 21, 447 (July 1, 1949).

[44]   Grossglauser, M. and Tse, D. N. C. Mobility increases the capacity of ad hoc wireless networks. *IEEE/ACM Transactions on Networking* 10(4), 477–486 (2002).

[45]   Grover, L. K. A fast quantum mechanical algorithm for database search. In *Proceedings of the 28th Annual ACM Symposium on Theory of Computing*, 1996.

[46]   McGuire, S. E., Deshazer, M., and Davis, R. L. Thirty years of olfactory learning and memory research in *Drosophila melanogaster*. *Progress in Neurobiology* 76(5), 328–347 (2005).

[47]   Gupta, P. and Kumar, P. R. The capacity of wireless networks. *IEEE Transactions on Information Theory* 46(2), 388–404 (2000).

[48]   Harrison, E. R. Why the sky is dark at night. *Physics Today* 27(2), 30–36 (1974).

[49]   Hartle, J. B. Generalized quantum theory in evaporating black hole spacetimes. *Black Holes and Relativistic Stars* (p. 195), 1998.

[50]   Hartley, R. V. Transmission of information 1. *Bell System Technical Journal* 7(3), 535–563 (1928).

[51]   Hauben, M. History of ARPANET. *Site de l'Instituto Superior de Engenharia do Porto* 17, 1–20 (2007).

[52]   Hawking, S. W. Breakdown of predictability in gravitational collapse. *Physical Review D* 14(10), 2460–2473 (1976).

[53]   Hawking, S. W., Perry, M. J., and Strominger, A. Soft hair on black holes. *Physical Review Letters* 116(23), 231301 (May 1, 2016).

[54]   Heisenberg, W. and Bohr, N. Copenhagen interpretation. *Physics and Philosophy* 16, 40–50 (1958).

[55]   Hendrikx, H., Bach, F., and Massoulié, L. Accelerated decentralized optimization with local updates for smooth and strongly convex objectives. In *The 22nd International Conference on Artificial Intelligence and Statistics*. PMLR, 2019.

[56]   Huffman, D. A. A method for the construction of minimum-redundancy codes. *Proceedings of the IRE* 40(9), 1098–1101 (1952).

[57]   Huffman, W. C. and Pless, V. *Fundamentals of Error-Correcting Codes*. Cambridge University Press, Cambridge, UK, 2010.

[58]   Jacquet, P. and Szpankowski, W. Asymptotic behavior of the Lempel-Ziv parsing scheme and digital search trees. *Theoretical Computer Science* 144(1–2), 161–197 (1995).

[59]   Jacquet, P., *et al.* Optimized link state routing protocol for ad hoc networks. In *Proceedings. IEEE International Multi Topic Conference, 2001. IEEE INMIC 2001. Technology for the 21st Century.* IEEE, 2001.

[60] Jacquet, P., *et al.* Performance analysis of OLSR multipoint relay flooding in two ad hoc wireless network models. Dissertation INRIA, 2001.

[61] Jacquet, P., Szpankowski, W., and Apostol, I. A universal predictor based on pattern matching. *IEEE Transactions on Information Theory* 48(6), 1462–1472 (2002).

[62] Jacquet, P. Shannon capacity in poisson wireless network model. *Problems of Information Transmission* 45, 193–203 (2009).

[63] Jacquet, P. Naissance de la théorie de l'information. Bibnum. *Textes fondateurs de la science* (2009).

[64] Jacquet, P., Mans, B., Muhlethaler, P., and Rodolakis, G. Opportunistic routing in wireless ad hoc networks: Upper bounds for the packet propagation speed. *IEEE Journal on Selected Areas in Communications* 27(7), 1192–1202 (2009).

[65] Jacquet, P. and Szpankowski, W. *Analytic Pattern Matching: From DNA to Twitter*. Cambridge University Press, Cambridge, UK, 2015.

[66] Jacquet, P. AI vs information theory and learnability. In *IHES Nokia 2nd Joint Workshop*, 2019. https://www.youtube.com/watch?v=Scdy5mJ83is-&list=PLx5f8IelFRgFY7udFWZRLkxtmYwGVgOBh&index=5.

[67] Jacquet, P., Shamir, G., and Szpankowski, W. Precise minimax regret for logistic regression with categorical feature values. *Algorithmic Learning Theory*. PMLR, 2021.

[68] Jacquet, P. Information theoretic study of COVID 19 genome. *Entropy* 26(3), 3 (2022).

[69] Jacquet, P. Is quantum tomography a difficult problem for machine learning? *Physical Sciences Forum* (MDPI), 5(1) (2023).

[70] Philippe Jacquet. Is Quantum Tomography a difficult problem for Machine Learning? MAXENT 2022. Corrected Version hal-03942607v2.

[71] Knuth, D. E. *The Art of Computer Programming* (Vol. 3). Reading, MA: Addison-Wesley, 1973.

[72] von Koch, H. Sur une courbe continue sans tangente, obtenue par une construction géométrique élémentaire. *Arkiv för matematik, astronomi och fysik* (in French), 1, 681–704 (1904). JFM 35.0387.02.

[73] MacWilliams, F. J. and Sloane, N. J. A. *The Theory of Error-Correcting Codes* (Vol. 16). Elsevier, Berlin, Germany, 1977.

[74] Barnsley, M. F., Devaney, R. L., Mandelbrot, B. B., Peitgen, H. O., Saupe, D., Voss, R. F., and Voss, R. F. *Fractals in Nature: From Characterization to Simulation* (pp. 21–70). New York: Springer, 1988.

[75] Bordenave, C., Lelarge, M., and Massoulié, L. Non-backtracking spectrum of random graphs: Community detection and non-regular Ramanujan graphs. In *2015 IEEE 56th Annual Symposium on Foundations of Computer Science*. IEEE, 2015.

[76] Maccone, L. Quantum solution to the arrow-of-time dilemma. *Physical Review Letters* 103, 080401 (2009).

[77] Chlamtac, I., Conti, M., and Liu, J. J.-N. Mobile ad hoc networking: Imperatives and challenges. *Ad Hoc Networks* 1(1), 13–64 (2003).

[78] Jacquet, P., Mans, B., and Rodolakis, G. Information propagation speed in mobile and delay tolerant networks. *IEEE Transactions on Information Theory* 56(10), 5001–5015 (2010).

[79] Jacquet, P. Capacity of simple multiple-input-single-output wireless networks over uniform or fractal maps. In *2013 IEEE 21st International Symposium on Modelling, Analysis and Simulation of Computer and Telecommunication Systems*. IEEE, 2013.

[80] Landauer, R. Information is physical. *Physics Today* 44(5), 23–29 (1991).

[81] de Laplace, P. S. *Essai philosophique sur les probabilités*. Paris: Madame Veuve Courcier, 1814.

[82] Mathur, S. D. The fuzzball proposal for black holes: An elementary review. *Fortschritte der Physik* 53(7–8), 793–827 (15 July, 2005).

[83] Mathur, S. D. The information paradox: A pedagogical introduction. *Classical and Quantum Gravity* 26(22), 224001 (21 November, 2009).

[84] Nguyen, D. and Minet, P. Analysis of MPR selection in the OLSR protocol. In *21st International Conference on Advanced Information Networking and Applications Workshops* (AINAW'07) (Vol. 2). IEEE, 2007.

[85] Busson, A., Mitton, N., and Fleury, E. Analysis of the multi-point relay selection in OLSR and implications. *IFIP Annual Mediterranean Ad Hoc Networking Workshop*. Boston, MA: Springer US, 2005.

[86] Mohri, M. Semiring frameworks and algorithms for shortest-distance problems. *Journal of Automata, Languages and Combinatorics* 7(3), 321–350 (2002).

[87] Muller, H. J. Radiation and genetics. *The American Naturalist* 64(692), 220–251 (1930).

[88] Myles, A. J., *et al.* An introduction to decision tree modeling. *Journal of Chemometrics: A Journal of the Chemometrics Society* 18(6), 275–285 (2004).

[89] von Neumann, J. Wahrscheinlichkeitstheoretischer Aufbau der Quantenmechanik. *Göttinger Nachrichten* 1, 245–272 (1927).

[90] O'shea, T. and Hoydis, J. An introduction to deep learning for the physical layer. *IEEE Transactions on Cognitive Communications and Networking* 3(4), 563–575 (2017).

[91] Papadimitriou, C. H. and Steiglitz, K. Some complexity results for the traveling salesman problem. In *Proceedings of the 8th Annual ACM Symposium on Theory of Computing*, 1976.

[92] Parrington, A. J. Mutually assured destruction revisited, strategic doctrine in question. *Airpower Journal* (Winter 1997). Archived 2015-06-20 at the Wayback Machine.

[93] Pennebaker, W. B. and Mitchell, J. L. *JPEG: Still Image Data Compression Standard.* Springer Science & Business Media, Berlin, Germany, 1992.

[94] Jacquet, P. and Popescu, D. Self-similarity in urban wireless networks: Hyperfractals. In *2017 15th International Symposium on Modeling and Optimization in Mobile, Ad Hoc, and Wireless Networks* (WiOpt). IEEE, 2017.

[95] Penrose, R. The basic ideas of conformal cyclic cosmology. *AIP Conference Proceedings 11* (American Institute of Physics), 1446(1) (2012).

[96] Popescu, D., Jacquet, P., Mans, B., Dumitru, R., Pastrav, A., and Puschita, E. Information dissemination speed in delay tolerant urban vehicular networks in a hyperfractal setting. *IEEE/ACM Transactions on Networking* 27(5), 1901–1914 (2019).

[97] Preskill, J. Do black holes destroy information? In *International Symposium on Black Holes, Membranes, Wormholes, and Superstrings*, 1992.

[98] Błaszczyszyn, B., Jacquet, P., Mans, B., and Popescu, D. Energy and delay trade-offs of end-to-end vehicular communications using a hyperfractal urban modelling. *Annals of Telecommunications* 78(5), 363–381, (2023).

[99] Qayyum, A., Viennot, L., and Laouiti, A. Multipoint relaying for flooding broadcast messages in mobile wireless networks. In *Proceedings of the 35th Annual Hawaii International Conference on System Sciences*. IEEE, 2002.

[100] Raju, S. Lessons from the information paradox. *Physics Reports* 943, 1–80 (January 2022).

[101] Riebe, M., Häffner, H., Roos, C. F., Hänsel, W., Benhelm, J., Lancaster, G. P. T., Körber, T. W., Becher, C., Schmidt-Kaler, F., James, D. F. V., and Blatt, R. Deterministic quantum teleportation with atoms. *Nature* 429, 734–737 (2004).

[102] Rivest, R. L., Shamir, A., and Adleman, L. A method for obtaining digital signatures and public-key cryptosystems. *Communications of the ACM* 21(2), 120–126 (1978).

[103] Rosen, E. C. Vulnerabilitiesof network control protocols: An example. *Computer Communication Review* (July 1981).

[104] Scarani, V., Iblisdir, S., Gisin, N., and Acin, A. Quantum cloning. *Reviews of Modern Physics* 77(4), 1225 (2005).

[105] Schrödinger, E. Die gegenwärtige Situation in der Quantenmechanik (The present situation in quantum mechanics). *Naturwissenschaften* 23(48), 807–812 (November 1935).

[106] Shannon, C. E. A mathematical theory of communication. *The Bell System Technical Journal* 27(3), 379–423 (1948).

[107] Shannon, C. E. A mind-reading machine. *Bell Laboratories Memorandum* (1953).

[108] Sierpinski, W. Sur une courbe dont tout point est un point de ramification. *Comptes Rendus de l'Académie des Sciences* (Paris), 160, 302–305 (1915).

[109] Jacquet, P., Malik, S., Mans, B., and Silva, A. On the throughput-delay trade-off in georouting networks. In *2012 Proceedings IEEE INFOCOM* (pp. 765–773). IEEE, March 2012.

[110] Shor, P. W. Polynomial-time algorithms for prime factorization and discrete logarithms on a quantum computer. *SIAM Review* 41(2), 303–332 (1999).

[111] Sklar, B. Rayleigh fading channels in mobile digital communication systems. I. Characterization. *IEEE Communications Magazine* 35(7), 90–100 (1997).

[112] Symul, T., Assad, S. M., and Lam, P. K. Real time demonstration of high bitrate quantum random number generation with coherent laser light. *Applied Physics Letters* 98(23) (2011).

[113] Teles, S., Lopes, A. R., and Ribeiro, M. B. Fractal analysis of the UltraVISTA galaxy survey. *Physics Letters B* 813, 136034 (2021).

[114] Shtarkov, Y. M. Universal sequential coding of single messages. *Problems of Information Transmission* 23(3), 3–17 (July-September 1987).

[115] Szpankowski, W. and Grama, A. Frontiers of science of information: Shannon meets Turing. *Computer* 51(1), 28–38 (2018).

[116] Tipler, F. J. Rotating cylinders and the possibility of global causality violation. *Physical Review D* 9(8), 2203 (1974).

[117] Regge, T. and Wheeler, J. A. Stability of a Schwarzschild singularity. *Physical Review* 108(4), 1063 (1957).

[118] Turing, A. M. On computable numbers, with an application to the Entscheidungsproblem. *Journal of Mathematics* 58(345–363), 5 (1936).

[119] Jacquet, P. and Viennot, L. Remote-spanners: What to know beyond neighbors. In *2009 IEEE International Symposium on Parallel & Distributed Processing*. IEEE, 2009.

[120] Weisberg, S. *Applied Linear Regression* (Vol. 528). John Wiley & Sons, Hoboken, New Jersey, 2005.

[121] Winters, J. H. Optimum combining in digital mobile radio with cochannel interference. *IEEE Transactions on Vehicular Technology* 33(3), 144–155 (1984).

[122] Winters, J. On the capacity of radio communication systems with diversity in a Rayleigh fading environment. *IEEE Journal on Selected Areas in Communications* 5(5), 871–878 (1987).

[123] Wootters, W. K. and Zurek, W. H. The no-cloning theorem. *Physics Today* 62(2), 76–77 (2009).

[124] Wright, R. E. Logistic regression (1995).

[125] Bouwmeester, D., Pan, J.-W., Mattle, K., Eibl, M., Weinfurter, H., and Zeilinger, A. Experimental quantum teleportation. *Nature* 390, 6660, 575–579, Nature Publishing Group, UK, London (1997).

[126] Ziv, J. and Lempel, A. Compression of individual sequences via variable-rate coding. *IEEE Transactions on Information Theory* 24(5), 530–536 (1978).

[127] Zhangozha, A. R. On techniques of expert systems on the example of the Akinator program. *Journal of Artificial Intelligence and Soft Computing Techniques*, 7(1), 26, MDPI, (2023).

[128] Zhihui, N., Lichun, W., Ming-hui, W., Jing, Y., and Qiang, Z. The fractal dimension of river length based on the observed data. *Journal of Applied Mathematics* 1–9 (2013).

[129] Jacquet, P. and Joly, V. (2000, June). Capacity of retro-information channels. In 2000 IEEE International Symposium on Information Theory (p. 181), RR-3836, INRIA. 1999. inria-00072821.

[130] Jacquet, P. and Joly, V. Retro-information implies quantum unitary violation even in absence of Closed Time Curves. Poster in 2024 International Symposium on Information Theory, hal-04819138 (2024).

Index

A

Additive noise, 20, 22–23, 27–28
Adelaide (34° 55 S, 138° 35 E), 145–146, 148
Akinator, 163
Arpanet. *see* Internet
Arpanet accident, 76–79
Artificial Intelligence (AI), 153–195
Artificial networks, 164
Aspect (Alain), 205, 225
Asynchronous Transfer Mode (ATM), 79, 81
Atomic physics, 204

B

Back to the future, 262–263
Bacteria, 157–158
Bees protocol, 112–113
Bekenstein (Jacob), 248–249
Bell (John), 205
Bell inequality, 225–229, 231, 235–236
Bell state, 222, 228–229, 233, 235, 261, 290
Bernoulli channel, 35–36, 57–58
Bernoulli source, 33, 58
Biomass, 157, 159–160
Black hole, 246–249, 266–269, 281
Black hole evaporation, 268
Bohr (Niels), 238
Boltzmann (Ludwig), 199

Boolean satisfiability, 217
Boson, 213
Brillouin (Léon), 5, 201
Broadcast network, 62

C

Canyon effect, 145–148
Causality, 23, 223–225, 261–266
Celestial mechanics, 124–125
Channel capacity. *see* Mutual
Channel coding, 34–40
Channel entropy. *see* Conditional
Chappe (Christian), 43
Clausius (Rudolf), 199
Closed Timelike Curves (CTC), 265–266, 279–280, 291
Coding, 34–51
Compression rate, 41, 56–58, 101
Conditional entropy, 22–23, 35–36

D

Dark night sky paradox, 139–140
Darwin (Charles), 156–157
Data. *see* Pyramid of information
Data compression, 31–59, 166
Deep learning, 164–166, 170, 174, 176, 188, 193
Definition of information, 209
Delay Tolerant Networks (DTN), 116, 118

Density operator, 215–217, 255–256,
  261, 271–276
Deoxyribonucleic acid (DNA), 156,
  160
Deterministic entropy, 18, 33, 39, 156
Dijkstra (Edsger), 75, 79
Distance vector protocol, 69
Distributed systems, 62–66
Divorzio all'Italina, 243
Dominating set, 90, 273–277, 280, 292
Drosophila melanogaster, 188–189

E

Einstein (Albert), 205, 223, 266
Einstein-Podolsky-Rosen (EPR),
  223
Electron, 204–205, 207–209, 212–214,
  221–222, 232–236, 239–240, 281
Energy differentiated theorem,
  126–128, 136–137
Entanglement, 222–233, 235–236,
  238, 260–261
Entropy, 18, 22–23, 31–36, 39, 44, 48,
  51, 54, 56, 81, 106, 156, 166,
  199–203, 215–217, 242–243,
  247–249, 285
Entropy of Markov chains, 59
Erdos–Renyi graph model, 90–92
Error correction, 29, 34–40, 93. *see*
  Channel
Euclidean geometry, 110–111, 128,
  132–133, 138, 144, 278
Euler Gamma function, 122
Event horizon, 246–248, 269

F

Fading, 121, 106, 112, 119–124, 126,
  128, 136, 149–150
Fermion, 213
Feynman (Richard), 217, 221
Flooding optimization, 96–99
Fourier (Joseph), 186
Fractal dimension, 132–133, 140,
  143–144
Fractal geometry, 129–140

G

Gateway Protocol (BGP), 74–76,
  78–79, 86, 88, 96, 98–99
Gaussian noise, 125, 198
Genetic heritage, 159
Geo-routing, 115–117
Geometry, 104, 107, 118–129, 136,
  140, 242, 246
Gödel (Kurt), 266
Grand father paradox, 263
Grover (Lov), 218

H

Hadamard (Jacques), 236
Halting problem, 162, 278
Hawking (Stephen), 154, 247
Heavyside function, 164, 172
Hidden variable hypothesis, 225
Huffman algorithm, 43–47, 59
Human brain, 52, 249
Hyperfractal. *see* Fractal

I

Information, 1–59, 61–101, 104,
  106–109, 117–118, 129, 153–195,
  197–283, 286, 291
Information loss paradox,
  266–269
Inria, 67, 88
Internet, 4, 34, 41, 56, 62–100
Internet addresses, 66
Internet Engineering Task Force
  (IETF), 66

J

Joint Complexity, 193–195

K

Kleinrock (Leonard), 66
Knowledge. *see* Pyramid of
  information
Knuth (Donald), 4
Kolmogorov (Andrei), 18
Kullback Leibler divergence, 167,
  174

L

Laplace (Pierre Simon), 204
Learning problem, 161, 164–165, 167, 169, 173–174, 181, 185
Lebesgue (Henri-Léon), 131, 135, 143–144
Lempel (Abraham), 48, 166
Life evolution, 156
Linear regression, 163–164
Link state protocol, 69, 74
Logistic regression, 164, 168
Lorentz (Hendrick), 260–261, 279
Lossless compression, 41–43, 47–48
Lossy compression, 32, 41, 56–58

M

Machine learning. *see* Artificial Intelligence (AI)
Mandelbrot (Benoit), 129–130
Many-worlds interpretation, 239–242, 244, 246, 256
Marconi (Guglielmo), 2–3
Markov (Andrei), 166, 206
Maxwell (James), 5, 200–201
Maxwell Daemon, 202
Microwave background radiation, 248
Mind reading machine, 52, 191–192
Minneapolis (44° 58 N, 93° 16 W), 141, 145
Mobile Ad hoc NETworks (MANET), 84–86, 88, 106, 146
Mobile networks, 84, 86, 115, 118
Morse (Samuel), 44
Multi-user networks, 62
MultiPoint Relays (MPR), 88–90, 99, 101
Mutual Assured Destruction (MAD), 64
Mutual information, 23, 56–58, 223

N

Neural networks, 160, 164–165, 170–171, 177–190
Newton (Isaac), 125, 204–205

No Cloning theorem, 210–211, 232, 234–235
Noise, 20, 22–24, 27–28, 106, 119, 125–128, 135, 175, 198, 221
Non unitary computer, 269–280
NP hard problem(s), 90, 166, 217, 219, 275–279
Nyon (46° 22 N, 6° 14 E), 145

O

Olfaction, 188
Open Shortest Path First (OSPF), 74
Optimized Link State Routing (OLSR), 88–90, 94–101
OSPF. *see* The Border

P

P *versus* NP problems, 272–275
Packets, 34–35, 63–64, 67, 69, 73–76, 78–83, 87, 89, 95–100, 104, 107, 109, 112–117, 120, 122–123, 146
Paris (48° 51N, 2° 21 E), 2, 73, 115, 130–131, 160
Paris Japonica, 160
Pattern matching, 47–55, 81, 191–193
Pattern matching compression, 47–51
Pattern matching predictor, 51–55, 192
Penrose (Roger), 269
Perceptrons, 164
Perfect codes, 24–28, 161
Photon, 63, 156, 173–175, 211–216, 221–223, 225–231, 233, 235, 238–239, 241, 267, 281
Physical signal. *see* Pyramid of information
Poisson shot model, 117–124, 139, 149–150
Polarization, 173–176, 211–216, 222–223, 225–226, 229–231, 241–242
Postman protocol, 113–115, 117
Predictor, 51–55, 161–162, 191–192
Probabilistic entropy, 31–34

Protocols, 4, 34–35, 62, 64, 66–83,
    86–87, 89, 107, 112–117, 191, 201
Proton, 208
Pyramid of information, 6–9,
    162

Q

Quantization, 54, 192, 205, 212
Quantum computer, 217–222, 234,
    270, 273
Quantum divorce. *see* Many-worlds
    interpretation
Quantum information, 197–282
Quantum measurement, 162, 173,
    176, 197, 209–212, 223, 253,
    257–258, 265, 270, 273,
    281
Quantum physics, 174, 203–217, 225,
    228, 238–239, 243, 266
Quantum state. *see* Wave function
Quantum teleportation, 232–235
Quantum tomography, 173–177

R

Rate distortion, 32, 56–58
Rayleigh fading, 106, 123, 150
Rectified Linear Unit (RELU), 164,
    172, 180, 190
Recursive process, 50, 130, 135,
    142–143, 146, 165
Reinforcement learning, 163
Remote spanner. *see* Topology
    compression
Retro-information, 254–266, 269,
    279–280, 291
Rivest Shamir Adleman algorithm
    (RSA), 219, 271
Routers, 63–80, 82–87, 89, 100, 112,
    116, 286
Routing Internet Protocol (RIP),
    66–83, 100
Routing tables, 67–70, 72, 74, 77–78,
    80–81, 89, 94–96, 100, 112–113,
    115, 286

S

San Francisco (37° 46 N, 122° 24 W),
    134
Schrödinger (Erwin), 206–207, 237,
    239, 241, 244
Schwarzschild (Karl), 246
Seattle (47° 36 N, 122° 20 W), 141,
    145
Second law of thermodynamics, 199,
    201–203
Self similar structures, 130–131, 139
Shannon (Claude), 17, 83, 154–155,
    200
Shannon half theorem, 25, 28
Shannon multi-user wireless capacity,
    27, 125
Shannon second half theorem, 27–28
Shor (Peter), 219–221, 270
Signal attenuation, 91, 105
Source coding. *see* Data compression
Space capacity paradox, 110–111
Spatial reuse, 107–109
Spin, 203, 211–215, 217, 221–222,
    232–235, 238–242, 244–245, 264,
    281
Star Trek. *see* Quantum teleportation
Stieltjes-Laplace transform, 119
Store and forward, 112, 116, 145
Store-hold-forward, 112
Suffix tree, 193
Supervised learning, 161–164
Swamp area. *see* Zero mean weight
    algorithms
Swiss flag, 133–136, 138
System components, 19, 33

T

Telecommunication, 1–4, 17, 22,
    24–28, 34, 41, 43, 62, 64–65, 76,
    79–83, 85, 116, 145, 175, 185, 223,
    228
Texas, 130
The Border Gateway Protocol
    (BGP), 74–76
Time arrow, 202–204, 216, 224, 257

Time capacity paradox, 103, 111
Time travel, 249–266
Timeline, 240–243, 246, 251, 255–256, 261, 263–264
Topology compression, 83–88, 95–96, 101
Transmission Control Protocol over Internet Protocol (TCP/IP), 82
Traveling salesman problem, 277
Trie, 81, 191–192
Turing (Alan), 169, 278

U

Unit disk graph model, 90, 93–94
Unitarity, 210–212, 216, 254–261, 266–267, 269, 291
Unitarity violation, 254, 256–258, 260–261, 266, 291
Unsupervised learning, 162–163

V

von Neumann (John), 200, 215–217

W

Wave function, 174, 205–215, 217–218, 221, 238–239, 241, 255–259
WiFi, 2, 63, 84–87
Wireless capacity, 27, 105, 148
Wireless internet, 69, 83–100

Z

Zero mean weight algorithms, 184–185
Ziv (Jacob), 47–48, 50–51, 53–54, 166